Advances in Planetary Science – Vol. 2

ORIGIN AND EVOLUTION OF COMETS

Ten Years after the Nice Model and One Year after Rosetta

Advances in Planetary Science

Series Editor: Wing-Huen Ip *(National Central University, Taiwan)*

Published

Vol. 2 *Origin and Evolution of Comets: Ten Years after the Nice Model and One Year after Rosetta*
by Hans Rickman

Vol. 1 *Nuclear Planetary Science: Planetary Science Based on Gamma-Ray, Neutron and X-Ray Spectroscopy*
by Nobuyuki Hasebe, Kyeong Ja Kim, Eido Shibamura and Kunitomo Sakurai

Advances in Planetary Science – Vol. 2

ORIGIN AND EVOLUTION OF COMETS

Ten Years after the Nice Model and One Year after Rosetta

Hans Rickman

Uppsala University, Sweden & PAS Space Research Center, Poland

World Scientific

NEW JERSEY • LONDON • SINGAPORE • BEIJING • SHANGHAI • HONG KONG • TAIPEI • CHENNAI • TOKYO

Published by

World Scientific Publishing Co. Pte. Ltd.
5 Toh Tuck Link, Singapore 596224
USA office: 27 Warren Street, Suite 401-402, Hackensack, NJ 07601
UK office: 57 Shelton Street, Covent Garden, London WC2H 9HE

Library of Congress Cataloging-in-Publication Data
Names: Rickman, H. (Hans), author.
Title: Origin and evolution of comets : ten years after the Nice model and one year after Rosetta / Hans Rickman (Uppsala University, Sweden & Polish Academy of Sciences, Poland).
Description: Singapore ; Hackensack, NJ : World Scientific, [2017] | Series: Advances in planetary science ; volume 2 | Includes bibliographical references and index.
Identifiers: LCCN 2017010483| ISBN 9789813222571 (hard cover ; alk. paper) | ISBN 9813222573 (hard cover ; alk. paper)
Subjects: LCSH: Comets--History.
Classification: LCC QB721 .R634 2017 | DDC 523.6/6--dc23
LC record available at https://lccn.loc.gov/2017010483

British Library Cataloguing-in-Publication Data
A catalogue record for this book is available from the British Library.

Desk Editor: Ng Kah Fee

Typeset by Stallion Press
Email: enquiries@stallionpress.com

Preface

In August 2005, the international research community concerning the small bodies of the solar system gathered for the triennial *Asteroids, Comets, Meteors* meeting in the small seaside resort of Buzios north of Rio de Janeiro, Brazil. I remember with pleasure the nice Brazilian winter and the interesting discussions we had around the hot topics of those days. The one thing that stands out in my recollections is the first news of the Nice Model — the term had not yet been coined, but it was immediately clear that this was extremely important. Something else, which did not attract so much attention while being a potential time bomb for the future, was the fact that ESA's Rosetta probe was on its way toward its asteroidal and cometary targets since more than a year.

Of course, the Nice Model is principally about planets and, at least for myself, its repercussions for cometary science took some time to grasp. But today it is practically impossible to discuss the origin and early evolution of comets without reference to the Nice Model. In addition, after all, according to this scenario, comets were the real arbiters of the planetary system in the early days, when they governed the dynamics of the giant planets by their collective gravitational effects. We call these icy planetesimals instead of comets, but it is the same objects that nowadays populate the Oort Cloud, the Edgeworth–Kuiper Belt and the Scattered Disk.

It is worth emphasizing that the central theme of the Nice Model remains unchanged, but there are many variations around this theme with different implications for the comets. To account for this was one of my goals when writing this book.

Concerning Rosetta, I was involved from the very beginning, and I played a part in the OSIRIS camera team largely by procuring the filters. I hence had a personal involvement and lived through the mission with my own personal hopes and worries. The mission concept was novel and daring, and people like myself who were not born optimists had a hard time realizing what was actually to come. But, eventually, the time bomb did burst, and it felt like a two-year long roller coaster ride. So many discoveries, so many stunning images, thanks to the sterling work by all involved! This is an experience that may rightly be called a privilege to have been part of.

Now, after the end of the mission, it is time to assess what we have learned. I am convinced that, one day, this will be done and our knowledge about comets will take a big step forward. However, I cannot reach such an ambitious goal in this book. For the moment, the big picture remains too dim, at least to my mind. In all modesty, I can only present some preliminary ideas and impressions that may be right or wrong. Hopefully, together with those of my international colleagues, they will be food for thought and discussion, awaiting the final breakthrough.

In the meantime, this book can be used as an introduction to cometary science from the perspective of origins and evolution. It is not intended to be a full account of the whole subject, and even important results may occasionally have been left out in the interest of brevity, to avoid boring the reader with too much details. While trying to be fair, I did not strive to give equal weight to all ideas or to downplay the natural bias toward my own opinions. The interested reader will find enough references to published literature in order to pursue any specific quest that may arise.

Let me finally express my heart-felt thanks to Tomek Wiśniowski for invaluable help with many of the illustrations, as well as Mike A'Hearn, Björn Davidsson, Gerhard Hahn, Alessandro Morbidelli and Giovanni Valsecchi for helpful discussions. Above all, this book could hardly exist without the patience of my wife Bożenna, who, practically without complaint, endured a year of hardship when I was busy writing all the time.

Arboga, 3 February, 2017,
Hans Rickman

Contents

Chapter 1

Introduction

In most parts of the world, it would be difficult to find people of age, who have no idea what a comet is. However, those ideas are generally different from the concepts that scientists have in mind when using the word comet. In fact, the scientific definition of a comet is a non-trivial issue, which we had better tackle before describing their physical properties and how they originated and evolved.

1.1. What is a Comet?

There are two interpretations of the word. Closest to the layman's impression is the one about the phenomenon observed on the sky, and according to this, a comet is a diffuse object whose technical term is *coma* (usually, including a bright spot called the *central condensation*), from which a *tail* may extend. This object is in orbit around the Sun. Referring to the orbit is often necessary to distinguish the comets from Galactic nebulae and external galaxies. In 1771, the first catalogue of such diffuse nebulae (the famous Messier catalogue) was in fact produced by Charles Messier in order to avoid wasting time on these objects when hunting for comets, because one sort of diffuse object was often difficult to distinguish from the other.

The second interpretation refers to a physical object belonging to the solar system. In 2006, the International Astronomical Union (IAU) at its 26th General Assembly adopted a classification of solar system objects in terms of *planets*, *dwarf planets* and *small bodies*.

In this definition, comets are counted with the small bodies together with, for instance, the asteroids. Here, the word comet means the solid object, orbiting around the Sun, which gives rise to the diffuse phenomena mentioned above. The mechanism whereby this occurs is another matter.

Generally, it is a question of ice sublimation due to heating by absorption of sunlight, which leads to an outflow of gas and dust into space. This outflowing material is seen as the coma with a possible, more or less anti-sunward extension, called the tail. The solid object, from which the coma and tail would emanate, is called the *nucleus*, and thus, the word comet is used as a synonym to the nucleus. It is fair to say that this usage dominates in recent scientific literature, and the same practice will be followed here. This is natural, because the concepts of origin and evolution always refer to the nucleus. The coma and tail evolve very rapidly and they typically come and go, as the comet moves around the Sun. But the nucleus persists and typically evolves on much more significant time scales encompassing many orbits.

The physical definition of a comet, as briefly sketched above, is in fact a bit ambiguous. If we consider ice sublimation as the cause of the outflow, it is obvious that this is strongly temperature dependent. Hence, it will depend on the distance between the object and the Sun. It is indeed a well-known fact that comets develop their comae and tails essentially in the innermost parts of their orbits. Thus, when comets are observed far from the Sun, they may appear starlike, and we may actually see the bare nucleus. Comets do not have to produce the diffuse cloud all the time, but reliable observations of such activity on at least one occasion are required. Thus, official status as a comet is recognized only after such observations, but it is not withdrawn if an established activity ceases.

Hence, the comet is in reality an object that has the potential to develop a coma and a tail under the right circumstances — essentially, when it comes close enough to the Sun. This means that the object has to contain ice but also that the ice must be found near the surface, so that sublimation may lead to an outflow of gas and dust. Moreover, the perihelion distance must be small enough for this to happen. One can easily see that this is problematic. If the orbit is perturbed because of a close approach to Jupiter, the perihelion distance can change appreciably. An object may thus be called a comet, if it is discovered before an increase of the perihelion distance but not if the discovery happens afterwards. In addition, even slight modifications of the surface layers without any orbital change may imply that the gas production subsides or resumes, and hence, what is essentially the same object may or may not be called a comet depending on when it is observed — if not, it would probably be called an asteroid.

The only reasonable solution to these problems would be to call an icy object a comet, even if its perihelion distance is too large, or the ice is too deeply buried beneath the surface. The essential property of a comet would hence be its ice content: comets are icy, while asteroids are rocky or metallic. This definition is attractive in theory, because the criterion used is intrinsic and more or less quantifiable. Moreover, it helps to convey an important message, namely, that the small bodies of the solar system belong together, even though there is a range of chemical compositions depending on their formation temperatures. Comets and asteroids are not fundamentally different — they are different incarnations of the small body population, representing objects that were formed at different distances from the Sun and thus have different ice content. However, the definition is not useful in practice as long as we cannot measure the ice content by probing the interior of the objects, and we therefore have to require observed cometary activity as an objective criterion when distinguishing comets from asteroids.

In any case, it is clear that we have to be open minded about the objects to discuss. The border between comets and asteroids is somewhat fluent, and there may be transitional objects that are difficult to classify. When limiting ourselves to "real" comets that have exhibited comae or tails, we must recognize that they have siblings that sometimes need to be discussed in the same context.

1.1.1. *The comet nucleus*

Let us now pay some more attention to what a comet nucleus is thought to be, as an introduction to all the recent findings to be described below. As mentioned, the starting point is the gas and dust forming the comae and tails in comets. In the early 20th century, the old concept of a solid object within this cloud was no longer a dominating idea. Comets had been seen to split and disappear, and prominent meteor streams had been shown to trace the orbits of well known comets. It thus seemed natural to imagine a comet as nothing but a concentration of grains moving together in space. There was also a theory that claimed to show, how such comets could be formed by interstellar material captured by the gravity of the Sun, as it travels through the denser regions of interstellar space.

However, computations had shown that the observed comets do not show a tendency to arrive along hyperbolic orbits. Moreover, comets had been found to approach Jupiter closely without being dispersed and losing their identity, as one would expect from large clouds without much internal gravity. It thus seems, in retrospect, that there was no physical basis for the picture of comets as loose clouds. However, this was clear to some but not to all.

One problem was how to explain the origin of the coma by ice evaporation. In the 1940s there was little information about the chemical composition of the coma, but some radicals had been identified in comet spectra and shown to provide much of the light that is observed. In 1948, the Belgian astronomer Pol Swings proposed that these radicals were produced by the release and dissociation of ices made of polyatomic molecules. One obvious way to release these molecules was proposed by Fred Whipple (1950) in the paper that introduced the modern concept of a comet nucleus, namely, a solid body consisting of an *icy conglomerate* composed of ices and refractories in an intimate mix. Sublimation of the H_2O-dominated ice in the solar heat would release the parent molecules, from which the radicals emanate.

Whipple's paper dealt with one particular comet. This is *Encke's comet*, which was known since the early 19th century. With an orbital period of only 3.3 years it had been observed on many returns, and scientists had noticed that each of these returns occurred a little too early, compared to the best predictions that could be made by integrating the orbit from the preceding apparitions. This *nongravitational effect* needed an explanation, and Whipple's solid nucleus offered a good explanation using the same concept as Bessel (1836) had used. This was a jet force acting on the nucleus due to the asymmetric outflow of material feeding the coma. Since observations of Encke and other comets indicated the outflow to occur mainly in the solar direction, this would mainly accelerate the nucleus in the radial direction outward from the Sun, and Bessel focused on this aspect. In Whipple's model, the asymmetry followed directly from the fact that the heating of the ice is strongest at the subsolar point, but a thermal lag due to the rotation of the nucleus could in principle add a transverse component to the radial acceleration. In principle, the latter could act persistently over time, if the rotation is markedly prograde or retrograde, and the effect would then be either too late or too early arrival at perihelion.

While Whipple's theory appeared to offer a good foundation for understanding the behavior of comets and thus seemed clearly preferable compared to its competitors, one problem would remain for decades. To explain the observed amounts of material in cometary comae, a km-sized nucleus was generally enough. The problem was that such a small object is very difficult to detect at the typical distances of observed comets, and the long-lasting absence of any clear observational verification of Whipple's nucleus caused some lingering skepticism by the proponents of alternative theories.

1.2. Comet Designations

In both media reports, popular descriptions and scientific literature, comets are referred to by names and designations. These are not always consistent and may appear confusing, so a brief guide may be helpful.

Comet orbits span an enormous range of revolution periods from just a few years to millions of years. Thus, comets can be subdivided into two categories: the single-apparition comets and the returning comets. The former often have so long orbital periods that, essentially, astronomers have only had a single occasion to observe them in connection with one perihelion passage. The latter, on the other hand, have periods short enough to present at least two such occasions. Those categories have traditionally been referred to as long-period versus short-period comets (see Sec. 1.4), and the limit has been placed at orbital period $P = 200$ years.

The present designation system dates back to a resolution passed by the IAU 22nd General Assembly in 1994.[1] Here, the returning category is referred to as *periodic comets*. These are defined to have revolution periods of less than 200 years or confirmed observations at more than one perihelion passage.[2] Upon discovery, all previously unknown comets get a designation, consisting of the year of discovery followed by an upper-case letter denoting the halfmonth in question and a numeral indicating the sequential order of this discovery announcement within the relevant halfmonth. Before this, a letter is applied, which indicates the category of the comet: "P/" denotes a periodic comet, and "C/" denotes a comet that is not periodic.

In 1999, the IAU Minor Planet Center decided to call single-apparition comets periodic only when their orbital periods are less than 30 years. Meanwhile, there is another way to designate returning comets, which is independent of the orbital period. This is a permanent, serial number followed by the letter "P", and it is assigned to comets that have been observed to return or have had their periodicity established otherwise. The name of the comet may be added, separated by a slash. The list of such comets is basically chronologic, starting with 1P/Halley. As of 1 January 2016, there were 330 comets listed as periodic in this way, and on 1 January 2017 this number had grown to 347.

A category of special interest for the evolutionary aspect of comets is those that have been deemed not to exist any more. In

---

[1] www.minorplanetcenter.net/iau/lists/CometResolution.html

[2] To date, there is only one comet of the second kind, namely, comet 153P/Ikeya-Zhang with perihelia in 1661 and 2002. The 1661 observations were carried out by Jan Heweliusz, while the independent discoveries in 2002 were made by Kaoru Ikeya and Zhang Daqing.

most cases, the reason is that they have not been found in spite of deep exposures of the sky area where they would certainly have been according to the ephemerides. For these comets, the letter P is replaced by D in the designation. The list of comets with permanent numbers currently hosts eight such members, the most famous of which is 3D/Biela.

In addition, there are single-apparition comets with "D/" designations. Among these, the most notable is the famous, very special comet D/1993 F2 (Shoemaker-Levy 9), which was discovered in a jovicentric orbit and collided with Jupiter in 1994. This is the only comet that is definitely deceased, even though the letter D generally stands for 'dead' or 'disappeared'. In other cases, one has to consider the possibility that the comet will reappear. A case in point is that the record of single-apparition comets used to contain two members with designations D/1783 W1 (Pigott) and D/1819 W1 (Blanpain), which were long lost comets with orbits known to be of short period and were thought to have disappeared. However, both were recently rediscovered and are now known as 226P/Pigott-LINEAR-Kowalski and 289P/Blanpain, respectively.

Comets are often referred to by names. For some of them, this is almost indispensible since the names are so deeply rooted in people's minds. For instance, it would be strange to discuss comet 1P without adding the name Halley or D/1993 F2 without clarifying that this is comet Shoemaker-Levy 9. However, in contrast to the designations, names are never unique. For instance, even though it seems quite unlikely, it cannot be excluded that one day somebody named Halley will discover a bright comet, and people will be tempted to call this "Halley's comet". For this reason, names are treated as of secondary importance in the official designation system.

There are many idiosyncracies and complications associated with comet names and designations, which cannot be covered here.[3] However, a few items are worth noting. The difficulty of

[3]An account may be found in the IAU-endorsed comet naming guidelines of 2003; see http://www.ss.astro.umd.edu/IAU/csbn/cnames.shtml.

distinguishing between comets and asteroids has already been dealt with, and one consequence is that many comets were designated as asteroids before their cometary nature was recognized. In such cases, the asteroidal designations are retained in conjunction with a cometary marker. One example is comet P/1999 DN_3, now known as 183P/Korlević-Jurić.

The names given to comets usually refer to their discoverers. Traditionally, these have been persons, but in recent times most comets are discovered due to team efforts within search programs or using space telescopes. Thus, many comets are named for the teams or space projects, like comets C/1999 S4 (LINEAR) or C/1997 B3 (SOHO). There are also a few periodic comets, which are named not after the discoverers but after the persons who investigated their orbits or established their periodic nature (1P/Halley, 2P/Encke and 27P/Crommelin).

At the time in 1994, when the current comet designation system was endorsed, one issue was especially controversial. In the 20th century, a practice had developed, whereby different periodic comets with the same discoverer were distinguished by adding a sequential number after the discoverer's name. With the introduction of the unique, permanent numbers, this secondary numbering was deemed redundant and thus abolished. However, in many cases those numbers have become an integral part of the name in the minds of comet scientists (and also, the general public as in the case of Shoemaker-Levy 9), and so they are still in frequent use. For instance, the target comet of the NASA Deep Impact mission is usually called 9P/Tempel 1, and the follow-up (EPOXI) target is usually called 103P/Hartley 2, even though the official names are 9P/Tempel and 103P/Hartley.

The last point to note here is that some objects have found their place in the permanent records of both comets and asteroids. Thus, they are recognized to have dual status. This group is heterogeneous as to both orbits and physical properties. The first object was asteroid (2060) Chiron — discovered in 1977 as the first among the so-called *Centaurs* (see Sec. 1.4). Observations made in 1988–89 showed that it exhibits cometary behavior and revealed its coma by imaging. This fact prompted an action and led to a parallel, permanent cometary designation as 95P/Chiron. So far, there is one more Centaur with a dual status: (60558) Echeclus or 174P/Echeclus. A quite different object is comet 107P/Wilson-Harrington, discovered in 1949 but later considered as lost. In 1992, it was securely identified with Apollo asteroid (4015) 1979 VA, which was then named (4015) Wilson-Harrington.

1.3. Discovery Bias

As mentioned, comets develop their comae and tails in the innermost parts of their orbits, where they receive sufficient energy from sunlight. Thus, they brighten up considerably. In fact, the persistent part of a comet — the nucleus — is generally only several kilometers in extent. Until recently, such small objects had to come very close to the Earth in order to be discovered, and yet due to the comae and tails, comets are known since many centuries, and some were seen by the naked eye in prehistoric times. Occasionally, they are still able to dominate the night sky, while most comets observed in recent times remain very faint all the time.

It is thus clear that individual comets may be extremely different in brightness and ease of discovery. But it is also clear that the threshold brightness for a comet to be discovered has developed with time, as astronomy itself has developed, instrumentation has improved, and more people have become engaged in research including the search for comets. Nonetheless, there must always be a *discovery bias* as long as some comets remain too faint to be discovered. One of the important tasks when looking for statistical properties of the comet population is to limit this bias by an optimal definition of the observed sample.

While differences in brightness between comets may partly be due to differences in internal properties, they may also come from the different orbits. Since closeness to the Sun is so important for cometary activity, the *perihelion distance* (smallest distance from the Sun) is an essential orbital parameter. Statistically speaking, the larger the perihelion distance of a comet orbit, the less active the comet will be, and the less likely its discovery. But, once again, this orbit-related bias has changed significantly over time.

Historically, the standard technique used for discovering comets has evolved dramatically. Centuries ago, it was a question of visual observations — by the naked eye or telescopic. Typical binoculars or telescopes used as comet finders had large fields of view and were thus very efficient for sky surveys, but their limiting magnitude was very bright so that only a minority of comets were within reach.

This situation gradually changed with the introduction of photographic techniques, using astrographs or Schmidt telescopes, for different purposes. Often the plates were aimed for star counts or probing extragalactic space, but they could also be scanned for moving objects like comets. Sometimes the aim of the observations was indeed to find comets or asteroids. The purpose would influence the sky coverage by the introduction of zones of avoidance or preference, depending on the nature of the targets. Galaxy counts would aim for high Galactic latitudes, while star counts would typically concentrate near the Galactic plane. Searches for asteroids would be made close to the ecliptic, avoiding the regions where the Galactic plane was crossed to avoid confusion with the rich stellar background.

Thus, an orbital bias could be introduced by an uneven sky coverage of the comet searches. This came on top of a fundamental source of bias due to the asymmetric distribution of observers between the northern and southern hemispheres of the Earth. The northern sky has been scanned much more efficiently than the southern sky — something that may, for instance, cause a bias in the distribution of arguments of perihelia.[4] Another obvious asymmetry arises from the fact that comets are most easily discovered on the night sky, favoring the opposition region. The discovery bias

[4]This is the angle measured along the plane of the comet orbit from a crossing point with the ecliptic (the ascending node) to the direction of perihelion. Together with the inclination, it determines the heliocentric ecliptic latitude of the comet, when it is at perihelion.

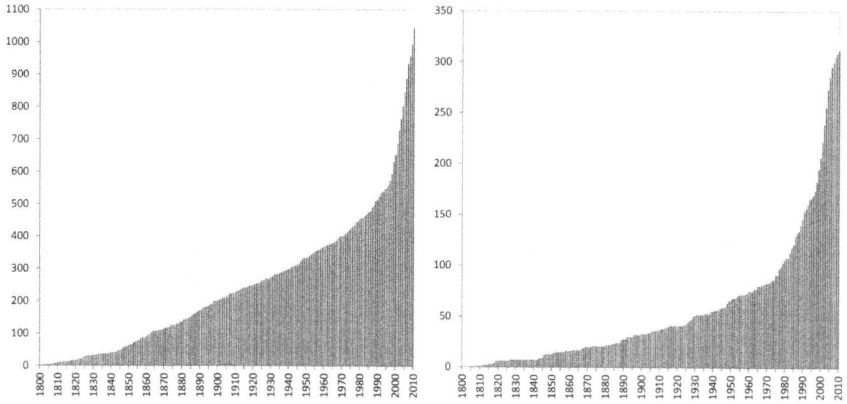

Fig. 1.1. Histogram representations of the cumulative number of discoveries during the years 1800–2010 referring to single-apparition comets (left panel) and numbered periodic comets (right panel). The list has been screened against comets with perihelion distances $q < 0.01$ AU or discovered by sun-staring space telescopes. The decrease of the discovery rate of periodic comets during the most recent years is due to the fact that, in general, two apparitions are needed before a comet is numbered. Hence, many of the recently discovered comets that will eventually be numbered are still counted as single-apparition comets. Courtesy T. Wiśniowski.

against comets passing perihelion on the side of the Sun opposite to the Earth is well known as the *Holetschek effect* since about a century ago.

Many of these biases have had a time dependence because of the changes in the ways comets are discovered. In particular, as illustrated by Fig. 1.1, the last few decades have seen a major rise in the rate of comet discoveries thanks to systematic search programs aimed at discovering Near Earth Objects (NEOs). This is obvious when glancing through the list of recent comet names, which is dominated by search projects like LINEAR, LONEOS, NEAT and Catalina or people associated with those. As a result, the total inventory of all ground-based comet discoveries is now dominated by those made recently by a small number of telescopes, located at a handful of sites with a certain geographical distribution and looking preferentially at low ecliptic latitudes. A special category is formed by the *sun-grazing comets* with perihelia close to the solar surface, which hosts an enormous harvest from sun-staring space telescopes — primarily, SOHO.

Of course, it is very important to correct the orbital statistics of the discovered comets for the biases involved in the discoveries. This has always been an issue in cometary science. Anomalies found in the distribution of comet orbits may lead to suggestions of underlying processes like, for instance, an ongoing comet shower (see Sec. 5.3.1), or a planet orbiting in the Oort Cloud, and to judge the reality of those processes requires an analysis of discovery biases. Nowadays, it is fortunate that bias corrections can be straightforwardly applied using discovery simulators for a majority of comets using the logs or observing routines of the contributing surveys similar to what is done for NEOs.

In any case, the strong discovery bias in perihelion distance and its relation to the intrinsic brightness distribution of comets remains a major issue. Figure 1.2 shows how the perihelion distances of newly discovered comets have evolved during the past two centuries. The trends seen reflect the improved sensitivity and efficiency of comet discoveries. It is obvious that the perihelion distance distribution

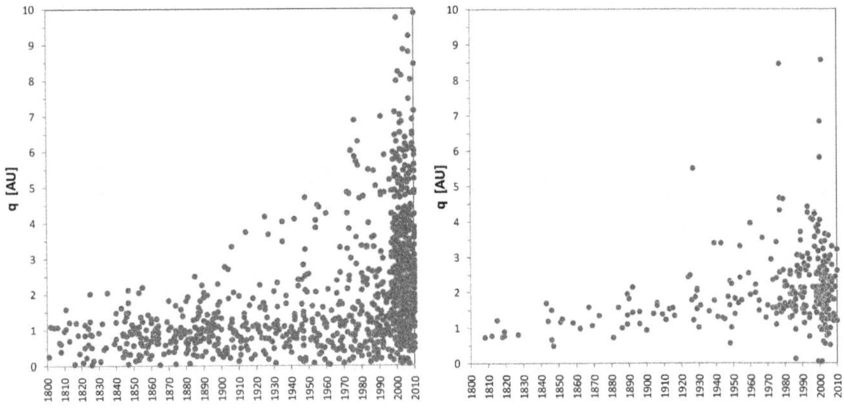

Fig. 1.2. Scatter plots of perihelion distance at discovery versus discovery year for comets discovered during the years 1800–2010: single-apparition comets (left panel) and numbered periodic comets (right panel) are plotted separately. The sample used is the same as in Fig. 1.1. A few comets with $q > 10$ AU are missing in each category. Courtesy T. Wiśniowski.

of the true comets has not undergone any major change in such a short time, so there have been effective limits in perihelion distance, beyond which comets escaped detection, or detections were limited by the brightness or activity of the comets. As time proceeded, these limits have moved outwards, and they are likely continuing to move.

A special case of discovery bias holds for the sungrazing comets. Before 1979, there were less than ten such comets on record throughout the history. These were extremely bright and overwhelming as a rule and frequently named "Great Comet" for the lack of only one or a few identifiable discoverers. Their brightness came from the fact that they passed perihelion at less than 0.01 AU from the center of the Sun. Only few such comets have been seen in the recent years, including the marginal case of comet C/2012 S1 (ISON). However, thousands of sun-grazing comets have been observed by a series of space observatories (SOLWIND, SMM, SOHO, Stereo), the most prolific of which was SOHO (SOlar and Heliospheric Observatory). These were seen passing perihelia or colliding with the Sun on coronographic images of the inner corona, but their faintness would have prohibited discovery by any other means. Most of them are tiny fragments of a common parent that also gave rise to the brightest sungrazers (Sec. 4.4.1). If included into the orbital statistics of comets, they completely skew the distributions of elements, and so they are excluded from the distributions presented here.

1.4. Comet Populations

Comets are often divided into separate populations. This of course concerns the observed comets, but we shall see that the observed comet populations tend to be associated with different, basically unobserved parent populations of objects with much larger perihelion distances. While the above-described categories like periodic comets and single-apparition comets are useful for book-keeping purposes, the populations are defined with an eye to the dynamical transfer routes that comets are expected to follow. One might ask if comets could be divided into different groups based on their morphologic appearance or physical-chemical properties as well. However, this has proved remarkably difficult. One case in point will be introduced in Sec. 2.6.3.

As mentioned, an early classification was made into *short-period comets* and *long-period comets*. The limit was arbitrarily put at $P = 200$ years. This would in principle give the short-period comets a chance of being observed at more than one apparition (or perihelion passage), while the long-period comets would not stand such a chance. However, this division does not have any obvious dynamical significance. The orbital period of a comet is a very unstable quantity due to perturbations, primarily by Jupiter. Thus, from early times, there was a concept of *comet capture*, whereby Jupiter in particular would transfer comets from the long-period into the short-period class by decreasing the periods. Of course, the opposite can also occur, but since some short-period comets were known to have disappeared, their number might be kept more or less constant by a predominance of captures. Long-period comets would hence be the source of short-period comets.

We now know that reality is much more complex. Of particular importance is the recognition that short-period comets can be divided into groups that are relatively stable in dynamical terms. This builds on theoretical results within the *circular restricted three-body problem* (see Sec. 3.1), which were derived more than a century ago. There is an algebraic function of the semi-major axis a, the eccentricity e, and the inclination i of a comet orbit, called the *Tisserand parameter*, which is quasi-conserved under the perturbations by a particular planet, to which it is referred. For most comets, the perturbations by Jupiter dominate over those due to all the other planets, and hence we use the jovian Tisserand parameter

$$T_J = \frac{a_J}{a} + 2\sqrt{\frac{a}{a_J}(1 - e^2)} \cos i, \tag{1.1}$$

where a_J is the semi-major axis of Jupiter's orbit. As long as the perturbations by other planets are insignificant, T_J remains approximately constant over considerable time intervals, even though each

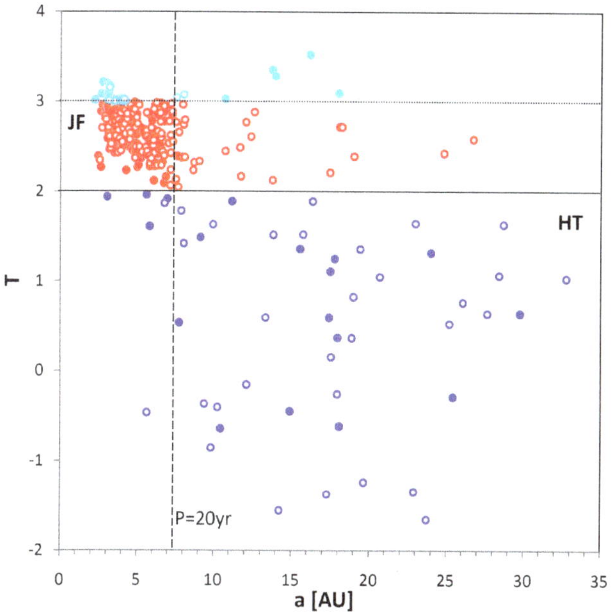

Fig. 1.3. Jovian Tisserand parameter versus semi-major axis of the short-period comets known in 2010, using their discovery orbits. Open circles denote single-apparition comets, and filled circles denote numbered periodic comets. The open circles are plotted on top of the filled circles in case of overlap. Blue symbols show Halley Type comets, red symbols show Jupiter Family comets, and sky-blue symbols are used for comets of Chiron or Encke type. The right-hand border of the diagram corresponds to an orbital period close to 200 years. Courtesy T. Wiśniowski.

of the elements entering into its definition may change dramatically as a result of close encounters with Jupiter.

Figure 1.3 shows how the comets with $P < 200$ years, known in 2010, are distributed with respect to a and T_J. By orbital evolution, these comets move essentially in the horizontal direction, so the diagram reveals that there are two, nearly disconnected main groups of short-period comets. The one with $2 < T_J < 3$ is in essence what we call the *Jupiter Family*, while the comets with $T_J < 2$ are usually called *Halley Type comets*. These two groups represent two distinct populations, which are not hermetically isolated but separated by a boundary ($T_J = 2$) that is difficult to cross. This boundary was introduced by Carusi and Valsecchi (1987) and is fundamental to the orbital taxonomy of comets introduced by Levison (1996) and further developed by Duncan *et al.* (2004). Here, all comets are considered irrespective of the orbital period, and those with $T_J > 2$ are called *ecliptic comets* due to their typically small inclinations, while those with $T_J < 2$ are called *nearly isotropic comets* to reflect the nearly isotropic distribution of their orbital poles.

In Fig. 1.3 we apply the taxonomy used by Duncan *et al.* (2004) by identifying as a special group the comets with $T_J > 3$. Those are of two different kinds. With $a > a_J$, they are referred to as Centaurs, of which there are many, but the objects in question have cometary designations due to their observed activity. They are also called *Chiron-type comets.* Those with $a < a_J$ and $T_J > 3$ are called *Encke-type comets* after their prototype, comet 2P/Encke, whose orbit is decoupled from Jupiter due to its small aphelion distance. This is quite remarkable, because in the current paradigm regarding the dynamical transfer of comets in the solar system (to be described in Sec. 3.6), comet orbits are either external to Jupiter's orbit in the case of remote reservoirs and the routes leading from these, or Jupiter-crossing in the case of observable comets. Orbits that fall entirely well inside that of Jupiter have to be seen as anomalous.

But nothing in our world is perfect. In particular, the term *Centaur* is used in a wider context for small bodies with orbits in the zone of the giant planets, but no exact definition has been agreed upon. The matter will be discussed later, but it is preferable to use the term Chiron-type comets rather than Centaurs for the comets in question. Moreover, the term Jupiter Family is usually thought to imply a dynamical control by Jupiter due to the possibility of close encounters. However, Fig. 1.3 shows a number of Chiron-type and Jupiter Family comets with similar semi-major axes, which straddle the dividing line at $T_J = 3$. The adherence to one group or the other is then a matter of orbital inclination and does not uniquely tell, if an object is under Jupiter's control or not. Finally, a recent category is formed by the so-called *Main Belt Comets*, whose orbits are typical of the asteroid main belt but which exhibit cometary activity. With the above definition, these are Encke-type comets but should really be treated as a distinct group.

A remarkable feature is that the two main groups (Jupiter Family and Halley Types) are almost disjoint also with regard to semi-major axis. Very few Halley Type comets have periods less than 20 years, while periods exceeding 20 years represent a minority in the Jupiter Family. A few decades ago the separation of the groups in orbital period was even more clearcut, and thus it was common to reserve the term short-period comets for those with $P < 20$ years and use *intermediate-period comets* for the range of periods between 20 and 200 years. This was almost equivalent to the distinction between Jupiter Family and Halley Type comets. However, as Fig. 1.3 demonstrates, the intermediate-period group nowadays has an important admixture of Jupiter Family and Chiron-type comets.

Figure 1.4(a) illustrates the major difference between ecliptic and nearly isotropic comets by means of the Jupiter Family and Halley Type comets. The former have inclinations that are almost always smaller than 30°, while the latter span the whole range up to 180° without any zone of avoidance. Among comets with periods

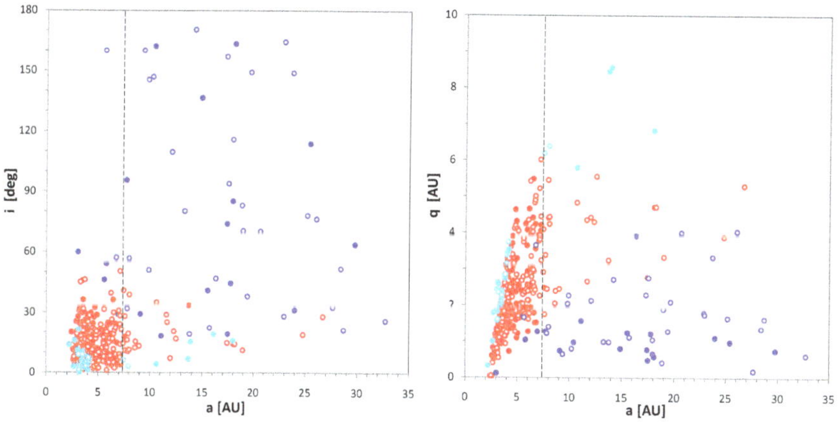

(a) (b)

Fig. 1.4. (a) Inclination versus semi-major axis, and (b) perihelion distance, versus semi-major axis for the comet apparitions plotted in Fig. 1.3. The symbol types and colors are the same. The vertical dashed lines indicate orbital period $P = 20$ years. One comet missing from the right panel is the Chiron-type comet 167P/CINEOS with $q = 11.8$ AU. Courtesy T. Wiśniowski.

$P < 20$ years, the few Halley Types have inclinations that are mostly close to the highest values of the Jupiter Family comets (30–60°), but one of them stands out by a strongly retrograde orbit. This is P/2006 R1 (Siding Spring).[5] At first sight, it looks like an extreme case of a Halley Type comet, while the others may rather be related to the Jupiter Family. However, the real nature of these objects cannot be judged without performing a comprehensive, dynamical investigation.

Regarding periods $P > 20$ years, the Jupiter Family and Chiron-type comets occupy the same range of low inclinations as some of the Halley Type comets. However, their orbits are obviously different because of the different values of T_J, and this difference is illustrated by Fig. 1.4(b). The perihelion distances of the Halley Types are mostly small, very rarely exceeding 3 AU, while the Jupiter Family comets show the opposite behavior. These follow the trend, seen among the smaller periods, for the perihelion distances to statistically increase with the orbital period — a natural consequence of the limited ranges of T_J and $\cos i$. It is now clear, why the number of Jupiter Family comets thins out for periods larger than 20 years. The large perihelion distances effectively counteract discovery of such comets.

Let us now consider the Encke-type comets. In Fig. 1.4(a), these mix perfectly with the Jupiter Family, and in Fig. 1.4(b) they appear to cling to the Jupiter Family on its left side. A better resolution of this orbital domain is provided by Fig. 1.5, which shows the (Q, q) plane of aphelion and perihelion distances with the line $Q = q_J$ indicated. This shows that the $T_J > 3$ comets are of three kinds. The Main Belt Comets are found at perihelion distances between 1.8 and 2.6 AU with aphelion distances less than 4.3 AU. They are well separated from the Jupiter Family, belonging to the orbital realm of the asteroid main belt. The only two comets that can indeed be called Encke-type are identified using their names: 2P/Encke (E) and 107P/Wilson-Harrington (W-H). These too are dynamically

[5] After 2010, two more comets have been discovered in rather similar orbits: 333P/LINEAR and P/2013 AL76 (Catalina).

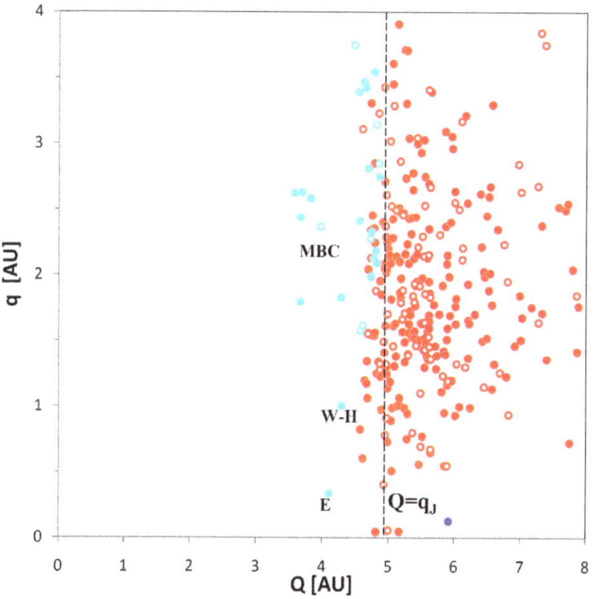

Fig. 1.5. Perihelion distance (q) versus aphelion distance (Q) for comets with $q < 4$ AU and $Q < 8$ AU. The meaning of the symbols is the same as in previous figures. The Encke-type comets with our definition ('E' and 'W-H'), and the Main Belt Comets ('MBC'), are identified (see the main text). Courtesy T. Wiśniowski.

decoupled from Jupiter due to their low aphelion distances. The rest of the comets are best counted with the Jupiter Family. In general, they mix with the low-Q border of the Jupiter Family, and at the highest q values (3.4–3.8 AU) there is a group of comets near the 3:2 mean motion resonance with Jupiter with a special dynamical history, which is considered a special subgroup of the Jupiter Family (see Sec. 3.2.2).

Finally, as will be detailed in Chaps. 3 and 5, the long-period comets form a more homogeneous group than the short-period ones. While the Jupiter Family and Halley Type short-period comets are considered to be derived from separate source populations in the outer solar system, no such subdivision can be made among the long-period comets. Their ultimate, distant source is considered as unique and common to all, namely, the *Oort Cloud*. This may also be the main source of the Halley Type comets, while the Jupiter Family mainly stems from the *Scattered Disk*. We shall not deal with the source populations here, but these will be central topics of the discussions in Chaps. 5 and 6. One group of closely related long-period comets stands out as very important in terms of numbers but does not represent a separate population. This is the *Kreutz group* of sun-grazing comets, which are fragments of one and the same parent comet that has split due to the tidal force of the Sun (see Sec. 4.4.1).

1.5. Lessons from Space Missions

Chapter 2 will deal with the physical and chemical properties of comets as a background to the discussion of how they may have originated. However, this will focus on our knowledge about comets in general. It is obvious that this knowledge has an essential input from the close-up studies of comets that have been performed within different space missions. Let us thus first briefly review what has been learned or glimpsed from these projects concerning a small number of individual comet nuclei. These descriptions are not meant to be comprehensive, and more information will be given in later chapters.

1.5.1. *The Halley flybys in 1986*

Comet 1P/Halley was observed in times that were prehistoric in most countries, where comet science is pursued today. Its periodic nature was established through orbit determinations by Edmund Halley in 1705. Largely by setting this historic example of scientific progress through Newtonian mechanics, but also by the fact that it has often been a remarkably bright comet, 1P/Halley has become legendary. At its long awaited return to perihelion in 1986, it was therefore a natural choice as target of the first space missions to the near-nucleus environment of a comet.

In fact, there was a whole armada of such spacecraft. The most important was ESA's first interplanetary mission, called *Giotto* after the Italian, early Renaissance artist Giotto di Bondone, who may have been inspired by comet Halley when painting the *Adoration of the Magi*, still to be seen in the Scrovegni chapel in Padova. Another pioneer achievement was made by the Japanese space institute ISAS, which launched two space probes called *Suisei* and *Sakigake* into the Halley environment. Last but not least, there were also two spacecraft of the USSR Academy of Science with French contributions, called *Vega 1 and Vega 2*, which first sent balloons into the Venus atmosphere and then were deflected toward Halley's comet by Venus' gravity (the name Vega comes from the Russian names of the targets, Venera and Galley).

Comet science was overthrown by the results of these encounters, mostly due to Giotto and the two Vegas. The most fundamental discovery was not unexpected, namely, that the comet has a solid nucleus. The most famous image of this nucleus, reproduced in Fig. 1.6, was taken by the Giotto HMC instrument (Halley Multicolour Camera). Even though the solid nucleus had been discussed since decades, this result was of paramount importance, since the hypothetical concept of the comet nucleus that Fred Whipple (1950) had introduced was now confirmed, and all lingering doubts could be disposed of.

Fig. 1.6. Composite image of the 1P/Halley nucleus acquired by the Halley Multicolour Camera in March 1986. Reproduced with permission from H. U. Keller.

More surprising was the size of the nucleus. Ever since the recovery of the comet in 1982, observations had been performed, which estimated the *photometric cross-section* of the nucleus. This is a measure of the product of the albedo and the geometric cross-section — the latter measuring the square of the average radius. In 1986, comet albedos were not known and models often assumed very high values, since the scientists' minds were fixed on a slightly contaminated snowball. With such an assumption, the radius of the Halley nucleus would be much smaller than the images revealed, and the result was an unexpectedly low albedo. In fact, this distinguished the only explored comet nucleus as one of the darkest solar system small bodies ever observed.

The large size of the nucleus had one more unforeseen consequence. The water production rate of comet Halley would have been much higher than observed, if its nucleus had been the expected snowball-like object. In fact, the flow of water vapor due to sublimation must be quenched over the whole surface or at least the majority of it. This could be consistent with the HMC image, which shows jet-like features of dust emerging from what seems like local, active spots. However, this interpretation has not been confirmed and may be an oversimplification. What is certain is that the surface has to be enriched in refractory substances, or the whole nucleus is made of a dusty matrix with included ice rather than the other way round. This is in agreement with the high surface temperatures observed on the Halley nucleus, reaching a level around 350 K.

In fact, thanks to the dust detection systems on board Giotto and Vega, a further important conclusion could be drawn. The grain size distribution turned out to fall off less steeply toward macroscopic chunks than the earlier models had predicted. As a consequence, the total mass of the ejected dust was inferred to be at least as large as that of the outflowing gas. Fragmentation of the larger grains into smaller pieces on the way out through the coma was also observed. Overall, the "Halley armada" showed comet nuclei — by inference from the single object under study — to be bigger, darker, more dusty, and possibly more porous and less coherent than earlier believed.

The elemental composition of the comet dust was studied by means of mass spectrometry of single, small grains. These turned out to be of different kinds: one dominated by silicates, a second dominated by organic molecules rich in carbon, hydrogen, oxygen and nitrogen (the so-called CHON grains), and a third being a mixture of the two. Combining this with what could be inferred about the gas composition led to a picture of the overall elemental composition of the material ejected from the 1P/Halley nucleus, which is illustrated in Fig. 1.7. Comparing this to the composition of the most primitive (i.e., solar-like) meteorites — the CI chondrites — shows the comet to be even more primitive.

This finding highlighted a property of comets that was not unexpected though extremely important. They appear to be unaltered, comprehensive collections of elements from solar nebula regions,

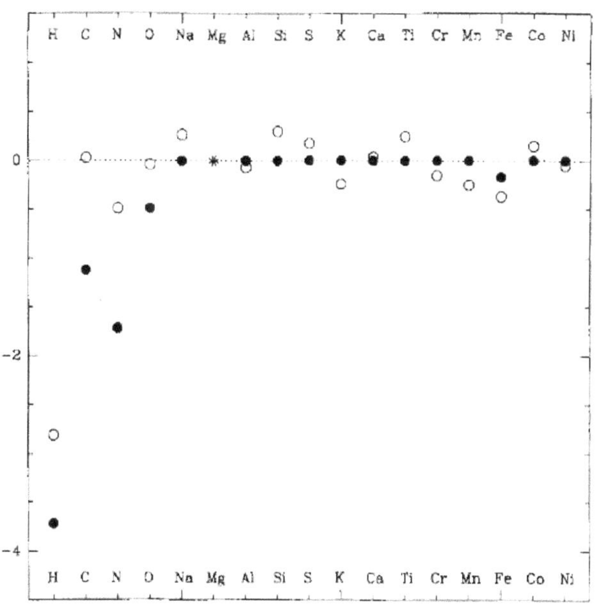

Fig. 1.7. For a suite of cosmically abundant elements, the ratios between abundances in CI chondrites and the Sun (filled circles), and in comet 1P/Halley and the Sun (open circles), are shown on a log scale. The reference element for the abundances is magnesium (Mg). From Festou, M. C. *et al.*, *Astron. Astrophys. Rev.* **5**, 37–163 (1993), © Springer-Verlag. With permission of Springer.

where all condensible material was indeed condensed, i.e., very cold regions. Further support for this came from the measured isotope ratios in grain material. Some isotope ratios in comet Halley (e.g., $^{12}C/^{13}C$ and $^{32}S/^{34}S$) are very close to the solar system average, presented by the Sun's photosphere and the most primitive meteorites. At the same time, large variations of isotope ratios in individual, microscopic grains exist, showing that pre-solar units may have been preserved.

1.5.2. *The flybys of 19P/Borrelly and 81P/Wild 2*

Comet 19P/Borrelly was discovered by French astronomer Alphonse Borrelly more than a century ago, and with a period of a little less than 7 years it has been observed at most returns since then. It is an average Jupiter Family comet that has not shown any remarkable behavior. However, it presented an attractive opportunity for NASA to steer its *Deep Space 1* technology test mission to a comet target after it had accomplished its prime scientific goal — the flyby of asteroid (9969) Braille. The flyby of 19P/Borrelly took place on 21 September 2001, when the comet was close to its perihelion at about 1.36 AU from the Sun.

The DS1 spacecraft carried much less instrumentation for comet exploration than Giotto and the Vegas did, but because this was the first visit to a comet after 1986 and the imaging conditions were much more benign than they had been for comet Halley,[6] the images taken of 19P/Borrelly had great significance. One of them is shown in Fig. 1.8. This nucleus is smaller than that of 1P/Halley, measuring about 8×3 km. The two nuclei are similar in their elongated shape, and the Borrelly nucleus was likened to a footprint. Just like the Halley nucleus, it was found to be extremely dark — in fact, it beat the record of darkness for surfaces of solar system objects with a mean Bond albedo of 0.01. Another similarity is that the global

[6]Due to the strongly retrograde orbit of 1P/Halley, the spacecraft encountered the comet with a speed of about 70 km/s, while the low-inclination orbit of 19P/Borrelly allowed the DS1 encounter to happen at only 16.5 km/s.

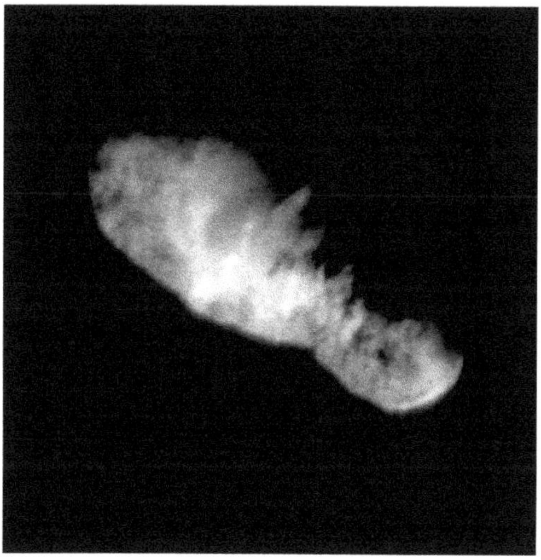

Fig. 1.8. Image of the 19P/Borrelly nucleus acquired from the Deep Space 1 spacecraft on 21 September 2001. Reproduced with permission from L. Soderblom. Credit: NASA/JPL-Caltech/US Geological Survey.

outgassing rate observed during the DS1 encounter was much smaller than expected for an icy surface of the same size.

The morphologic appearance of the surface is seen to be variegated. Several different units have been identified, and the local albedo varies by a factor of almost four. While the overall topography is rough, some local units appear relatively smooth. No impact structures have been identified. A possible indication of a binary nature from the overall shape remains speculative, and there is no sign of the "sole" and "heel" of the footprint having different morphologic characters.

Near-infrared spectroscopy of the 19P/Borrelly nucleus was also performed by DS1. The most noteworthy conclusion was that no trace of surface ice was seen in these spectra, although this had been an easy matter, if the surface had been rich in ice.

The next comet mission performed by NASA would be much more scientifically ambitious. This was the *Stardust* mission, launched in February 1999. Stardust was aimed to collect coma grains in comet 81P/Wild 2 in silica aerogel during a close approach to the comet nucleus on 2 January 2004, and bring them back to Earth for laboratory analysis. All went well, and the Stardust harvest of more than 10 000 comet particles represents one of the most important sources of information available on the formation conditions of comets.

Comet 81P/Wild 2 was discovered in 1978 by Swiss astronomer Paul Wild. It is a Jupiter Family comet with an orbital period of 6.4 years, but as such it is very young, and this is its distinguishing property. In September 1974, resulting from a close encounter with Jupiter, it was captured into the current orbit from a larger perihelion distance and much longer period. This may mean that it is less physically evolved and therefore relatively fresh, compared to most Jupiter Family comets — an attractive property for a space mission target.

Images of the Wild 2 nucleus were acquired by the Stardust navigation camera, and one of them is shown in Fig. 1.9. Contrary to what is often stated, its shape is not markedly oblate, but nor is it as prolate as the Halley and Borrelly nuclei. Its dimensions are

Fig. 1.9. Image of the 81P/Wild 2 nucleus acquired by the Stardust camera. Reproduced with permission from D. E. Brownlee. Credit: NASA/JPL-Caltech/University of Washington.

$5.5 \times 4.0 \times 3.3$ km. Similar to Halley and Borrelly, it has a low albedo estimated to be close to 0.03.

The surface features on the Wild 2 nucleus include elements that were not seen on the preceding mission targets. These may indeed provide insights into the evolution of the nuclei. The most characteristic features are circular depressions. These are of two kinds: pit-halo features and flat-floor features, ranging in size from 0.5 to almost 2 km. According to Brownlee *et al.* (2004), they are most likely impact structures. Their morphologies do not coincide with those of impact craters on rocky solar system objects, but the nature of the target and its material are also very different. If the inference is correct, one may conclude that the lack of impact features on the Borrelly nucleus is due to their destruction on a time scale similar to the typical residence time of Jupiter Family comets in orbits like those of Borrelly and Wild 2.

However, the most remarkable results from the Stardust mission concern the composition of the sampled coma grains. These will be discussed in later chapters and are only briefly summarized here. An unexpected result is a strong heterogeneity of the olivine and pyroxene compositions between individual grains, which indicates very different formation conditions. Thus, on a micrometer scale, comet Wild 2 appears to host particles from all over the solar nebula, which necessitates efficient radial mixing from the innermost parts into the trans-planetary region, where comets were likely formed.

In particular, high-temperature minerals like forsterite and enstatite are present along with much less refractory components like iron-magnesium sulfides and organics. Even a CAI-like[7] composition has been observed for one grain. Another remarkable finding is that the compositional evidence seems to exclude acqueous alteration in comet Wild 2, contrary to the case of most carbonaceous chondrite meteorites — see Sec. 8.1.2. Finally, even though the isotope ratios of the most abundant elements show a wide scatter between individual

[7]CAI stands for Calcium–Aluminum-rich Inclusions. These occur in many chondritic meteorites and consist of minerals with extremely high condensation temperatures, often involving calcium and aluminum.

grains, the absence of extreme anomalies indicates that presolar material is rare in comet Wild 2.

1.5.3. *The study of comet 9P/Tempel 1*

The next comet to be investigated by spacecraft flybys was 9P/Tempel 1. It was discovered 150 years ago by the German astronomer Wilhelm Tempel. After the discovery apparition in 1867 it was observed at its next returns in 1873 and 1879 but was then lost for nearly a century. During that time it was sometimes deemed to have disappeared, never to be found again, but the history of comet science knows several examples of a memorable fact: comets should not too easily be declared dead! Comet Tempel 1 was photographed again in 1967 and definitely recovered in 1972 after painstaking calculations by British astronomer Brian G. Marsden and has never been missed since then.

In this case, the reason for the long interruption of the observations has to do with the orbital evolution of the comet. It repeatedly undergoes moderately close approaches to Jupiter in connection with a libration around the 2:1 mean motion resonance with the planet (see Sec. 3.5). This means that the comet spends intervals of about a century on either side of this resonance, and the approaches occur during the transits. When Tempel 1 is outside the resonance (semi-major axis larger than one half that of Jupiter), the perihelion distance is close to 2 AU, and this was the case from the 1880s to the 1950s. In recent times, the comet has had $q \simeq 1.5\,\mathrm{AU}$ like it had, when it was discovered. This librating motion can be traced with reasonable confidence back to a close encounter with Jupiter in 1703 (Yeomans *et al.* 2005), before which the perihelion distance was apparently much larger.

Comet Tempel 1 was chosen as the target of NASA's *Deep Impact* mission. This carried a spacecraft designed to perform close-up observations of the comet nucleus and an impactor, which was disconnected from the main spacecraft a little before the encounter and steered onto collision course with the nucleus. The aim was to excavate nuclear material from some depth to observe its composition and thereby to better understand the relationship between the well-observed coma abundances in many comets and the make-up of the nuclear material from which the coma species originate. The operations were basically flawless. On 4 July 2005, this man-made little cosmic impact took place, delivering a kinetic energy of 19 GJ to the selected place on the comet nucleus.

The Deep Impact (DI) spacecraft would not return to comet Tempel 1. However, on 14 February 2011, this comet was revisited by the Stardust spacecraft, which had been steered to this goal using an Earth swing-by in 2009 and several other trajectory correction maneuvers in what was called the *Stardust-NExT* mission (Veverka *et al.* 2013). The Tempel 1 nucleus is so far the only one that has been visited by spacecraft more than once.

Figure 1.10 shows two pictures of the Tempel 1 nucleus — one from each of the visits. The combination of the two sets of images was important to determine the dimensions of the nucleus as 7.9×4.2 km, since the spin period is relatively long (40.7 hours). Once more, like for the preceding targets, this object was seen to be of low albedo and very large for its observed gas production rate.

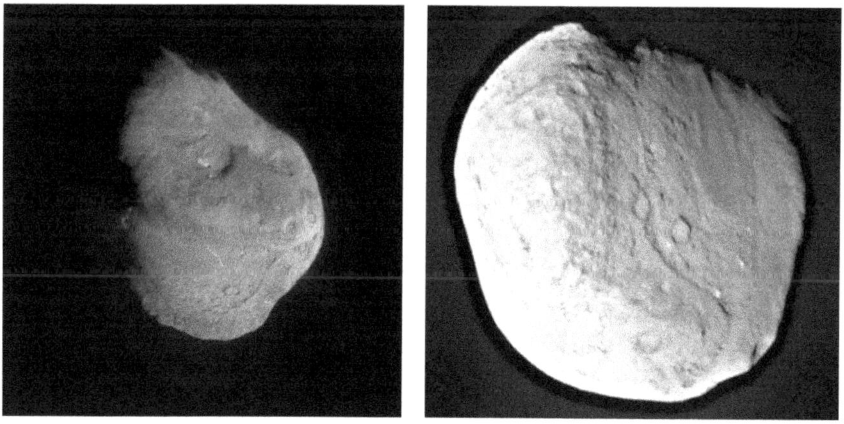

Fig. 1.10. Left panel: Image of the nucleus of comet 9P/Tempel 1, acquired by the Deep Impact spacecraft. Reproduced with permission from M. F. A'Hearn. Credit: NASA/JPL/University of Maryland. Right panel: Image of the same nucleus, acquired more than five years later by the Stardust NExT spacecraft. Credit: NASA/JPL/Cornell University.

It is natural to compare the surface morphologies between the two well-imaged comets, Wild 2 and Tempel 1. Concerning the circular features that are so prominent on Wild 2, such features are also seen on Tempel 1. In this case they appear more subdued and eroded, but an origin as impact craters cannot be excluded. One may wonder, if the surface erosion that has occurred on Tempel 1 since 1703 is on its way to wipe out the memory of a past collisional evolution, which is still prominent on Wild 2.

Let us not forget that round depressions, or pits, can have an endogeneous origin by local, jet activity as well as being impact craters. Time scales are of relevance for this discussion. Can the pits observed on the two comets be formed by collisions with interplanetary boulders in the time available? This time may be limited by both the dynamical residence time in a collisional environment and the surface erosion of comets with small perihelion distances. For Tempel 1, Belton *et al.* (2013) found the impact origin to be unlikely by comparing the number of pits with an expected dynamical lifetime of ~30 000 years or less for Jupiter Family comets. However, the dynamics in question allows for the possibility of extended stays in orbits with large perihelion distance and minimal erosion, so the issue remains open.

Even more attention has been paid to a morphological feature that was discovered on the Deep Impact images of Tempel 1, while evidence could *a posteriori* be traced on the Borrelly and Wild 2 images too. This is referred to as *layering*. Two different manifestations were identified (Thomas *et al.* 2007). One is a set of linear outcrops running more or less in parallel across the nucleus over a length of 3.5 km, forming bands of width ~10–100 m. These are interpreted as surface manifestations of layers penetrating deep into the interior of the nucleus. The other is a region of flat surface with scarps of ~1–10 m height. These scarps are interpreted as erosional features, formed when surface mass loss proceeds along thin layers rather than perpendicular to them.

The near-IR spectroscopic studies made by Deep Impact have brought important information about the water production activity of comet Tempel 1. Temperature maps of the nucleus have been constructed, showing that the surface temperature is far above the level expected for sublimating H_2O ice (like earlier observed for the 1P/Halley nucleus). This means that the flux of water molecules leaving the nucleus in the near perihelion part of the orbit must be maintained essentially by subsurface sublimation. A spectral search for icy patches on the Tempel 1 nucleus by Sunshine *et al.* (2006) revealed only very small areas, far too small to sustain the observed H_2O outgassing. This confirmed the DS1 results from 19P/Borrelly concerning the near absence of surface ice. Thus, active regions on comet nuclei, as often used in theoretical models, do not appear to be identifiable with icy areas of considerable extent.

The above-mentioned temperature maps have been derived using the short-wavelength part of the thermal emission from the Tempel 1 nucleus as captured by the DI imaging spectrometer together with modeling of the physical processes behind this emission. A good fit to the observational data may then imply not only a charting of the surface temperature but also a determination of the thermal inertia and surface roughness, if the observing geometry is favorable (Davidsson *et al.* 2015). The best effort in doing this for Tempel 1 (Davidsson *et al.* 2013) led to the conclusion that significant variations of the thermal inertia occur across the nucleus surface — some regions being measurably rough with extremely low inertia (less than 50 MKS units) and others reaching 3 to 4 times larger inertia values depending on roughness (though still very low). A low thermal inertia essentially means a low conductivity, which in turn implies a loose, fine-grained material with little contact between the grains.

The impact experiment brought several interesting results. Volatiles like H_2O and CO_2 were released in large amounts in the earliest ejecta, showing that they exist close to the surface. Organics were also abundant in the ejecta. The large volume of very fine particles that was ejected shows that these grains are an integral part of the nuclear material — either as individual units or as building blocks of loose aggregates. Moreover, this fine-grained material must exist at considerable depth as well as at the very surface. Finally, the analysis of the ballistic motion of the ejecta revealed that the material had almost vanishing strength, and that the bulk density of the nucleus was very low, estimated at $\sim$400 kg/m^3. While of great importance, both the last mentioned results essentially confirmed already published findings (see Sec. 2.5.3).

1.5.4. *Observations of comet 103P/Hartley 2*

Comet 103P/Hartley 2 was discovered by English astronomer Malcolm Hartley in March 1986. He was using the Uppsala Schmidt telescope at Siding Spring Observatory in Australia within a program forming part of the US-based Catalina search for Near Earth Objects. Comet Hartley 2 is indeed such an object, whose perihelion distance at the time was 0.95 AU. Soon before its discovery, it had completed the third perihelion passage in such an orbit, but in 1971 it underwent a close encounter with Jupiter, which brought it closer to the Sun from a preceding perihelion distance of about 1.5 AU. Before another encounter in 1947, the perihelion distance was more than 2 AU (Carusi *et al.* 1995).

The main spacecraft of the Deep Impact mission had enough instrumentation left in good shape after the passage of comet Wild 2 that it was worth pushing it toward a second cometary target. Thus an extended investigation was approved under the name *EPOXI* (first called Dixi), and comet 103P/Hartley 2 was the target within reach. Among the cometary space mission targets thus far, it was probably the least explored, but its relative freshness from a dynamical point of view was of course a positive aspect. The encounter took place on 4 November 2010, one week after perihelion, and the main results came from the imaging and spectroscopic studies.

A picture of the Hartley 2 nucleus is shown in Fig. 1.11. This nucleus was known in advance to be quite small, and a good size determination using the Spitzer Space Telescope had been published (Lisse *et al.* 2009). Thc EPOXI imagery showed a bi-lobed shape of dimensions 2.33×0.69 km. The albedo is only 0.04 in spite of its fresh appearance. Ejection of material is prominent from the Sun-facing, minor lobe.

When comparing the water production rate of comet Hartley 2, as measured during the 2010 apparition, with the size of the nucleus, a

Fig. 1.11. Image of the nucleus of comet 103P/Hartley 2, acquired by the Deep Impact spacecraft within the EPOXI mission. Reproduced with permission from M. F. A'Hearn. Credit: NASA/JPL/University of Maryland.

remarkable feature stands out. This comet is hyperactive, producing significantly more H_2O than an equivalent sphere of pure H_2O ice at the same distance from the Sun. In this regard, it is quite different from all the other cometary spacecraft targets. The reason for the hyperactivity was revealed by the EPOXI observations. The prominent activity emanating from the small lobe is driven by sublimation of CO_2, which consequently is abundant in the near-surface material, at least in this part of the nucleus. This causes the release of large grains ("chunks") of nearly pure H_2O ice, which subsequently sublimate — thereby contributing a large part of the observed H_2O production rate (A'Hearn *et al.* 2011).

Hence, it is obvious that the outgassing activity of the Hartley 2 nucleus is not primarily indicated by the water production rate. Water was found to sublimate from the nucleus too, but mainly from the waist region and probably resulting from icy grains that were released from elsewhere and fell back to this area. Some degree of chemical heterogeneity, apparently between the two lobes, has been suggested with a variation of the CO_2/H_2O ratio by a factor two. The large abundance of CO_2 is remarkable, especially relative to CO. The latter molecule is less than 1/60 as abundant as the former in comet Hartley 2, indicating that the material is much more oxidized than what has been observed in other comets (A'Hearn *et al.* 2012).

The satellite-based exploration of comet Hartley 2 during its 2010 apparition led to another remarkable result. This was the discovery, using the Herschel Space Observatory, of an ocean-like deuterium/hydrogen ratio in the water molecules of the inner coma. The submillimeter line of two isotopic variants of H_2O (HDO and $H_2^{18}O$) were observed simultaneously, and a value of 500 ± 50 was assumed for the $H_2^{16}O/H_2^{18}O$ ratio. From this, the D/H ratio was found to be consistent with the Vienna Standard Mean Ocean Water (VSMOW) value (Hartogh *et al.* 2011). Such ratios had been observed in several other comets, starting with 1P/Halley in 1986, and these had shown a more or less common behavior with an average D/H ratio close to a factor two larger than VSMOW. Since all the earlier observed comets belonged to the groups associated with the Oort Cloud, while Hartley 2 may rather be referred to the Scattered Disk (see Sec. 1.4), the difference in D/H ratio might reflect a difference in formation conditions at different distances from the Sun — a fundamental issue when discussing comet origins. We shall return to this in Sec. 8.1.3.

1.5.5. *The scrutiny of comet 67P/ Churyumov-Gerasimenko*

The ESA Rosetta mission to comet 67P/Churyumov-Gerasimenko represents a quantum leap in cometary exploration. Some of its results will be referred to in the following chapters, and here we only give a brief introduction. Comet 67P was discovered on photographic plates in 1969 by Soviet astronomers Klim I. Churyumov and Svetlana I. Gerasimenko, who were active at Kiev University. This is in most respects an average Jupiter Family comet except for the fact that it has been scrutinized by the Rosetta instruments, yielding information far beyond what has been obtained for any other such comet. Its perihelion distance at discovery was 1.29 AU and is now slightly smaller.

A close encounter with Jupiter in 1959 had transformed the orbit from a preceding perihelion distance of 2.76 AU. In this sense, the comet reminds us a bit of 81P/Wild 2. The latter case was more extreme, but comet 67P may also be seen as a relative newcomer in its present orbit. In view of the official aim of the Rosetta mission — to read the early history of the solar system from its imprint on the comet nucleus like the Egyptian hieroglyphs were interpreted due to the trilingual inscriptions on the Rosetta stone — this kind of freshness is definitely an attractive property. In this sense, comet 67P is an improvement over the original Rosetta target,[8] comet 46P/Wirtanen, whose recent dynamical history is more quiescent.

The mission scenario in itself is quite remarkable. After launch in March 2004 the spacecraft had to travel for ten years through interplanetary space, using three gravity assists at close encounters with the Earth and one with Mars. Thus, its heliocentric orbit was made similar to the comet orbit, and a very slow approach was performed relatively far from the Sun in 2014, after which a rendezvous followed, lasting about two years. Science data were obtained at the planetary encounters and especially at two main belt asteroid flybys: with (2867) Šteins in September 2008 and (21) Lutetia in July 2010.

The rendezvous with comet 67P started with arrival in August 2014, when the comet had passed its aphelion and was slowly approaching the Sun. The heliocentric distance was then 3.7 AU. Orbit insertion was finished within a few weeks. In November 2014, at 3.0 AU from the Sun, a lander called *Philae* was dropped onto the nucleus surface. The comet was then still of low activity, and the gradual increase until perihelion passage in August 2015 and decline thereafter were monitored from the orbiter with multiple instruments. The scientific use of Philae, however, was severely hampered by an unfortunate failure at touchdown, which made it

[8]The reason for the change of target comet was an Ariane 5 launch failure shortly before the planned Rosetta launch, which was thus delayed by about one year.

bounce a few times while travelling a long distance across the nucleus and come to rest without radio contact with the orbiter. For almost the whole remaining part of the mission, its exact location remained unknown. The mission ended spectacularly on 30 September 2016, as the orbiter was steered into a hard landing on the nucleus in a controlled manner.

One of the most remarkable properties of this nucleus is its binary nature, which is clearly displayed in Fig. 1.12 and will be a topic of discussion in Sec. 7.3.1. The most commonly used analogue from real life experience is a rubber duck — thus, the small and large lobes are often called the head and the body, respectively, while the connecting region is called the neck. Moreover, an intricate and varied surface morphology is seen to characterize the whole object. Some of these features, including temporal and episodic changes, will be presented

Fig. 1.12. Image of the nucleus of comet 67P/Churyumov-Gerasimenko, acquired by the OSIRIS camera onboard the ESA Rosetta spacecraft on 6 August, 2014. Credit: ESA/Rosetta/MPS for OSIRIS Team MPS/UPD/LAM/IAA/SSO/INTA/UPM/DASP/IDA.

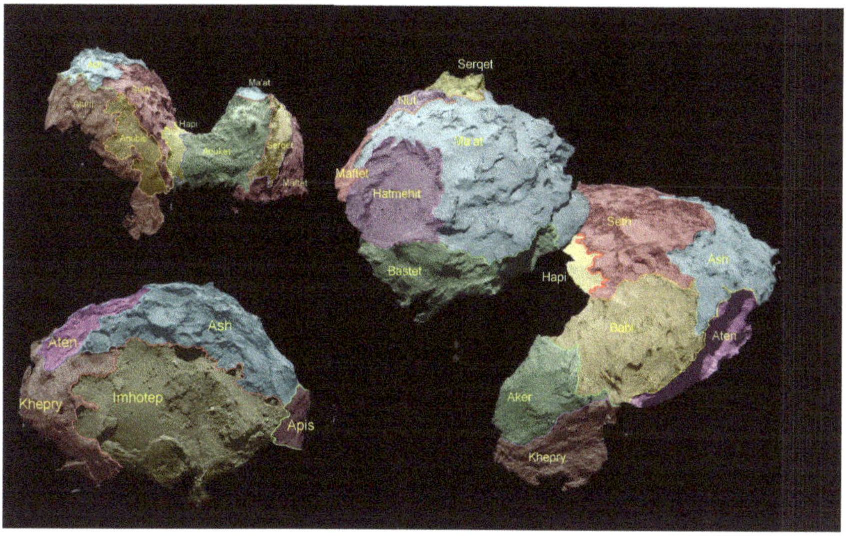

Fig. 1.13. Regional definitions for the 67P nucleus, corresponding to large-scale morphologic provinces. From Thomas, N. *et al.*, *Science* **347**, aaa0440 (2015). Reprinted with permission from AAAS.

in Chaps. 2 and 4. To prepare for those discussions, Fig. 1.13 shows the morphologic provinces defined by Thomas *et al.* (2015a), which are used to describe the different geologic formations and localize the individual features. This picture comes from the initial exploration by the OSIRIS cameras and thus shows only the part of the nucleus that was sunlit at that time.

Chapter 2

Physical and Chemical Properties

The first topic to discuss concerns the observational evidence regarding comet nuclei. Physical modeling is of importance to elucidate some of the properties and determine the parameters in question. However, the observations are of prime importance and will be dealt with in the first place. These are often difficult to perform or vulnerable to biases and systematic errors, so the results have to be discussed with a critical mind. It is natural for this to be reflected in the following descriptions, which sometimes tend to look a bit pessimistic. However, when we deal with the foundations of the ambitious and comprehensive theories on the role of comets in solar system formation and evolution, it is of essence to distinguish what we actually know from what appears to be favored by indirect evidence, and what we simply have to guess.

2.1. Size Distribution

One of the most important statistical properties of any comet population is the size distribution of the nuclei. This was initially shaped by the processes whereby the nuclei were formed, and it may later have been changed by the physical evolution of these nuclei, thus providing a record of these processes. Clearly, if we knew the size distributions of all kinds of comets with great precision, we would have an invaluable tool to unravel and understand the origin and evolution of comets.

Unfortunately, this is not the case. There are several reasons and the most important is that the detailed observations, from which the sizes and shapes of comet nuclei can be inferred, are very demanding. Hence, these are available only for a minority of comets. In particular, from close-up images of comet nuclei obtained using space missions (Sec. 1.5), the sizes and shapes are reasonably well known without ambiguities, but there are only a handful of comets that have so far been visited this way. This leaves remote sensing observations as the only way to deduce the sizes of all the other nuclei.

2.1.1. *Visual nuclear magnitudes*

Since the size distributions essentially rely on such remote observations, let us now see how these are performed. First of all, given the size range involved, it is hopeless to wait for comets to pass, by chance, close enough for their nuclei to be resolved by Earth-based imaging. We thus have to deal with unresolved objects. The usual way to infer the sizes of such objects in the solar system is by means of photometry at visual wavelengths. We need to observe the *nuclear magnitudes* of comets.

Specifically, the input is photometric data referring to the nucleus, obtained in a standard system and thus expressible as apparent visual magnitudes (m_V). Treating these magnitudes as truly nuclear, they can be used to infer an *absolute magnitude* H_N from the formula

$$m_V = H_N + 5\log\Delta + 5\log r + \delta m, \tag{2.1}$$

where Δ is the distance from the Earth and r is the distance from the Sun, both expressed in AU, and δm is the phase correction of the magnitude. Its dependence on the phase angle ϕ is sometimes called the *phase function*. Due to measurement errors, inaccuracies in the assumed phase function, and the fact that the nucleus may spin and have an irregular shape, the values of H_N obtained at different times for the same comet may differ, and thus it is preferable to use an average over many observations.

Lacking information about the actual shape, this average can then be assumed to correspond to an equivalent sphere[1] with radius R_N. The relation between R_N and H_N can be expressed as

$$\log\left(p_V \pi R_N^2\right) = 16.85 + 0.4(m_\odot - H_N), \tag{2.2}$$

where R_N is expressed in km. Here, $m_\odot = -26.77$ is the apparent visual magnitude of the Sun, and p_V is the visual geometric albedo of the nucleus (Tancredi *et al.* 2000).

A potential problem with the above procedure is that there is no guarantee that the photometry refers to the actual nucleus. Under many circumstances it is quite likely that the light emanates from both the nucleus and a surrounding cloud of dust. If so, the derived value of R_N will be an overestimate.

The problem is unavoidable, when we deal with real comets that experience activity in parts of their orbits. Usually, the visible activity (i.e., the coma) subsides as the comet moves away from the Sun, and it may even disappear completely. However, when this happens, the comet is mostly far away. The nucleus is a very faint object, and with the limited angular resolution of the imaging, even a sizeable dust cloud would appear starlike. This leaves open the possibility that such a cloud exists in the form of residual grains orbiting around the nucleus since the last period of activity. Such grains were positively identified in the Rosetta target comet (67P/Churyumov-Gerasimenko), as the probe explored the comet in the outer part of its orbit (Rotundi *et al.* 2015).

A second approach to the problem has been successfully pursued in many cases. This is to use the excellent angular resolution of space-borne telescopes and image the inner parts of the dust coma including the central condensation, where the nucleus dwells. For visual observations, the Hubble Space Telescope has been used (Lamy *et al.* 2004) taking advantage of the occasions, when comets pass relatively close to the Earth and the linear measure of each imaging pixel is minimal. By extrapolating the observed brightness of the

[1]This is a sphere, whose cross-section πR_N^2 equals that of the observed nucleus.

dust coma, one can estimate its contribution to the brightness of the central pixel, where the nucleus is situated. When subtracting this, one has to assume that the dust does not obscure the nucleus, but such obscuration would only occur in extremely active comets. Here, comets of low activity are preferentially selected. The apparent nuclear magnitude thus derived is potentially the most reliable of all remote data. However, the phase angle can be substantial, and the accuracy of the absolute magnitude may be limited by the uncertainty of the phase function as well as that of the extrapolated central dust brightness.

Especially when space telescopes are utilized, a detailed advance planning of the observations is of essence. Therefore, short-period comets are much easier to target than the long-period ones. While some observations of the latter have been made, they do not suffice for a study of the size distribution. In fact, the lack of data concerning the nuclear sizes of long-period comets is a serious problem, since it makes it much more difficult to constrain the masses of these comets and hence to estimate the total mass of their source population, the Oort Cloud. More or less, the same problem holds for the Halley Type comets, and the only group that has been well studied is the Jupiter Family.

2.1.2. *Thermal radiation*

This group was recently targeted for nuclear photometry in the mid-IR wavelength region using the Spitzer Space Telescope (Fernández *et al.* 2013). The wavelengths used were near 16, 22 and 24 μm, and thus the detector arrays received the thermal radiation of the comet nucleus and the surrounding dust. Coma corrections were applied, but the comets were at $r > 4$ AU to facilitate this procedure, and residual dust in the central pixel could not be ruled out. It was, however, concluded that the nuclear brightnesses were not overly affected by this.

As seen from Eq. (2.2), for visual photometry the absolute nuclear magnitude constrains the product of the visual albedo and the square of the radius. Since it is rarely possible to determine the albedo independently (see Sec. 2.4), the estimate of the radius generally depends on an assumed albedo. The albedos that have been determined for Jupiter Family comets are very low, so an error in the assumed value can have a large effect. But fortunately, these albedos are rather similar, so using an average (0.04 is the most commonly used value) leads to radius estimates that should be fairly good in terms of internal consistency. However, for the mentioned mid-IR photometry the situation is better, since the quantity involved in that case is the emissivity (ϵ), which should have a higher relative accuracy than the albedo. This is because, as indicated by Kirchhoff's law, a low value of p_V implies a value close to unity for ϵ. Therefore, even if the error in p_V is quite significant in relative terms, the error in ϵ should be less important.

The constraint from mid-IR observations is obtained as follows. The thermal continuum flux density at frequency ν can be written

$$F_{\rm th}(\nu) = R_N^2 \epsilon(\nu) \frac{\Phi_{\rm th}}{\pi \Delta^2} \oint_{2\pi} B_\nu(T_s) d\omega, \tag{2.3}$$

where $\Phi_{\rm th}$ is the phase function at the relevant frequency. The integral over cometographic solid angle $d\omega$ is taken over the Sun-facing hemisphere of the nucleus. Here, $B_\nu(T_s)$ is the Planck function at frequency ν evaluated for the surface temperature T_s. The expression of $\Phi_{\rm th}$ in terms of thermal infrared magnitude $m_{\rm th}$ would be $\delta m_{\rm th} = -2.5 \log \Phi_{\rm th}$.

Accordingly, the solution for R_N requires a thermal model, from which a surface temperature map is obtained. Inherent in this are several additional parameters: at the very least, the Bond albedo, the bolometric emissivity, and the beaming factor that expresses the anisotropy of the thermal emission from a rough surface in terms of concentration to the solar direction. Many papers have been written on this topic, and recent works often use more sophisticated models than Eq. (2.3). A thorough discussion of the effects of roughness and thermal inertia is found in (Davidsson *et al.* 2015). While Fernández *et al.* (2013) were able to circumvent some of the modeling uncertainties by independent fitting of beaming parameters, these authors also noted the possible problems caused by the unknown shapes — since the model uses a sphere, it may be inaccurate for non-spherical nuclei.

2.1.3. *Statistical analysis*

There is nowadays a good set of nuclear magnitudes and radii for Jupiter Family comets due to several surveys and data collection efforts. However, the data set of radii is not homogeneous. The close-up imaging provides the basic reference data. Visual absolute magnitudes in general are of varying quality depending on amount of data, orbital coverage, and observational techniques used. Only rarely have they been obtained together with simultaneous IR data for determination of radius and albedo together. The most prominent cases are comets 28P/Neujmin 1 and 49P/Arend-Rigaux, which are known since long ago for their very low activity. For a Table summarizing the basic physical properties of all such comets, see Sec. 2.4.2.

Judging from comparisons of the different data sets on nuclear radii, there are still large uncertainties over the actual radii of most individual comets. This became evident, when Fernández *et al.* (2013) made such comparisons between their own results and those from compilations of data from visual observations. While the average offsets between mid-IR and visual radii are not remarkable, the scatter around the 1:1 line is much larger than expected from non-sphericity or the spread in individual albedos.

Figure 2.1 shows the comparison of nuclear radii from Tancredi *et al.* (2006) and Fernández *et al.* (2013) for 38 comets as an illustration. The worst quality estimates according to the former paper are identified, and these do not stand out as the major culprits for the scatter. Moreover, looking separately at the seven comets of the two best quality classes, these do not stand out as especially good fits. The material does not include any comets visited by spacecraft, so there is no independent verification. It appears that the source of the scatter may be problems with the modeling involved or residual dust contributing to the nuclear brightness.

Both the visual and the mid-IR data sets have been used to derive the size distribution of Jupiter Family comet nuclei. The method used

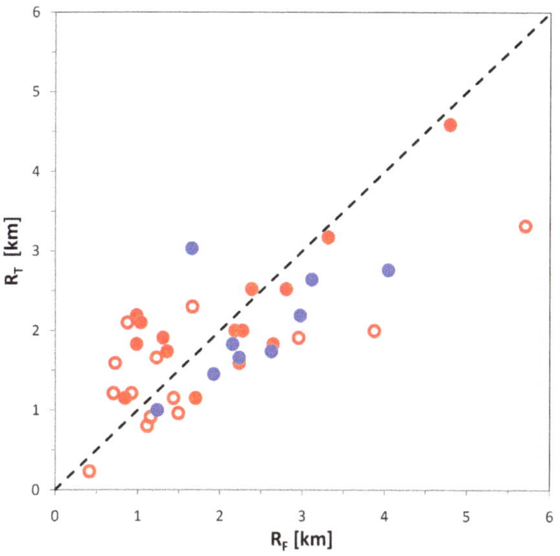

Fig. 2.1. Comparison of radii for Jupiter Family comets, determined independently by Tancredi *et al.* (2006) (R_T) and Fernández *et al.* (2013) (R_F). The dashed line indicates a 1:1 regression. Filled red circles show comets with R_T of quality class 3, and open red circles concern comets of quality class 4 (the lowest quality). The blue circles indicate quality class 1 or 2 (the best quality). Courtesy T. Wiśniowski.

is always basically the same. The cumulative distribution of nuclear radii is imagined to be well fitted by a power law, and the task is to fit the index (α) of this function according to

$$N(R > R_o) \propto R_o^{-\alpha}. \qquad (2.4)$$

The straightforward way to do this is to plot $\log N$ versus $\log R_o$, fit a straight line to these data, and measure its slope.

However, there are problems with this method too. One problem is that the data set of nuclear radii may be inhomogeneous and of varying quality, causing errors in the horizontal location of the points to be fitted. Another problem is caused by incompleteness of the observed sample of comets, possibly causing a radius dependent discovery bias. A third problem is that, at the level where N is just a few (i.e., for the largest observed comets), the sampling of the parent distribution is obviously insufficient. Yet another, more fundamental problem is that the parent distribution may not be an exact power law. If there was a very rich set of data points of high accuracy, this might not be a problem, since the actual shape including wiggles or waves could then be observed. But with a rather small amount of low-quality data, this may actually be a serious source of error.

Keeping in mind all these caveats, it is not very surprising that the record of published α values is rather discordant. In short, from visual data bases used during the last 20–25 years, the range of α extends from about 1.5 to 2.5, and there is no telling where the truth is. From their mid-IR survey, Fernández *et al.* (2013) made an ambitious search for the best power-law index (while acknowledging some evidence for deviations from a power law), which resulted in a preferred value of 1.9.

An important piece of indirect evidence was recently obtained from the NASA *New Horizons* mission. From studies of the impact craters on Pluto and Charon, Singer *et al.* (2015) concluded that the small projectiles (about km-sized) causing such craters have a size distribution with $\alpha \approx 2.3$. These projectiles belong to the trans-neptunian population, which includes the likely source of most Jupiter Family comets (see Sec. 5.4). Hence, there is good reason to assume that these comets start out with this value of the size distribution index. The index pertaining to the actual Jupiter Family comets, after some aging and erosion is probably not much different.

This is where the issue stands regarding the observational constraints on the size distribution of comets. Related issues concerning the number of comets in the solar system and their total mass, and the evolution of these quantities, will be discussed later (Chaps. 6 and 7).

2.2. Brightness and Gas/Dust Production

2.2.1. *Apparent total magnitude*

The most common measure of the activity of comets is found in the sets of visual magnitudes observed at different times during their apparitions. These represent the integrated brightness of the coma including the central condensation, to which the nucleus may also contribute. The wavelength coverage is generally broad, and typically, the eye is the detector, unaided or using binoculars or the eyepiece of a telescope. Photographic and CCD data also exist in large quantities, but these typically fail to incorporate the whole coma. An advantage of these data sets is that the activity is often easily monitored, as the comet moves through the near-perihelion part of its orbit. For a periodic comet, it is also possible to compare its behavior on successive orbits and look for evolutionary changes.

However, there are also drawbacks. In contrast to other extended astronomical objects like gaseous nebulae or galaxies, which generally stay the same all the time, comets keep changing as a result of their orbital motions. They move across the sky, and they brighten or fade — sometimes quite rapidly. Therefore, in a sense, it is like observing different objects from month to month or night to night, and one cannot go back and reobserve to improve the accuracy of a measurement. This is particularly troublesome, since the comet magnitudes are susceptible to uncertainties caused by observational circumstances and observer-specific biases.

The reason for the problem is that the total magnitude should ideally encompass a rather large, diffuse object whose surface brightness varies strongly from the center to the outskirts. An important fraction of the light may be missed, if the comet is observed with too high magnification, or the outer coma is lost in the sky background. These effects would always lead to underestimates of the integrated brightness, but in reality there are other problems too, affecting the measurements by most individual observers. Hence, even if lots of data are available (which is not often the case), deriving an accurate light curve in terms of total magnitude versus time during a cometary apparition is far from being a trivial task.

Nonetheless, these data provide the only way to determine quantitatively, how complete the observational record is, and to set a standard for counting the numbers of comets in different populations. This is because it is essentially the total magnitude of a comet that determines its chances to be detected or discovered. Consequently, the data have to be carefully selected and seriously discussed. The light received remotely from a comet coma consists of two parts: sunlight scattered from the dust grains, and what is traditionally interpreted as fluorescent emission from a few radicals (molecular fragments) populating a region of radius $\sim 10^5$ km around the nucleus. The most prominent of these are C_2, C_3 and CN, each shining in its particular wavelength bands. The so-called Swan bands of C_2 are often dominant. The ratio between the dust and gas contributions is variable, but generally they are of similar magnitude.

2.2.2. *Absolute and heliocentric magnitudes*

It is obvious that the total apparent magnitude of a comet should depend on its distances from the Sun and the Earth. Correcting for this dependence will lead to an *absolute magnitude*, revealing the intrinsic brightness of the comet. As expected, the flux received on Earth turns out in most cases to fall off as the inverse square of the geocentric distance, corresponding to a term $5 \log \Delta$ in the expression for the total magnitude like in Eq. (2.1) for the nuclear magnitude. However, for comets approaching the Earth within 1 AU it may happen that the angular extent of the coma increases so much that the photometry suffers from the same problem as when using a too high magnification. The apparent brightness then varies with Δ more slowly than the second power. This is sometimes called the *Delta effect.*

The dependence on the heliocentric distance is less obvious. A standard correction that is often applied to the total magnitude uses a term $10 \log r$, which corresponds to a fall-off of the brightness as r^{-4}. This has some support in the following argument. We consider the comet to be so close to the Sun that the entire solar energy input to the nucleus (decreasing as r^{-2}) goes into sublimation of ice with associated outflow of gas and dust. Within a short time, the parent molecules of the luminous radicals get dissociated, so thc amount of dust grains and radicals in the visible coma also drops as r^{-2}. But the flux of radiation emanating from a given amount of such material is proportional to the flux of sunlight that it receives — at least in the case of fluorescent emission by the gas. Thus we have two factors r^{-2} that should be multiplied.

The resulting formula for the total magnitude (m_1) of a comet is:

$$m_1 = H_{10} + 10 \log r + 5 \log \Delta, \tag{2.5}$$

where r and Δ have to be expressed in AU. Formally, this would mean that H_{10} is a *total, absolute magnitude* of the comet, expressing its intrinsic brightness reduced to a distance of 1 AU from both the Sun and the Earth. Contrary to Eq. (2.1), there is no phase term in Eq. (2.5). This is natural, because a phase effect would only be present in the dust contribution, and it would be difficult to characterize without detailed knowledge of the grain properties.

However, the H_{10} magnitudes are of dubious value. On the one hand, the argument for using the r^{-4} dependence is weak at best. Several points can be made. The use of r^{-2} for the production rate of gas and dust is a serious oversimplification for any individual comet, as evidenced over and over again by the shapes of the actual, accurate light curves and the direct observations of gas production curves. After a lot of progress during the last decades and, in particular, the exploration of comet 67P by Rosetta, we have some insight into the mechanisms behind the real production curves, and the use of a r^{-2} law should now be regarded as obsolete.

On the other hand, the second factor r^{-2} is suspect too for different reasons. The dust contribution to the coma brightness must depend on the scattering phase function of the optically dominant grains. When comets are observed near opposition, this is not very important, but many times bright comets are observed not very far from solar conjunction, and their brightness may be enhanced by a tendency for forward scattering by the smaller grains. The gas contribution, according to recent Rosetta results (Bodewits *et al.* 2016) may not be realistically modeled as pure fluorescence, since the excitation of the radicals is often dominated by electron impact. Thus, the question remains, whether the number density of electrons in a typical coma varies as r^{-2} or not.

Moreover, the bulk of experience on comet light curves has shown that the more general formula

$$m_1 = H + 2.5n \log r + 5 \log \Delta \tag{2.6}$$

may rather be used with $n \neq 4$. In this case, H is still an absolute magnitude referring to the total brightness of the comet extrapolated or interpolated to $r = \Delta = 1$ AU. Of course, the predictive power of Eq. (2.6) is limited by the need to determine the *photometric index* n individually for each comet or estimate it by some other means.

There is another quantity that is of more direct use, namely, the *heliocentric magnitude* (m_h):

$$m_1 = m_h + 5 \log \Delta, \qquad (2.7)$$

which measures the brightness of the comet at the respective position in its orbit, only corrected for the geocentric distance. This should somehow be related to the gas production rate of the comet and may therefore be used as an indicator of the gas production curve. In particular, Jorda *et al.* (1992) showed that there exists a correlation between the heliocentric magnitudes of comets and their measured H_2O production rates ($Q[H_2O]$) with a regression formula

$$\log Q[H_2O] = 30.74 - 0.24 m_h. \qquad (2.8)$$

The error bars on the numerical parameters are small. However, the sample of comets was small (13 objects), and water production rates were only generally known from the OH lines at 18 cm observed with the Nançay radio telescope. Equation (2.8) implies that the water production rate varies as the 0.6 power of the intrinsic coma brightness, in rather good agreement with the value of 0.5 suggested by the above theoretical argument.

Naturally, to provide the outgassing rates of gas molecules and the release rates of dust, direct observations are much superior to the visual magnitudes. The problem with the gaseous species, i.e., the parent molecules of nuclear origin that give rise to the observed radicals, is the opacity of the Earth's atmosphere for most of the critical transitions at infrared and sub-mm wavelengths. To this comes the limited access to space telescopes with the necessary cryogenic equipment, especially for monitoring purposes. Monitoring can often be made for the radicals by ground-based narrow-band photometry or spectroscopy, but it is often difficult to relate the observed production rates to those of specific nuclear parents. In practice, the production rates of the radicals are often translated into water production rates by a simple, empirical correlation coefficient.

2.2.3. *Dust production rates*

As regards the dust, the situation is even more complex. The coma can be imaged using filters that transmit the continuum with minimal contamination by gas emission, so that the flux is attributed to the totality of dust grains within the field of view through scattering of sunlight. In this case, the $Af\rho$ parameter (A'Hearn *et al.* 1984) is generally used to constrain the integrated cross-section of all these grains via the filling factor f.[2] This parameter is easily derived from the observed quantities, where ρ is the radius of the field of view at the comet. A good estimate of the grain albedo A thus allows us to find the integral of the grain area over the entire population. Clearly, this is somehow related to the dust mass production rate, but conclusions about the latter are hampered by strong model dependence.

There are two features that would need to be taken into account but cannot be straightforwardly assumed. One is the *dust size distribution* (DSD), and the other is the expansion velocity of the dust as a function of grain size. Observational estimates of the DSD are generally made by analyzing the brightness distribution over cometary dust tails, in either visible or infrared light. As a model for the differential DSD, a power law with index β is used for the radius interval under consideration. Here, the equation used is analogous to Eq. (2.4) in Sec. 2.1.3 with $\beta = \alpha + 1$.

This issue has been reviewed by Fulle (2004) with the following conclusions. Oversimplification of the dynamical model used is extremely risky, as was seen during the ESA/Giotto flyby of comet 1P/Halley in 1986, when the spacecraft was hit by a grain far too large (its mass was ~1 g) to be compatible with the advance predictions. Nowadays, Monte Carlo methods are preferred to invert the brightness distribution with the use of a full description of the

[2]This is defined to be the fraction of the solid angle in the field of view, where the line of sight crosses a grain.

dust dynamics. Resulting values of β tend to be in the range between 3 and 4, implying that the total dust mass is dominated by the largest grains, which may actually be meter-sized boulders. Even the scattered, visual light from parts of the dust tails may involve large chunks to an extent that previously seemed hard to imagine.

The actual dust mass loss rate will of course depend on the maximum size of the ejected particles. In comet 1P/Halley, during the Giotto flyby, the dust impact detection system (DIDSY) detected the mentioned 1 g particle, thus showing that chunks of at least this size were ejected. Thus, the ratio of dust to gas mass loss rate from the nucleus — the so-called *dust/gas ratio* (DGR) — was at least of order unity, which came as a surprise in view of earlier, much lower estimates. Since that time, the signatures of large coma particles in some comets have been measured by both radar at cm wavelengths and radio observations at mm wavelengths. High dust mass loss rates have thus been derived, implying large DGRs in the observed comets. In addition, similar conclusions have come from studies of *dust trails* associated with several comets, observed at infrared wavelengths. These trails consist of large grains that have left their parent comet at low speeds and slowly drifted away from it counter to the direction of orbital motion. Sykes and Walker (1992) concluded that a mass ratio of about 3 between ejected refractories and gas was indicated by the dust trail observations.

Recently, dust measurements by Rosetta instruments in the coma of comet 67P were analyzed (Rotundi *et al.* 2015), revealing that at 3.4–3.6 AU inbound, the ejected grains had diameters up to $\sim$2 cm, and the larger chunks in orbit around the nucleus since a time near the preceding perihelion were sometimes meter-sized. These estimates translate into a DGR of $\simeq$3 near perihelion and $\simeq$4 for the material ejected in the mentioned part of the comet orbit. The DWR (dust/water ratio) would be $\simeq$6, but Rosetta/ROSINA data for CO and CO_2 had shown these molecules to contribute $\sim$50% in mass relative to H_2O (Hässig *et al.* 2015).

These results may imply a paradigm shift in comet science. Before the Halley exploration, comet nuclei were described in popular wording as "dirty snowballs", and the "dirt" — i.e., the dust — was generally considered a minor contamination. Afterwards, the term "snowy dirtballs" has occasionally been used in a somewhat provocative way, and this may be more to the point. But two issues should be kept in mind. First, the number of individual comets for which large DGRs seem confirmed remains small, and there is no way yet to say how the DGR varies between different kinds — say, between periodic comets and fresh, long-period comets. Second, the surface layers of comet 67P are heavily processed as a result of insolation, ice sublimation and grain redeposition (so-called *airfall* — see Sec. 4.2.1) during many orbits, and the same is likely true of most periodic comets. This makes it risky to conclude that a large DGR of the material currently leaving the surface can be interpreted as a large ratio of refractories to ice in the bulk of the nucleus.

2.3. Light Curves and Gas Production Curves

One of the most obvious ways to study the nature of comet nuclei is to observe in detail, how the activity of the comet responds to the changing insolation, as the comet approaches the Sun before perihelion and recedes afterwards. This was the basic, direct goal of the Rosetta mission: to learn "how comets work" by escorting comet 67P from far out through perihelion and thereafter, observing from close range how the activity developed and the nucleus surface changed. As often emphasized, the Rosetta results are truly remarkable and deeply impressive, but strictly speaking, they refer to one comet only. The major scientific harvest of Rosetta calls for a way to generalize the findings, to put 67P into a wider context, and thus to understand much better, not only how this comet works but how all comets work. An Earth-orbiting telescope with UV and IR detectors primarily aimed to monitor the activity variations of many other comets is an example of a facility that could provide a good match to Rosetta but has not yet been realized.

Therefore, we have to make the best use of what we have, and this means in particular to study comet light curves along with evidence from imaging and measurements of production rates of gas species and dust grains. Figure 2.2 illustrates, as an example, the heliocentric

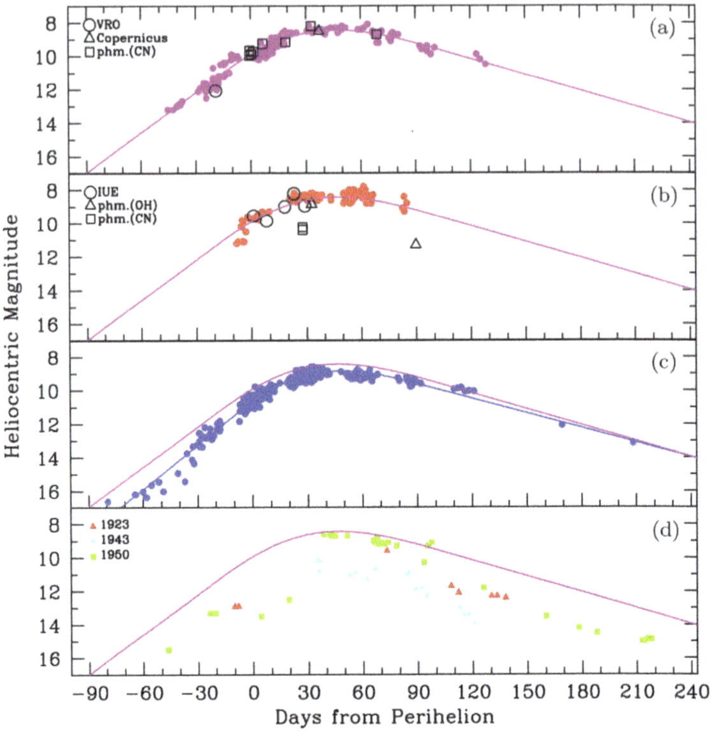

Fig. 2.2. Observed light curve for comet 6P/d'Arrest as observed in 1976 (panel a), 1982 (panel b), 1995 (panel c) and during three earlier apparitions (panel d). The magenta and blue curves show polynomial fits to $m_h(t)$ in 1976 and 1995, respectively. Reprinted from Szutowicz, S. and Rickman, H., *Icarus* **185**, 223–243 (2006), with permission from Elsevier.

magnitude of comet 6P/d'Arrest during three favorable apparitions together with some direct observations of its gas production rates. Similar data are available for many comets although rarely of comparable quality. As seen from Eq. (2.8), the plotted curve directly translates into the variation of the log of the H_2O production rate with time.

2.3.1. *Isothermal model*

To see how this relates to the properties of the nucleus, let us first consider a very simple physical model. We assume that the nucleus is spherical and that its surface is smooth and covered with H_2O

ice. Placing it at a certain distance (r) from the Sun, it is clear that different local elements of this surface receive different fluxes of solar radiation, corresponding to the local zenith distance (ζ) of the Sun. However, we disregard this circumstance and use a common solar zenith distance all around the nucleus. The projection factor that governs the energy absorption rate by the surface is $\cos\zeta$, and its spherical average is 1/4. This is sometimes inaccurately called the rapid rotator approximation, because one may imagine a nucleus spinning so fast that its thermal inertia causes the whole surface to have the same temperature (T). Calculating this temperature, one should then use the average of $\cos\zeta$. Another, more accurate name for the procedure is hence the *isothermal approximation.*

The energy absorption rate should be balanced by the loss rates. One of these is the thermal radiation of the surface, which is found from the Stephan-Boltzmann law. Another is the latent heat (H) consumed by the sublimating H_2O molecules, and this must be combined with an expression for the sublimation rate (Z) — i.e., the flux of molecules leaving the surface — as a function of temperature. The third loss rate (either positive or negative) would correspond to the heat exchange between the surface and the sub-surface layer, but in the isothermal approximation one also neglects the vertical temperature gradient, so this term vanishes.

According to the simplest gas-kinetic sublimation theory, the flux of escaping molecules can be written in terms of the product of the number density in a saturated gas (n_s) and the average velocity (v_{th}) of the molecules. From this, one derives

$$Z(T) = \frac{p_s(T)}{\sqrt{2\pi mkT}}, \tag{2.9}$$

where p_s is the saturation pressure of the gas, m is the molecular mass and k is Boltzmann's constant. To a good approximation, p_s is an exponentially increasing function of T, and hence, $Z(T)$ shares the same qualitative behavior. Using a constant value for H, the energy balance equation for the surface is

$$\frac{1}{4}F_{\odot}r^{-2}(1 - A_v) = \epsilon\sigma T^4 + H \cdot Z(T), \tag{2.10}$$

where $F_{\odot}$ is the solar constant,[3] A_v is the hemispherical visual albedo and ϵ is the thermal emissivity of the surface. Using appropriate values for A_v and ϵ, and taking the thermophysical data (H and p_s) for H_2O, one can easily solve for $Z(r)$, and the same can be done for other molecules too, if the comet should be composed of other ices.

Examples of such solutions are seen in Fig. 2.3. The exponential behavior of $p_s(T)$ is common to all molecules and leads to a common feature of the curves. Close enough to the Sun, where T is high enough, the sublimation loss term dominates over the thermal radiation. This means that the solar energy flux goes essentially into sublimation, and the sublimation rates tend to fall off as r^{-2}. On the other hand, when the comet is far from the Sun, the thermal radiation dominates over sublimation. Consequently, T falls off nearly as $r^{-1/2}$, causing an exponential drop of the sublimation rate. The transition

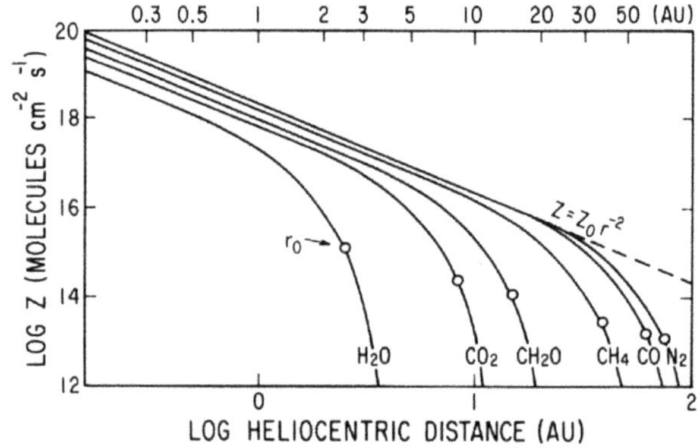

Fig. 2.3. Sublimation rates for a selection of cosmically important molecules versus heliocentric distance, plotted in a log–log diagram. A dashed line indicates a fall-off as the inverse square of the distance. The open circles marked r_0 indicate the knees, where the sublimation and thermal radiation heat losses are equal. See Delsemme (1982). From *Comets*, edited by Laurel L. Wilkening. © 1982 The Arizona Board of Regents. Reprinted by permission of the University of Arizona Press.

[3]The solar constant is the amount of energy passing per unit time through a unit surface perpendicular to the solar direction due to solar radiation at all wavelengths, at a distance of 1 AU from the Sun.

between these two regimes is marked by a knee on the curve, and the location of this knee depends on how volatile the substance is, i.e., how large its saturation pressure is at a reference temperature.

As one can see, H_2O is less volatile than all the other selected, cosmically abundant substances. A comet made of H_2O will start sublimating appreciably at a distance around 2–3 AU from the Sun, while comets made of the other species would become active much further away. Here is the reason why comets were considered to be made mostly of water ice long before this could be verified by spectra or in situ exploration. Their activity tends, in a statistical sense, to commence and subside at heliocentric distances close to the knee of the H_2O curve.

Thus, one may imagine that this H_2O curve may yield at least a crude fit to the observed H_2O production curves of different comets. For each comet, it would be possible to multiply $Z(r)$ by the surface area of a sphere with the same radius as the comet nucleus to get an idealized production curve for a fully active nucleus. This would be shifted by some amount from the observed curve, and the shift could be interpreted to mean that the activity is limited to some constant fraction of the surface. Hence, the fitting would mean to find the relevant value of the active fraction. However, reality is not so simple.

2.3.2. *Seasonal variation*

The thermal model of Eq. (2.10) is of course extremely crude and may only be useful to arrive at general conclusions like the importance of H_2O. However, for interpreting real light curves, it is inadequate. This is obvious from Fig. 2.2 already at first sight. As seen, the light curve of comet 6P/d'Arrest — and, therefore, the H_2O production rate according to Eq. (2.8) — peaks about 40 days after perihelion, while Eq. (2.10) predicts Z to be a monotonously decreasing function of r, and thus the light curve has to peak at perihelion and to rise and fall on either side in a symmetric way. Even if comet d'Arrest is a rather extreme example of perihelion asymmetry, the phenomenon is quite general, and it would be hard to find any comet exhibiting a light curve in full agreement with Eq. (2.10).

How should this fact be explained? Which underlying assumption shall we abandon? Obviously, comet nuclei cannot be absolutely spherical, and their non-sphericity is verified by all the images acquired to date. But it is not obvious that such large disagreements as seen can arise from shape effects only. The idea of a thermal lag, whereby the surface is preferentially cooled by heat conduction toward the interior before perihelion, can also be discarded, since the light curve peaks are often found on the pre-perihelion orbital branch. Another obvious conclusion is that real nuclei cannot be isothermal. However, by simply introducing the temperature contrasts of a day-night cycle, one will not solve the problem of perihelion asymmetry, even though the temperature maps may become more realistic.

A different improvement is much more promising and also verified by imaging: replace the isotropic, more or less icy surface by one where ice exists only locally if at all, and gas production from the remaining area is quenched by an overlying, refractory or ice-depleted layer. In this case, depending on the orientation of the spin axis, seasonal variation in the insolation of the most active (least quenched) spots may cause a strong perihelion asymmetry in the light curve, leading to a maximum either before or after perihelion. Seasonal variations in comet activity were first considered by Paul Weissman (1986).

It is clear that the sources of comet activity are to some extent local and thus in general subject to seasonal effects. Some models of comet nuclei have involved *active spots* of constant size with a freely sublimating icy surface, while the rest is completely inactive. The observed perihelion asymmetry may then be explained by a proper choice of spin axis and spot latitudes. The fit to the observed gas production curve then consists of choosing the best areas of the active spots, and the sum of these areas divided by the total surface area of the nucleus then yields the active fraction.

However, reality is certainly more complicated. On the one hand, the black and white picture of such models with surface elements that are either 0% or 100% active is oversimplified, as shown by the Deep Impact and Rosetta missions (Sec. 1.5). Most of the gas production of the target comets comes from regions with activity levels larger than zero but less than 100%, as one may expect for ice sublimation at shallow depths below the surface, involving partial quenching of the outflow.

On the other hand, given that we can speak of such an activity level as a continuous variable between 0 and 1, there is no reason to think that it should be everywhere the same or that it should stay the same as time proceeds at any given place. Many surface changes on comet 67P during its 2015 apparition have been recorded by the Rosetta/OSIRIS imaging, indicating likely changes in activity level as well. Thus, it appears that comet light curves and gas production curves may reflect a complex interplay of surface phenomena in addition to the fundamental variation of the ice sublimation rate due to the varying heliocentric distance. To some extent, these curves may hence be beyond the predictive capability of simple models, but the bulk features should clearly be understood. For instance, the observed perihelion asymmetries are often quite repeatable from one orbit to the next and should contain important information about the activity distribution and nucleus spin. Other features of similar importance include the width of the light curve peak, and the fall-off slopes on either side as indicated by the photometric index — see Eq. (2.6) in Sec. 2.2.2.

2.3.3. *Remote activity*

One more feature remains to be described. This is the gas-producing activity at large heliocentric distances, which may call for other explanations than H_2O ice sublimation.[4] There are many examples among both short-period and long-period comets. The latter include at least one comet that made the newspaper headlines, namely, C/1973 E1 (Kohoutek). At discovery on 7 March, 1973, this comet was more than 4 AU from the Sun, and its perihelion distance was only 0.14 AU. Considering the heliocentric distance, the discovery magnitude of 15 was quite bright, and extrapolation to perihelion by a standard photometric index led to the prediction of a magnificent

[4]Other species than H_2O can be important close to the Sun as well, as shown by the EPOXI observations of comet 103P/Hartley 2 (see Sec. 1.5.4).

display, which triggered the use of the term "comet of the century" in news media. When comet Kohoutek came close to perihelion in December and January, it was a respectable object reported to be of second magnitude with a long tail, but this did not live up to the early expectations. As a result, most people who remember comet Kohoutek think of it as a failure.

Although we cannot tell for sure what happened to comet Kohoutek, in retrospect its brightness evolution does not appear enigmatic. The activity observed at discovery may have been caused by the release of species more volatile than H_2O. If so, there was no reason for a very rapid increase as the comet approached the Sun. At some point, closer to the Sun, the H_2O outgassing likely became dominant, but the brightness then took off from a lower level than predicted. This behavior might have been similar to the one observed in the famous comet C/1995 O1 (Hale-Bopp) as illustrated in Fig. 2.5 below, even though there is an important difference between the orbits of the two comets: comet Kohoutek was a new Oort Cloud comet, and Hale-Bopp was a returning long-period comet.[5]

When Jupiter Family comets pass through the aphelion part of their orbits at heliocentric distances $\simeq$5–6 AU, they often appear totally inactive but not always. For instance, Licandro *et al.* (2000) reported on a program to observe such comets far from the Sun to determine the photometric cross sections of their nuclei, but seven out of the 18 targets were deemed active, whereof six at $r >$ 4 AU. Figure 2.4 shows two examples of the images obtained: comet 37P/Forbes showing no activity at $r = 3.6$ AU and comet 65P/Gunn exhibiting a large coma at $r = 4.3$ AU. Other observing programs too have reached similar conclusions. In their above-mentioned mid-IR photometric study of Jupiter Family comets (see Sec. 2.1.2), Fernández *et al.* (2013) found extended dust emission in 35 out of 89 detected comets — all at $r > 4$ AU.

The question remains, what does it mean that many comets produce dust comae at large heliocentric distance? Is the dust

[5]The concepts of new and returning comets are explained in Sec. 5.1.

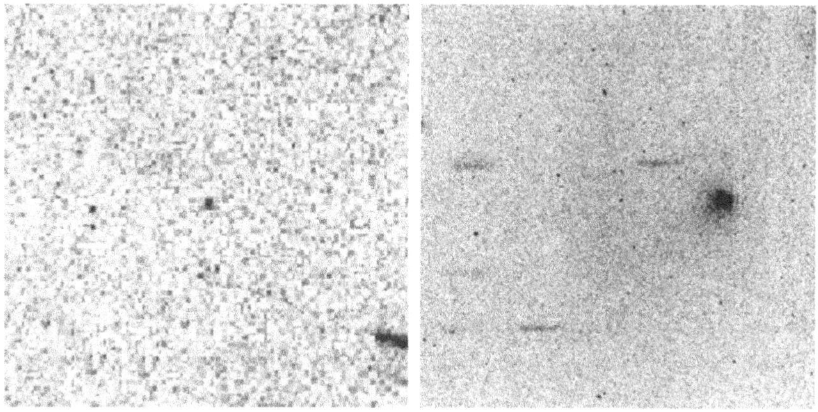

Fig. 2.4. CCD images of comets 37P/Forbes (left) and 65P/Gunn (right), obtained at the ESO/La Silla and Pic du Midi observatories, respectively. Reprinted from Licandro, J. *et al.*, *Icarus* **147**, 161–179 (2000), with permission from Elsevier.

production necessarily a direct consequence of gaseous outflow? If so, how much gas is required? Quite possibly, there may not be a unique relation between the strength of the gas outflow and the amount of dust particles forming the coma. Too little is yet known about how the refractories are lifted off the surface of the nucleus and how the grains are derived from what is likely a coherent, refractory matrix. In any case it is clear that some gas is needed to accelerate the grains away from the nucleus into the observed large, roundish cloud.

The amount of data on gas production rates of distant comets is quite small. Figure 2.5 shows the only impressive data set, obtained from radio astronomical observations of C/1995 O1 (Hale-Bopp) by Biver *et al.* (2002). Of the molecules included, CO was clearly the dominant one, when the comet was at $r > 3.5$ AU before perihelion and at $r > 3$ AU afterward. The $Q(CO)$ production curve at such distances is much shallower than the corresponding curve for $Q(H_2O)$, but closer to perihelion, the CO/H_2O ratio was more constant. This so-called Christmas tree diagram contains lots of information on how various gases may have been produced from this comet in different parts of its orbit. At this point, let us note that one likely, major molecule is missing for lack of observations, namely, CO_2.

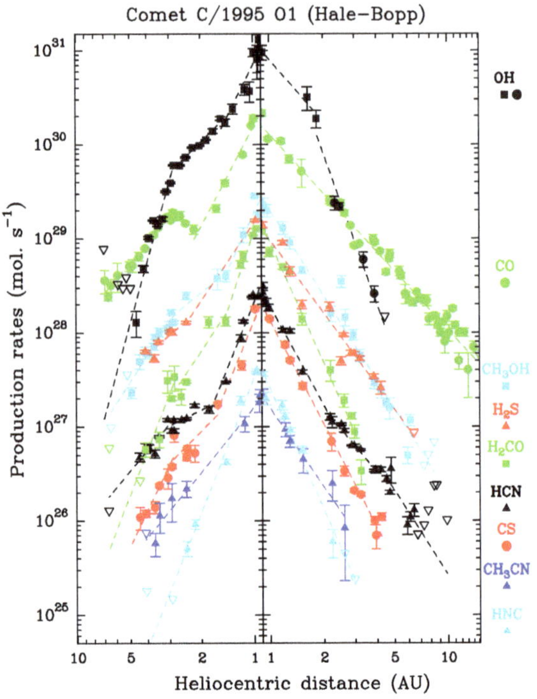

Fig. 2.5. Observed production rates of different molecules in comet C/1995 O1 (Hale-Bopp), plotted versus heliocentric distance in a log-log diagram. The perihelion passage at $q = 0.914$ AU is marked by the vertical line. Pre-perihelion observations are to the left and post-perihelion observations to the right. From Biver, N. *et al.*, *Earth, Moon, and Planets* **90**, 5–14 (2002), © Kluwer Academic Publishers. With permission of Springer.

The CO molecule has been observed in other distant comets too — notably, comet 29P/Schwassmann-Wachmann 1. This comet orbits around the Sun at distances of about 6 AU and is known to exhibit a persistent grain coma. In this case the word "grain" should be preferred to "dust", because the grains are likely composed partly of H_2O ice. This may in fact be a general phenomenon in comets too far from the Sun for such grains to sublimate at an appreciable rate, while gases like CO may drag them into the coma. Comet 29P is also famous for its intermittent outbursts when, apparently, huge quantities of grains are suddenly ejected from the nucleus and the comet brightens up by five magnitudes or more. The reason for these outbursts remains unknown, while there are some clues as to the mechanism of CO outgassing, to be discussed in Sec. 2.6.6.

2.4. Albedo and Activity

A good size determination for a comet nucleus in principle opens up for estimates of two important physical parameters — the albedo and the level of activity. The albedo is then found from Eq. (2.2), using the radius and the visual absolute magnitude. Of course, a good size determination then means a direct one that does not depend on an assumed albedo! This limits the sample to a handful of comets that have been visited by spacecraft or were subject to observing campaigns involving both visual photometry and thermal radiometry, allowing determination of both radius and visual albedo simultaneously. While spacecraft targets can have any amount of activity, the second method is limited to comets with minimal activity, so that the observations approximately refer to the bare nucleus.

In Sec. 2.3 we saw how the word activity can be used to describe the relation between the actual water production rate of a comet and the production rate corresponding to a model surface with the same size as the nucleus, made up by H_2O ice under some assumptions. Some such assumptions were discussed and deemed not to be realistic. This means that the derived parameter, which may be called *active fraction* or *activity level*, is model dependent. It is linked to a physical model for the sublimation rate from the surface — a model that should be chosen so as to imitate the observed H_2O production curve as closely as possible. However, even if a very good fit is found, there is no guarantee that the chosen model is a true description of the nucleus.

In other words, the activity level has no prescribed physical interpretation. Quenching of the gas flow is a reasonable expectation in the case of H_2O production beneath the surface, but this holds only for local outgassing. The global activity level of an entire nucleus depends on other things as well, as we shall see for the case of comet 67P below. The only message conveyed by this parameter is that the comet produces a certain fraction of the predicted H_2O flux from an idealized standard nucleus with ice on its surface.

2.4.1. *Activity and thermal modeling*

Concerning the model that underlies the definition of activity, there are two ways to go. If the comet is very well observed and its H_2O production curve is known in some detail, and if its nucleus is very well imaged so that a digital terrain model of reasonable quality is available, one may pursue the thermal modeling to a very high degree of sophistication. Within the framework of such a model, the resulting activity parameter has a well-defined physical meaning. It thus tells something useful about the comet in question, but it cannot be generalized. However, for most comets, the observational constraints are far from adequate to develop such a model. We must then use a suitably crude model that is reasonably realistic and applicable to the sparse data typically available for the average comet.

The isothermal model is too extreme by underestimating the sublimation rate, unless the comet is so close to the Sun that the insolation is anyway balanced by sublimation and thermal radiation is unimportant. Most Jupiter Family comets do not come close enough for this to hold true. For increased realism, the simplest model (Weissman and Kieffer 1981; Rickman and Froeschlé 1983a) introduces the instantaneous value of $\cos\zeta$ and features another term in the energy balance, corresponding to the heat exchange between the surface and the interior:

$$\cos\zeta F_{\odot} r^{-2}(1 - A_v) = \epsilon\sigma T^4 + H \cdot Z(T) - K\left(\frac{\partial T}{\partial z}\right)_0, \quad (2.11)$$

where K is the thermal conductivity at the surface, and the partial derivative expresses the temperature gradient with respect to depth z counted downward from the surface. This is to be treated as a boundary condition to the heat diffusion equation, which is solved numerically for the temperature $T(z, t)$ using also an interior boundary condition. The latter is usually taken to be isolating, i.e., the temperature gradient is zero at the lower boundary of the computational grid. However, when solving explicitly for the diurnal heat flow that occurs in a very thin surface layer, it may be advantageous to use a non-zero thermal gradient at the bottom, resulting from the orbital heat flow that extends to larger depths.

These equations may be solved for local spots at different latitudes on the nucleus surface. For the diurnal heat flow, the factor $\cos\zeta$ then depends on the orientation of the nucleus spin axis, the latitude and the rotational phase. Seasonal effects generally appear, if attention is focused on certain latitudes. This may thus be useful for investigating comets whose gas production curves have perihelion asymmetry. However, for the definition of activity in the general case, it is better to neglect seasonal effects by assuming the spin axis to be at right angles to the orbital plane.

For an arbitrary spin axis orientation, the poles are often close to the orbital plane, and it is worth considering a model predicting the water flux from a pole facing the Sun as an extreme case (opposite to the isothermal one). In this case, Eq. (2.10) can be used with only one modification — the factor $\frac{1}{4}$ is replaced by 1. In Fig. 2.6 the two extreme models are compared. We see that the difference

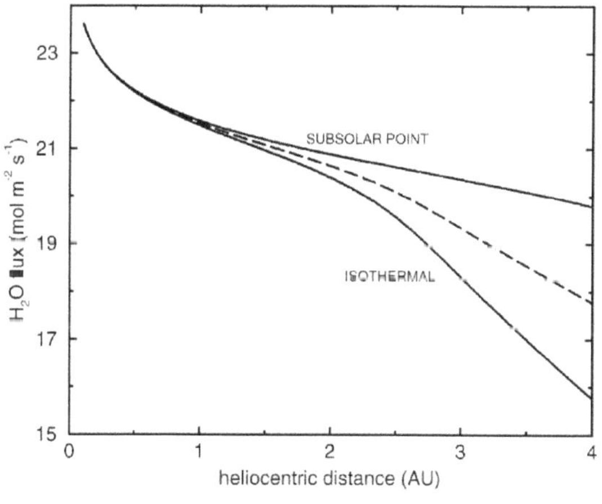

Fig. 2.6. H_2O sublimation flux versus heliocentric distance for an icy surface, computed for a pole facing the Sun (subsolar point) and the average surface of an isothermal sphere, represented by solid curves. The dashed curve is an average of these two models. Reprinted from Tancredi, G. *et al.*, *Icarus* **182**, 527–549 (2006), with permission from Elsevier.

in H_2O flux is moderate or small for $r < 2\,\mathrm{AU}$, while it grows to orders of magnitude for $r > 3\,\mathrm{AU}$. Since the heat diffusion model just described will always be intermediate between the two extremes, we may replace it by a simple average between the latter, if we restrict our attention to gas production rates observed at $r < 2\,\mathrm{AU}$ (also shown in the figure).

Tancredi *et al.* (2006) used this average to derive what they called active fractions for a sample of Jupiter Family comets with estimated water production rates from individual observations (single or in small groups) made at $r < 2\,\mathrm{AU}$ with the aid of their own values for the nuclear radii (see Sec. 2.1.3). These fractions f came from the identity

$$Q(H_2O) = 4\pi R_N^2 \cdot fZ, \tag{2.12}$$

where Z is the average sublimation flux at the heliocentric distance of the observation. Their results are shown in Fig. 2.7. Here we prefer the term activity rather than active fraction to avoid giving the impression of a 100% active area on an otherwise inactive surface.

2.4.2. *Statistical trends*

The most remarkable feature of the above diagram is that the values of f span the whole range from very close to zero up to more than 1. The largest values (and also the longest "error bars") are found for sub-km nuclei, and for $R_N > 2\,\mathrm{km}$ all the activities are less than 0.4. Only three comets have $R_N > 3\,\mathrm{km}$, and all of these have extremely low activities. It hence appears that there may be a correlation such that smaller comets have higher activities, and models of the physical evolution of comets should probably aim to explain this phenomenon.

While most comets that do not belong to the lowest quality class for nuclear radius have $f < 0.2$ (Tancredi *et al.* 2006), there are also a few such comets with much higher activities. Those for which f is consistently larger than 0.8 are: 21P/Giacobini-Zinner, 22P/Kopff and 46P/Wirtanen. The observational material seems solid, except that the radius of 21P (1.0 km) has a rather large error bar, being of quality class 3. There is hence no reason to doubt the existence of

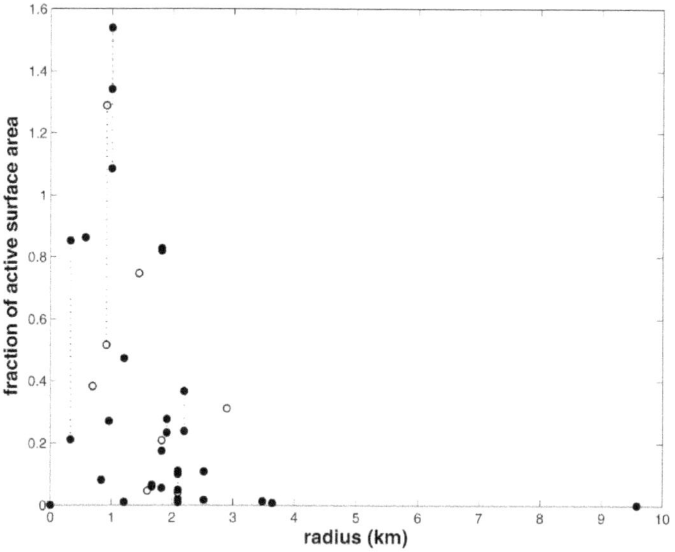

Fig. 2.7. Activities observed for Jupiter Family comets at $r < 2$ AU, using mainly the production rates from the near-UV OH band and the [O(^{1}D)] line according to A'Hearn *et al.* (1995). Open circles denote comets with nuclear radii of the lowest quality class, while filled circles are used for the other three classes. When more than one production rate is used, the individual symbols are joined by a vertical dashed line. The data point at the origin is erroneous and should be disregarded. Reprinted from Tancredi, G. *et al.*, *Icarus* **182**, 527–549 (2006), with permission from Elsevier.

comets that sometimes emit H_2O in quantities similar to what they would do, if their nuclei were icy spheres subject to sublimation due to solar heating.

Another comet of the same kind can be noted from recent observations including the EPOXI mission — comet 103P/Hartley 2 (see Sec. 1.5.4). While the physical mechanism behind the high water production rates cannot be clearly discerned for the three first mentioned comets, the situation is better in the last case, because it was established that much of the water came from sublimating icy grains in the inner coma (A'Hearn *et al.* 2011). In turn, as suggested by these authors, these grains might have been released from the nucleus by the very important CO_2 outgassing that the mission discovered. Of course, it cannot be excluded that other

Table 2.1. Nuclear parameters for comets that have been visited by space missions (sm) or have been subject to simultaneous visual and IR photometry (ph) in a state of very low activity, allowing accurate size determinations. The activities and albedos are global averages around the whole nuclei or their observed parts.

Comet	Nucleus dimensions (km)	Activity near perihelion	Geometric albedo	Source of information
1P/Halley	15.3 × 7.2 × 7.2	0.10	0.04	sm
9P/Tempel 1	7.9 × 4.2	0.03	0.06	sm
19P/Borrelly	8.0 × 3.15	0.3	0.03	sm
28P/Neujmin 1	19.2	0.001	0.03	ph
49P/Arend-Rigaux	9.2	0.007	0.04	ph
67P/Chur.-Geras.	4.3 × 4.1	0.06	0.06	sm
81P/Wild 2	5.5 × 4.0 × 3.3	0.25	0.03	sm
103P/Hartley 2	2.33 × 0.69	1.3	0.04	sm

Jupiter Family comets share the same behavior in the near-perihelion parts of their orbits.

In Table 2.1 we summarize the current estimates of physical properties for the best studied comets. These are not of homogeneous quality. Moreover, the thermal models used for individual comets are not the same. We will discuss comet 67P separately below, since this is by far the best studied case.

The most noteworthy information conveyed by these data is that the different comets have very different activities (like we saw in Fig. 2.7) but very similar albedos. The two quantities are not correlated, which verifies the impression that free surface sublimation is not the prime explanation of comet activity. Had it been so, the high-activity comets would also have had higher albedos due to the presence of ice on much of the surface, even if the albedo of the ice may be lowered by dust contamination.

The reason for the low albedo has not been fully elucidated. Since surface ice is rarely seen, we have two possibilities. Either the surface layers on all comets are made up of very dark material, or the light is trapped in the porous surface by multiple scattering together with absorption. Both effects are likely present. The organic refractories, forming an abundant constituent in comet material, may well contribute importantly. The absence of the red spectral slope seen on small bodies that are considered to be related to comets, like Jupiter Trojans and D-type asteroids, may be explained by the freshness of the continually eroded comet surfaces, in case the other objects are reddened by long-term cosmic ray sputtering.

2.4.3. *The activity of comet 67P*

Dust jets in the inner coma of comets have frequently been used as indicators of local activity and sources of information about the spin of the nucleus. Such an analysis was carried out before Rosetta on comet 67P/Churyumov-Gerasimenko by Vincent *et al.* (2013) and led to the identification of three active regions at latitudes $+60°$, $\simeq 0°$ and $-45°$. These source regions were later localized more precisely by Lara *et al.* (2015) using Rosetta data. However, observing dust features is not the same as tracing gas sources on the nucleus surface.

The body of information on comet 67P from Rosetta is enormous. Activity was observed already during the approach phase in 2014. The MIRO instrument first detected water vapor on 6 June, when the comet was situated 3.92 AU from the Sun. The OSIRIS cameras observed that the nucleus was surrounded by dust already at the end of April and in July detected dust jets emanating from the nucleus at a distance of 3.7 AU from the Sun. Somewhat later, ROSINA detected H_2O, CO_2 and CO molecules in situ in the vicinity of the nucleus.

It soon became clear that the source of this activity was situated in the neck. Figure 2.8 (left panel) illustrates this by showing an early picture of the dust jets. Specifically, the Hapi region was identified. This is an area of smooth-appearing surface running along the neck with big boulders strewn along its central part — see the right panel of Fig. 2.8. It is close to the north pole and experienced a period of relatively high illumination at the time in question, though other areas on the nucleus were receiving even stronger insolation. In fact, the equinoxes of the 67P nucleus occur at $r = 1.7$ AU before perihelion and at $r = 2.6$ AU on the outbound branch. The northern hemisphere experiences a long summer, mostly in the outer parts of

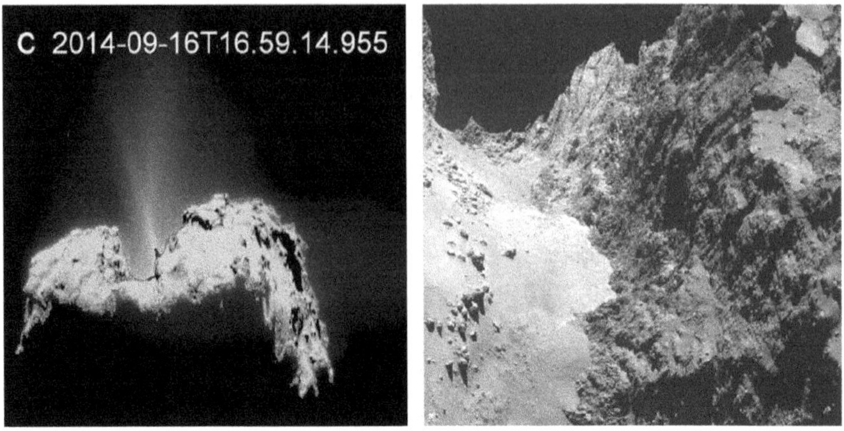

Fig. 2.8. Images of the 67P nucleus taken by the OSIRIS cameras. Left panel: Global view from 16 September 2014, showing a dust jet emanating from the neck region (wide angle camera). Credit: Z.-Y. Lin *et al.*, *A&A* **583**, A11 (2005), reproduced with permission © ESO. Extracted from Fig. 2 in Lin *et al.* (2005). Right panel: To the right, the Hathor wall, and to the left, the Hapi region (narrow angle camera). From Thomas, N. *et al.*, *Science* **347**, aaa0440 (2005). Reprinted with permission from AAAS.

the orbit, while the southern hemisphere is strongly insolated during perihelion passage and is therefore potentially more strongly eroded by sublimation.

A key to explaining the localization of the early activity is found in the shape of the nucleus. During August and September 2014, the Sun shone into the Hapi valley for part of each rotation, while the region was in shadow for most of the time. Thus, the integrated amount of direct insolation was rather low, but the surrounding walls of Hathor and Seth on the head and body, respectively, made an important contribution to the heating of Hapi by their thermal radiation and scattering of sunlight. This has been clarified by sophisticated thermal modeling, performed by Keller *et al.* (2015). The comprehensive model in question shows that the highest surface temperatures were mainly located in a band on the side of the Hapi valley closest to the Hathor wall under the assumption that the whole surface is ice covered. These temperatures are high enough to explain the activity by H_2O sublimation.

The fact that the activity of 67P started exactly where it should start assuming an isotropic surface, i.e., in the region of highest temperature, may indicate that the surface is indeed rather isotropic. This impression is supported by ROSINA measurements of gas density in the inner coma and their interpretation by gas dynamic simulations starting with outflow from the surface (Bieler *et al.* 2015; Fougere *et al.* 2016). From these it appears that the coma structure is best explained, if the sources of activity are widespread around the nucleus rather than restricted to a few active areas. Even so, the activity level of the surface was clearly non-uniform. Moreover, after the comet had approached the Sun close enough for the activity to become global (around New Year 2015), dust features were seen from all the illuminated parts of the northern hemisphere.

Meanwhile, the study of dust jets during the approach of the comet to perihelion (Vincent *et al.* 2016) has offered interesting, complementary information. With the exception of the broad, diffuse feature arising from Hapi, these jets could be resolved into very narrow structures, whose three-dimensional orientation could be discerned using series of images from the moving spacecraft. The footpoints of these jets were found to be concentrated to special topographic features, including nearly vertical cliffs or scarps. Roundish pits that are abundant in the Seth region form one category of features that jets arise from.

At least during the time in question, it appears that much of the distributed sources of outgassing were situated in such steep slopes — not only in the vicinity of Hapi. As a physical interpretation, Vincent *et al.* (2016) proposed a scenario, where flat areas are dust covered and therefore poor sources of outgassing, but cliffs expose fresh material. The latter would also be vulnerable to cracking by thermo-mechanical stresses, which facilitates the outgassing and also undermines the terrain behind the cliff. This may lead to collapse and mass wasting, whereby a layer of talus is deposited on the neighboring surface. Hence, as time proceeds, the topography levels out and the whole area may become dust mantled and thus inactive.

In the meantime, surface erosion proceeds in a lateral sense rather than the vertical erosion imagined by standard theories. This fits well with results from other comets like 81P/Wild 2 and 9P/Tempel 1 (see Sec. 1.5). Formation of new pits, e.g., by sinkhole collapse (Vincent *et al.* 2015), may allow the process to continue and thereby save the comet from complete deactivation.

Whether the surface distribution of such local activity sources is actually uniform is a justified question. In particular, the gas production curve of comet 67P offers a chance to check. Keller *et al.* (2015) found that the uniform surface of their standard model does not produce a good fit to the observed curve. It fails to predict the amount of perihelion asymmetry and the fall-off slopes far from perihelion. Therefore, they considered a second thermal model that was computationally more efficient by sacrificing some details of the standard model but also more realistic from other points of view. This could be used in a search for an optimal distribution of activity over the surface to represent the observed H_2O production curve. Figure 2.9 shows that a reasonable fit to the data requires a spotted model instead of a uniform one. In the spotted models, a small number of directions from the nucleus center were selected as

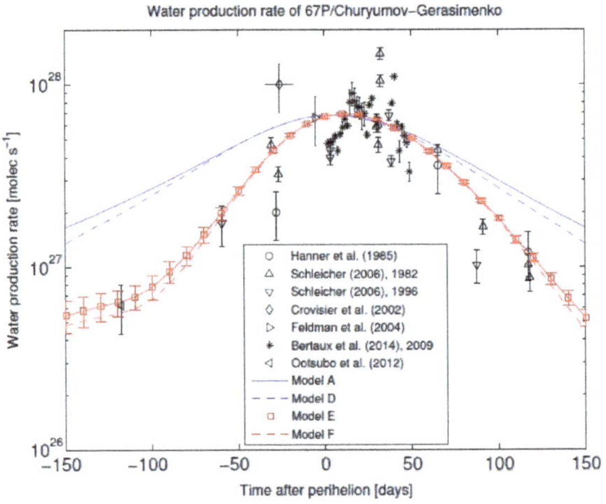

Fig. 2.9. H_2O production curve of comet 67P/Churyumov-Gerasimenko. The symbols indicate observations during preceding apparitions of the comet. The blue curves correspond to the model with a uniform surface, and the red curves illustrate the results of an optimal, spotted surface. Credit: H. U. Keller *et al.*, *A&A* **583**, A34 (2015), reproduced with permission © ESO.

centers of an active solid angle, while the rest of the surface had no activity.

Twenty-nine random surface activity maps were deemed successful in their fits to observations. From a combination of these it was seen that the local potential for activity has a maximum in a region joining the neck (opposite to Hapi) with an extended area on the large lobe. This coincides with the part of the nucleus, which is in permanently high illumination during the perihelion passage. Keller *et al.* (2015) found that there should be a physical explanation for this fact, and further discussion is given in Sec. 4.2.1.

The mentioned study of localized outgassing goes some way toward a complete activity model for comet 67P. Such a model should rather be unique and consistent with all the observations. However, the time may not yet be ripe for this endeavor. So far, detailed thermal models of cometary outgassing have not been coupled to hydrodynamic simulations of the coma, but this would be necessary for a full understanding of the activity whenever there is relevant information about the gas distribution in the inner coma.

The standard model of Keller *et al.* (2015) may not be ideal for determining the global activity parameter of comet 67P, since it does not yield a very good match to the H_2O production curve. However, it is the best as yet available and it was used for this very purpose. The result, based on gas production rates near perihelion, was about 6% — this is the value reported in Table 2.1. Alternatively, one may derive a value for the time integrated activity referring to the whole apparition as follows. The integrated mass loss of H_2O has been estimated as 2.7×10^9 kg (Bertaux 2015), while the standard model of Keller *et al.* (2015) with an icy surface produced an integrated loss amounting to 6.5×10^{10} kg. The ratio of these leads to a time integrated activity of about 4%.

Finally, we come to the issue of exposed ice on the surface of the nucleus. A very small amount of such ice was detected on the nucleus of comet 9P/Tempel 1 (Sec. 1.5.3), but the pre-perihelion search for H_2O ice by its infrared spectral signature by the VIRTIS instrument on board Rosetta only led to an upper limit of 1% of the area (Capaccioni *et al.* 2015). However, it takes only an extremely thin layer of overlying dust to hide this signature, and thus, ice might be generally present very close to the surface, and small patches might even be ice-covered. Indeed, meter-sized icy patches were identified by OSIRIS as bright spots (Pommerol *et al.* 2015; Barucci *et al.* 2016) and by VIRTIS as small exposures of surface frost (DeSanctis *et al.* 2015; Filacchione *et al.* 2016a).

The situation changed, as the comet came to perihelion and the southern hemisphere was targeted by the Rosetta investigations. The surface of the nucleus gradually became more neutral (less red) in color on approach to perihelion and then again, took a redder hue on the way out (Fornasier *et al.* 2016) — a variation that reveals an increased presence of near-surface ice in the innermost part of the orbit. We note that a full thermal model aiming to reproduce the gas production curve of 67P should also incorporate this feature. In addition, two large ice-rich patches (about $1\,500\,\mathrm{m}^2$ each) were observed by OSIRIS shortly before perihelion (Fornasier *et al.* 2016). They lasted only for about 10 days, after which the ice had apparently sublimed. The patches were situated in the southern Anhur and Bes regions.

About three weeks earlier, VIRTIS had observed exposed CO_2 ice in one of these areas (Filacchione *et al.* 2016b) — again, only of short duration. The nature of this ice, whether it is a recondensed frost or a newly exposed patch of pristine material, remains to be explored.

2.5. Structure, Density and Porosity

A fundamental question about comet nuclei is how they came together. It is usually thought that this is part of the broad picture of planetesimal formation, which starts from micron-sized grains and leads to macroscopic objects the size of comets or even larger (see Chap. 7). The structural properties of comet nuclei may hold important information about this process. In particular, the way a typical nucleus is composed of sub-units of various sizes, the strength of the bonds that hold these together, and the porosity of the assemblage are items of prime interest. Let us now review the observational evidence that has a bearing on these.

2.5.1. *Morphologic evidence*

The first feature to strike the viewer of a picture showing the nucleus of comet 67P/Churyumov-Gerasimenko is that it looks like two nuclei glued to each other (Sec. 1.5.5). Such objects are called *contact binaries.* The elongated nuclei of 1P/Halley, 19P/Borrelly and 103P/Hartley 2 have all been put forward as candidates for a contact binary structure based on the close-up images, but only the last mentioned case can be deemed nearly as clearcut as that of comet 67P. Among other comets there is only one unquestionable case, namely, the nucleus of comet 8P/Tuttle. This showed its binary nature by radar imaging (Harmon *et al.* 2010) after Hubble Space Telescope observations by Lamy *et al.* (2008) had suggested such a nature.

It is important to note that the absence of evidence for contact binaries among the rest of the comets is no evidence of absence of this structure. Lacking close-up images, it is hard to conclude against a contact binary nature for any individual object. From the imaged cases, if one accepts that 9P/Tempel 1 and 81P/Wild 2 do not have contact binary nuclei, the extant small number statistics still indicates that contact binaries may be at least as common as monolithic nuclei. It is thus reasonable to include this feature among the constraints, when studying the formation of the nuclei.

Concerning other large-scale structural features on comet nuclei, we may note the layers observed on the 9P/Tempel 1 nucleus and those, possibly of similar nature, on the 67P nucleus (Massironi *et al.* 2015). These might be remnants of the accretion process, if the latter involved cometesimals that were flattened by the pressure exerted by the impacts, as in the "talps" model by Belton *et al.* (2007). On the other hand, evidence of non-flattened cometesimals is rare. One possible case on the 67P nucleus was mentioned by Davidsson *et al.* (2016). This consists of three convex surface features on the small lobe with diameters of a few hundred meters, which may be cometesimals that preserved their structural integrity to some extent during their accretion — see Fig. 2.10.

A small-scale feature of possible relevance seen on the Rosetta/OSIRIS images of 67P was named "goosebumps" by Sierks

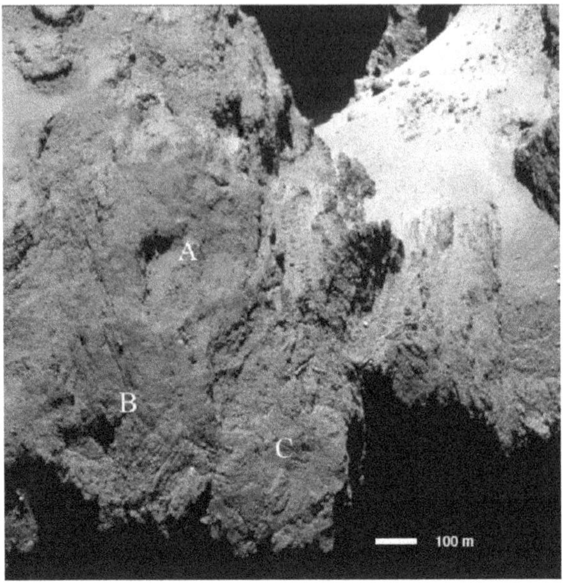

Fig. 2.10. Bastet region on the small lobe of the 67P nucleus (lower left), imaged by the OSIRIS narrow-angle camera. The letters denote three "positive relief features" that may be outcrops of individual cometesimals. The Hapi region is seen to the right. Credit: B. Davidsson *et al.*, *A&A* **592**, A63 (2016), reproduced with permission © ESO.

et al. (2015). An example is shown in Fig. 2.11. The typical size scale is a few meters, and they appear as more or less consolidated clods sitting side by side in large numbers but are only visible under favorable conditions like on the vertical walls of pits. There are several examples of goosebump fields in different regions of the nucleus (Davidsson *et al.* 2016), and the first question is how they were formed. Whether or not thermal cracking (El-Maarry *et al.* 2015) was involved, an intrinsic lumpiness of the comet material on the meter scale may be indicated with obvious consequences for understanding its origin.

2.5.2. *Strength and splitting*

When listing the phenomena that comets are known for, one cannot ignore the fact that they are prone to split apart. This has been known for 150 years, and recent time has witnessed several very

Fig. 2.11. Circular pit in the Seth region of the 67P nucleus, imaged by the OSIRIS narrow-angle camera. The grainy structure of the opposite wall is caused by meter-sized clods called goosebumps. From Sierks, H. *et al.*, *Science* **347**, aaa1044 (2015). Reprinted with permission from AAAS.

interesting observations of splitting events. Comet splits fall into two categories. In one case, they occur in the close vicinity of a massive, gravitating body (the Sun or a planet), and these are caused by the tension applied to the nucleus by the tidal force. In the other case, their occurrence seems random in space and time, and very little is known about the underlying reason. However, in both cases detailed observations and theoretical modeling have yielded information about the strength and buildup of the nuclei.

In March 1993 a strange comet, later to become one of the most renowned comets ever, was discovered in Arizona by Carolyn and Eugene Shoemaker together with David Levy. It is nowadays officially referred to as D/1993 F2 (Shoemaker-Levy 9). Upon discovery, the comet appeared like a string of about a dozen comets, which had to be physically related and thus signaled the recent split of a common parent. Soon enough, orbit determinations verified this hypothesis by showing that the comet was orbiting around Jupiter and had passed very close to the planetary surface in July 1992. This would allow the comet to be torn apart by Jupiter's tidal force.

The attention that comet Shoemaker-Levy 9 caught among the public and in media was mostly due to the fact that the following

Fig. 2.12. A mosaic of images from the Hubble Space Telescope, showing the whole train of fragments that together make up comet D/1993 F2 (Shoemaker-Levy 9) on 17 May 1994, two months before these plunge into Jupiter's atmosphere. Photo Credit: H. A. Weaver and T. E. Smith (STSci), NASA.

perijove passage in July 1994 implied a series of collisions with Jupiter, which were accurately predicted and very well observed. The comet was shown to have orbited around the planet since several decades before the tidal breakup, and it provides one of the most prominent examples of temporary jovian satellite captures among observed comets (see Sec. 3.2.2).

A fundamental concept when discussing tidal splits is the *Roche limit*, which was introduced in the mid-19th century by French astronomer Édouard Roche. In the present context, the Roche limit is the smallest distance from the center of a planet, where a strengthless object can exist without being disrupted by the tidal force. Such an object is held together by its self-gravity, while the external gravity field in which it is placed tends to pull its surface away from its center along the line connecting the bodies. The latter is the tidal force, and it falls off rapidly (inverse cube dependence) with increasing distance from the planet. The distance at which the two forces are in equilibrium is the Roche limit (r_R). This is a small number of planet radii, but there is no universal value, since it depends on the density of the object. A standard expression is

$$r_R = 2.45 R_p \left(\frac{\rho_p}{\rho} \right)^{1/3}, \tag{2.13}$$

where R_p and ρ_p are the planet radius and density, respectively, and ρ is the density of the object.

If a comet approaches Jupiter very closely without ever passing inside the Roche limit, it will not undergo any tidal split. If it does penetrate within the Roche limit, it may survive intact due to its internal strength. If the strength is much smaller than the gravity forces and thus negligible, a split may or may not occur. Whether it does occur depends on detailed circumstances that may differ from case to case, but an important criterion is how far inside the Roche limit the perijove is situated.

There is a certain time around the perijove passage, during which the surface of the comet nucleus is accelerated away from the center in the two directions facing and opposing the planet. If this results in some material reaching the escape speed, the nucleus splits. After exiting through the Roche limit, the nucleus may reaccrete material moving at lower speeds, and concentrations of shattered material may fall together under self-gravity. According to Sridhar and Tremaine (1992), the shedding of material requires a perijove distance smaller than $0.69 r_R$.

The encounter of comet Shoemaker-Levy 9 with Jupiter in 1992 was the closest known for observed comets with a minimum distance of 1.3 jovian radii (20 000 km from the surface). This is deep inside the Roche limit for any reasonable density, so the split is no surprise as long as the strength is very low. It is of interest to compare with other comets, which are known to have made very close encounters with Jupiter. The only previous case of a tidal split is that of comet 16P/Brooks 2, which was discovered in 1889 as a multiple comet with several components. The background turned out to be an encounter with Jupiter in July 1886 with a perijove distance of 2.0 jovian radii. Sekanina and Yeomans (1985) modeled the separation of the components as a tidal split due to fracturing of a parent nucleus, following Aggarwal and Oberbeck (1974), and concluded that both the bulk density and the tensile strength of the Brooks 2 nucleus had to be very low.

For comet Shoemaker-Levy 9 the situation was much better, since much more observations were available to constrain the modeling of the breakup. In this case Erik Asphaug and Willy Benz (1996) made the most comprehensive study, in which they first showed that the observed train of comets cannot be explained by any mechanism involving fracture in a monolithic precursor nucleus. They then applied hydrocode modeling of the tidal breakup of a strengthless precursor consisting of an aggregate of uniform, frictionless, spherical grains followed by the reaccumulation under self-gravity into a number of sizeable clumps. The outcome was studied as a function of the basic properties of the precursor: its size and density.

Assuming a non-rotating precursor, the best results were found for a diameter close to 1.5 km and a bulk density of $0.6 \pm 0.1\,\mathrm{g/cm^3}$. This would reproduce the basic characteristics of the train of comets: its length and the similar sizes of most of the members. With a higher density, a large central clump would form in contrast to the observations. However, a rapid prograde rotation of the precursor would offset this tendency, and a good fit could then be obtained with a 1 km diameter and a density of $1\,\mathrm{g/cm^3}$. Since the state of rotation is not constrained by any other observations, and the total mass of the precursor cannot be determined accurately enough from observations and modeling of the 1994 impacts, there is hence a range of possible densities from about 0.5 to $1\,\mathrm{g/cm^3}$.

In summary, the experience from comet Shoemaker-Levy 9 tells us that this comet had an essentially strengthless nucleus of low density. Its temporary jovian satellite capture does not suggest that its dynamical evolution could have bestowed special properties on it, different from most Jupiter Family comets. Therefore, one has good reason to believe that strengthless, low density nuclei are the rule, but of course, support for this picture from other comets is essential. In addition, care needs to be exerted concerning the term "strengthless". In the Asphaug–Benz model there is no strength whatsoever, but for real comets this only means that strength should be negligible compared to the already very small forces of the planetary tide and self-gravity.

The vast majority of comet splits are non-tidal. These occur at large distance from any massive object, covering a very wide range of heliocentric distance. Unfortunately, it has not been possible to reveal any particular mechanism as the main culprit for these phenomena, and thus it is difficult to draw conclusions about physical properties based on their occurrence. However, one thing that is certain is that such random splits are quite common, possibly playing an important role in comet evolution. There are hence many cases of well-observed splits, from which valuable data can be derived. The most recent, general review was written by Boehnhardt (2004), who presented the relevant statistics and discussed their implications.

In particular, in many cases, the separation velocities, lifetimes, and accelerations of fragments relative to the primary component have been estimated. At first impression, there is no uniformity among these parameters — they all vary within a few orders of magnitude between different events. One thing that stands out (Boehnhardt 2004) is that the largest separation velocities (tens of m/s) are often associated with periodic comets. Among several splitting mechanisms that are often discussed, *rotational splitting* is perhaps the most popular due to its relative lack of problems explaining the observations. This involves a spin-up of the nucleus by a torque due to asymmetric outgassing until the centrifugal force exceeds the self-gravity at the equator, so that shedding of material may occur.

If there is virtually no cohesion in the nucleus, the shed material would leave at the equatorial spin velocity $V_{\rm eq}$, which in centrifugal equilibrium would be proportional to the nuclear radius R and amount to 0.36 m/s for $R = 1$ km with a density of 0.5 g/cm^3. Thus, without strength, the larger separation velocities would require unreasonably large nuclear radii. In fact, if the separation instead occurs by yielding of a cohesive material under the centrifugal force, the tensile strength can be estimated to reach the order of 10^5 Pa (Sekanina 1982; Boehnhardt 2004).

Even though this may appear a bit speculative, it is supported by the fact that the largest separation velocities are restricted to periodic comets. These are the ones that make a large enough number of orbital revolutions that the outgassing torque can build up a very fast spin. Long period and new comets would only be subject to rotational splitting, if they have very low strength so that a slower spin may do the job.

One may also obtain some information from the observed spin rates of comets concerning the minimum strength to hold them together. For the usual, km-sized nuclei the situation is rather inconclusive, since spin periods short enough to endanger their stability have not been observed. However, the very big nucleus of comet C/1995 O1 (Hale-Bopp) needs a strength similar to the above estimate to survive intact with its spin period of 11.5 hours (Boehnhardt 2004).

It is not entirely clear, what this means for the comparison with comet Shoemaker-Levy 9. One must keep in mind that the split of the latter involved the whole nucleus, while the rotational splits would only affect layers close to the surface. From the findings concerning comet 67P/Churyumov-Gerasimenko it appears that a surface layer of possibly considerable thickness is characterized by high-strength material, while the deep interior may well be strengthless.

2.5.3. *Density and porosity*

There is only one generally available method to determine the density of a comet nucleus, i.e., to determine its mass and volume and divide the two. However, in practice this approach is realistic only for a minority of the short-period comets, as far as individual densities are concerned. To see why this is so, we first consider the mass determination. Except in peculiar situations, this rests on observation and analysis of the nongravitational effects in the orbital motion together with the gas production curve. These are well known only for some of the short-period comets. To this comes the problem that the average radius of the nucleus has a large observational error bar for many of those, making volume estimates extremely uncertain.

Naturally, spacecraft targets are the preferential objects for density determinations. Not only do these present the most direct and accurate volume estimates, but their mass determination is also facilitated by the fact that, for preparation and follow up of the space missions, they are well observed both ground-based and from satellites. This is a guarantee for availability of good information about the gas production curve, which is necessary to interpret the nongravitational effects.

The basis for these mass determinations is Whipple's (1950) theory for the solid nucleus, in which the nongravitational effects observed in comets were interpreted as results of the jet force caused by asymmetric outgassing. To illustrate how this has been put into practice, we list some historical facts. The nongravitational effect that was first established was an offset of the time of perihelion passage from the ephemerides for returning periodic comets. In orbital solutions, this was first represented as an empirically determined correction to the mean motion that would make the orbital period longer or shorter as needed. By itself, this correction involved no physics.

The same can be said about the improved treatment introduced by Marsden *et al.* (1973). In Sec. 2.3.1 we discussed an extremely simplified thermal model for a comet nucleus, called isothermal, and in Fig. 2.3 we presented theoretical gas production curves based on this model for various sublimating molecules. As noted, the curve for H_2O is of particular interest, since it comes closest to portraying the observed activity variations of comets. An analytical representation of this curve, usually denoted $g(r)$, forms part of the Marsden *et al.* model.

The model implies that a nongravitational acceleration is introduced into the equations of motion for the comet. In comparison to the empirical corrections to the mean motion, this was a great step forward. However, there was no physical theory available to parametrize the expression for the nongravitational force, and hence an empirical approach was once again used. Here, the acceleration vector is described as $\mathbf{A} = (A_1, A_2, A_3)\,g(r)$, and A_1, A_2 and A_3 are constants representing the orthogonal radial, transverse, and normal components of the acceleration at $r = 1\,\mathrm{AU}$. These constants can be solved for along with the osculating orbital elements to fit the observed astrometric positions of the comet at various times. The influence on the orbital period means a perturbation of the orbital energy, and the A_2 term yields such a perturbation, while in a first approximation the influence of the A_1 term cancels out before and after perihelion, and the A_3 term has no influence at all.

That this model practically involves no physics can be seen from the fact that it is based on the isothermal model of sublimation from the nucleus. In this approximation there is no asymmetry of the outgassing and hence no jet force. Moreover, there is no physical basis for assuming the constancy of the three parameters, and it can even be surmised that these would not be constant, if the acceleration vector was derived from a more realistic thermal model. Even so, the accuracy of the observations is limited, and as long as the fit is satisfactory, the model can be deemed satisfactory too. In fact, it quickly became the preferential model in almost all nongravitational orbit determinations and is usually called the *standard model.*

Calculations of non-uniform distributions of surface temperature and sublimation rate from a thermal model were first applied to the problem of the nongravitational force by Rickman and Froeschlé (1983b). Huge variations of A_2 as derived from the instantaneous force vector were found, but upon closer inspection it was concluded that the standard model cannot be replaced by an equally comprehensive, more realistic model (Froeschlé and Rickman 1986).

As mentioned above, the parameter A_2 is closely related to the perturbation ΔP of the orbital period during one revolution. Comets 1P/Halley and 22P/Kopff were the first, for which the mass determination was carried out (Rickman 1986). Halley's comet is known to arrive at perihelion about four days too late due to its nongravitational perturbation, so $\Delta P \simeq 4$ days. Using the Gauss equations (McCuskey 1963), the following formula can be derived for calculating the nongravitational perturbation of the orbital period from the radial (F_r) and transverse (F_t) components of the jet force in the orbital plane:

$$\Delta P = \frac{6\pi\sqrt{1-e^2}}{Mn^2} \cdot \left\{ \frac{e}{p} \int_0^P \left(F_r \sin f + \frac{F_t}{r} \right) dt \right\}, \tag{2.14}$$

where M is the mass of the nucleus, t is time, e, n and p are the eccentricity, mean motion and semilatus rectum of the comet orbit, f is the true anomaly, and r is the distance from the Sun.

The jet force can formally be expressed as

$$\mathbf{F} = -Q\, m\, \mathbf{u}, \tag{2.15}$$

where Q is the outflow rate of gas molecules from the nucleus, m is the mean molecular mass, and $\mathbf{u}$ is the effective outflow velocity, which means the vectorial average of all the velocities of individual molecules as they leave the nucleus. This definition was given by Rickman (1989), but a few comments are in order. It is not obvious that Q can be identified with the observed gas production rate of a comet, because some of this gas may have been produced in the coma by decomposition or sublimation of solid grains. The averaging of the velocities is a complicated procedure, since active comets near the Sun have a so-called *Knudsen layer* next to the surface, where the individual molecules get together into the bulk outflow of an equilibrium gas by collisional mixing. This involves a back flow to the nucleus, which adds to the momentum exchange. Moreover, if Q refers to the observed H_2O production, there may be other, unobserved gases that also contribute to the jet force.

The mass M can be obtained from Eq. (2.14), if F_r and F_t are derived from a combination of observations and thermal modeling of the nucleus. In Eq. (2.15), the gas production curve $Q(t)$ essentially comes from observations, while u_r and u_t come from modeling. For comet Halley, it turned out that the accuracy of the mass determination is mainly limited by the lack of consistency between different data sets concerning the H_2O production rate (see Sec. 4.1). Depending on which data is preferred, the analysis by Rickman (1989) gave masses from about 1 to 3 times 10^{14} kg.

The volume of the 1P/Halley nucleus was determined from the *in situ* imaging with thus far unprecedented accuracy as 365 km^3 (Merényi *et al.* 1990), and thus the bulk density of the nucleus could be crudely estimated as 0.3–0.7 g/cm^3. An update was made by Skorov and Rickman (1999), using new modeling of the Knudsen layer, whereby the density range was raised to 0.5–1.2 g/cm^3.

In the 21st century, new thermal models of comet surfaces were developed by Björn Davidsson, and within this framework new mass determinations for spacecraft targets were made by Davidsson and Pedro Gutiérrez in a series of papers (2004, 2005, 2006). Here the modeling was made self-consistent in that activity maps of the nuclei were made to fit the observational data on H_2O production rates and spin axis orientations as well as to predict the nongravitational effects (not only the perihelion delay) for comparison with the available orbit determinations. For 19P/Borrelly the density came out as 0.2–0.3 g/cm^3, for 67P/Churyumov-Gerasimenko they found $<$ 0.5–0.6 g/cm^3, and for 81P/Wild 2 the result was $<$ 0.6–0.8 g/cm^3. In a following paper, Davidsson *et al.* (2007) found 0.2–0.7 g/cm^3 for 9P/Tempel 1.

There is a fairly good consistency within this set of density determinations, although there is of course some model dependency due to the assumptions used for the nature of cometary outgassing. However, there is also independent support from a few other cases. One of these is comet Shoemaker-Levy 9, which was discussed above. For comet Tempel 1, the mass of the nucleus was estimated from the Deep Impact observations of the expansion of the base of the ejecta plume as 0.4 g/cm^3 with an uncertainty range of 0.2–1.0 g/cm^3 (Richardson *et al.* 2007), and the new volume estimate from Stardust-NExT led to a slight revision upward (Thomas *et al.* 2013). The most solid result so far has recently been obtained for comet Churyumov-Gerasimenko from the mapping of the gravity field by the Radio Science Investigation and the OSIRIS imaging, and the value is 0.533 ± 0.006 g/cm^3 (Pätzold *et al.* 2016).

Hence, it now seems well established that comet nuclei have low densities, but of course, if the density should be called low depends on what material the nuclei are made of. In fact, the porosity is the interesting quantity, meaning the fraction of the volume that is not occupied by solid material (gas can be present in the pores, if the temperature allows the sublimation of volatiles from the pore walls). Assessing the porosity of comet nuclei involves considering two critical issues that are not yet fully understood. The first is the question just raised: which substances contribute to the comet material, and in which proportions? Answering this question means providing a value for the compact density of the solids in the nucleus, and the porosity is then the ratio between the actual bulk density and this compact density.

The second issue deals with the extent of the voids causing the porosity. It may be a question of pores on the microscale, like those occurring when micron-sized grains are packed in a loose structure. But there may also be macroscopic voids, if the nucleus is a rubble pile where large cometesimals sit together without much compaction. Finally, one needs to distinguish the bulk density from the local density of surface units or the average bulk density of the surface layer. The latter can possibly be both larger and smaller than the bulk density of the whole nucleus. Surface layers can be compacted by the effects of cooking or sintering, at least locally, and this may contribute to the elevated range of surface densities (~ 0.5–$1.5\ \mathrm{g/cm^3}$) found from radar observations by Harmon *et al.* (2004).

Concerning the issue of the abundance of refractories versus ices in the comet material, the Rosetta exploration has brought some preliminary insight. The dust/gas ratio in the outflowing material of comet Churyumov-Gerasimenko before perihelion was reported by Rotundi *et al.* (2015) as 4 ± 2. Using their accurate bulk density determination, Pätzold *et al.* (2016) found, assuming the measured dust/ice ratio to hold throughout the nucleus, that the porosity of the comet would be 72–74%. The material would be quite dusty, since the inferred dust/ice ratio by volume is about 2.

Concerning the large-scale structural homogeneity of the nucleus, Pätzold *et al.* (2016) found that the C_{20} and C_{22} coefficients of the comet's gravity field agreed quite well with those inferred from the shape model assuming a constant density. Therefore, it seemed that any internal voids large enough to cause departures in these coefficients could be excluded, and this places significant limits to the amount of macroporosity in the nucleus. The CONSERT measurements (Kofman *et al.* 2015) are in agreement with both the homogeneity and the dust/ice ratio in the upper layers of the minor lobe of the nucleus.

2.6. Chemical Nature of Comets

The material of comet nuclei can be divided into three categories: ice, organics and rock minerals. The ice is fundamental to the picture of the nuclei as introduced by Whipple (1950), and the other two contribute to the solid particles observed in the comae and tails.

Silicate minerals have been taken for granted since long ago and supported by spectroscopic observations of Sun-grazing comets, while the organics were discovered more recently. They can be analyzed one by one in terms of chemical and mineralogical composition.

2.6.1. *Cometary volatiles*

Being the most abundant, H_2O is the natural reference molecule for cometary volatiles. But the abundances of other volatiles relative to H_2O in the ice cannot be directly observed. In the best case, the observations refer to simultaneous production rates from the nucleus, but even so it is not obvious that different molecules are produced by the same mechanism at the same place. This would be the case, if the other molecules existed in the nucleus as guest molecules in clathrate hydrates formed of hexagonal water ice. Sublimation of the ice at the surface would then yield the other molecules in the same proportions as they occupy as guests in the ice. However, the truth is not so simple.

The Christmas tree diagram in Fig. 2.5 brings out this fact very clearly. The production curve of H_2O (traced by OH) is not parallel to the other curves in comet Hale-Bopp. For CO one question is evident: Is this more abundant or less abundant than H_2O? To answer this question, we may consider that the integrated productions of all molecules during the whole apparition correspond to the erosion of one and the same surface layer, so that the ratios of different integrals reveal the abundance ratios in that layer. However, the assumption made is not necessarily correct. Alternatively, we should have a good idea about the actual processes behind the release of the different molecules, allowing to estimate their efficiencies quantitatively.

Neither of these alternatives is generally available, however. Therefore, the best remaining option is to simply compare the observed production rates in comets close enough to the Sun for all molecules to be released into space. Such a compilation, made by Dominique Bockelée-Morvan and co-authors (2004), is shown in Fig. 2.13. We note the high abundance of oxidized carbon (CO and CO_2) relative to the reduced (hydrogenated) species. The molecules

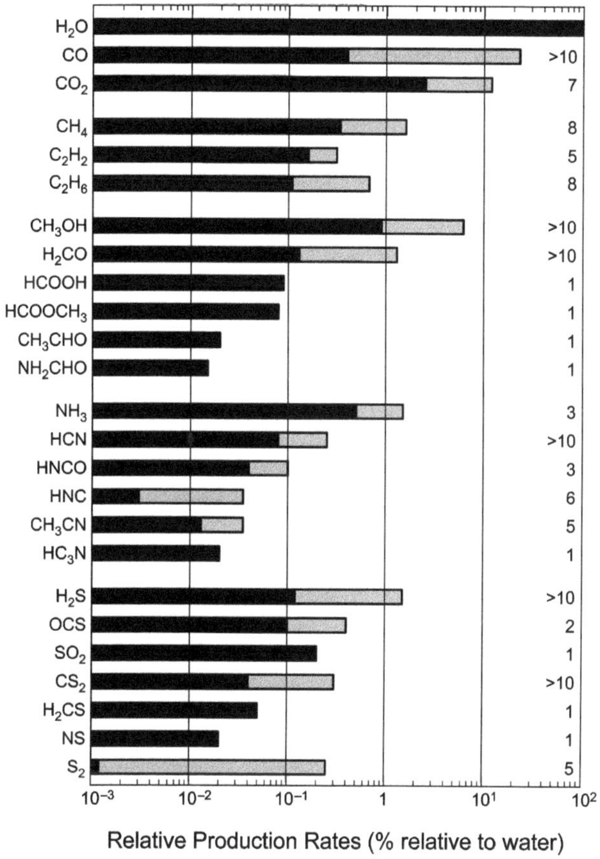

Fig. 2.13. Observed abundances of molecules relative to H_2O as found from spectral detections in various comets. The number of individual comets used for each specific molecule is shown at the right edge of the panel. From D. Bockelée-Morvan *et al.*, "The Composition of Cometary Volatiles," in *Comets II*, edited by M. C. Festou *et al.* © 2004 The Arizona Board of Regents. Reprinted by permission of the University of Arizona Press.

are generally outgassed from the nuclei, but for CO and H_2CO there is often an important distributed source in the form of organics, and HNC is produced by coma chemistry.

Historically, CO_2 observations have in general been made more recently than those of CO. Of special importance was the Japanese *AKARI* satellite (Ootsubo *et al.* 2012). An analysis of CO and CO_2 abundances in all types of comets relative to H_2O was presented by

Mike A'Hearn and co-workers (2012). This showed a large scatter of individual values for both molecules as well as of the ratios between CO and CO_2 in the same comet. No correlation between the two is evident, and no clearcut difference between Jupiter Family comets and other types is found.

Both CO and CO_2 were observed along with H_2O in comet 67P/Churyumov-Gerasimenko by the ROSINA mass spectrometer on board the Rosetta orbiter. These three molecules were clearly identified as the dominant ones in the cometary coma during the early part of the investigation. The instrument observed gases coming from the nadir direction on the nucleus, which changed with latitude and longitude due to the orbital motion of Rosetta and the spin of the nucleus. Strong temporal variations of the relative detection rates of both CO and CO_2 with respect to H_2O were observed (Hässig *et al.* 2015), while both molecules were on the average very important in this remote part of the cometary orbit. In particular, a pronounced dichotomy of the nucleus was identified, whereby the outgassing from the northern hemisphere (specifically, the Hapi region) was H_2O dominated, while the other molecules predominated near the unilluminated south pole. Is this a sign of chemical heterogeneity in the 67P nucleus? This question will be discussed in Sec. 7.4.2.

When several comets are available for comparison, the abundances often differ substantially. The thermophysical properties of the different molecules span a wide range of saturation pressure, or volatility. An extreme case is the S_2 molecule, which is highly variable but reached 0.25% of water in comet C/1983 H1 (IRAS-Araki-Alcock) according to Budzien and Feldman (1992). This molecule has a very low equilibrium condensation temperature. Comparing carbon and nitrogen is of interest. Their hydrides, methane and ammonia, are often detected at about 1% of H_2O, but while pure carbon in the form of graphite is common in the refractory component, molecular nitrogen — like molecular oxygen — took a long time to be detected in any comet.

This happened with the *in situ* Rosetta exploration of 67P, specifically by ROSINA. Both molecules were detected but at quite different abundances. While O_2 was found to be produced at a ratio of 3.8% with respect to H_2O (Bieler *et al.* 2015) and was subsequently found to have been produced at 3.7% also in comet 1P/Halley (Rubin *et al.* 2015a), the production rate of N_2 in comet 67P was only ~1% with respect to CO (Rubin *et al.* 2015b). Moreover, argon was for the first time identified in a cometary coma. Its production rate was ~1% with respect to N_2 (Balsiger *et al.* 2015). How to best interpret these results will be discussed in Sec. 7.3.2.

2.6.2. *Cometary silicates*

A real insight into the nature of the cometary silicates had to await the invention of relevant spectral IR-observing facilities, in particular the Hawaii-based NASA Infrared Telescope Facility and the *Infrared Space Observatory* (ISO). Fortunately, several bright comets including Hale-Bopp were thus observed with good spectral resolution in the region of silicate emission around 10 μm. Later on, equally important observations have been made by the *Spitzer Space Telescope*, and the *Stardust* sample return mission brought fundamental information on the composition of the most refractory coma grains in comet 81P/Wild 2.

From early IR data, specifically 7.8–13 μm spectra of seven comets, Hanner *et al.* (1994) concluded that amorphous silicates of both olivine and pyroxene type were very common, but that three comets also showed features indicative of crystalline olivine. The former type dominates in interstellar space and is likely the native form of the silicates. The latter type, on the other hand, requires high temperature processing — for instance, the evaporation and subsequent recondensation of the silicates according to the common scenario for equilibrium models of solar nebula chemistry. However, such processes would only occur close to the Sun, while comets should have formed much further out.

The presence of both amorphous and crystalline silicates was confirmed by spectra of comet Hale-Bopp (Wooden *et al.* 1999). Finally, the Stardust samples from comet 81P/Wild 2 showed that the grains emitted by this comet contained both amorphous and crystalline silicates (Zolensky *et al.* 2006) — the latter very often in the form of almost pure magnesium olivine called forsterite, which is a high-temperature condensate. One of the tracks formed in the silica aerogel was especially remarkable, since the responsible particle seems to have had a composition similar to the extremely refractory, meteoritic *Calcium–Aluminum-rich Inclusions* (usually called CAIs).

Of the low-temperature hydrated silicates that in meteorites appear as products of *acqueous alteration*, none were seen. This simply means that the comet material never saw liquid water, which is not surprising. But the intimate mix of unaltered silicates with variants formed at very high temperatures is a fascinating discovery. The altered material appears to have been transported from the place near the Sun, where it was formed or the alteration took place, all the way into the cold regions where the comet nucleus was aggregated. We shall return to this in Sec. 7.3.2.

2.6.3. *Cometary organics*

Each of the Giotto and Vega missions to comet Halley carried on board a time-of-flight ion mass spectrometer for roughly micron-sized dust grains. The results mentioned in Sec. 1.5.1 are illustrated in Fig. 2.14, from Kissel *et al.* (1986). Three single grains are shown in terms of the mass distributions of their atoms. They represent the three basic kinds of grains that were identified. One kind is obviously made of silicate minerals, being dominated by the rock-forming elements Si, Fe and Mg. Another type features the light elements H, C, O and N, and these are the *CHON grains.* The third kind is a mixture of the two components. The mixed kind was the most abundant followed by the CHON grains, while the pure silicate grains were rare.

Just like the silicate grain atomic compositions cannot be uniquely synthesized into specific minerals, the CHON grains can only vaguely be referred to as "organics". Furthermore, having survived expulsion from the nucleus in the solid state, they are obviously more or less refractory, so more precisely they have been called *organic refractories.* Their exact nature could not be established from the Giotto data. Since they mostly are not refractory enough to have survived the entries into the Stardust aerogel (Sec. 1.5.2),

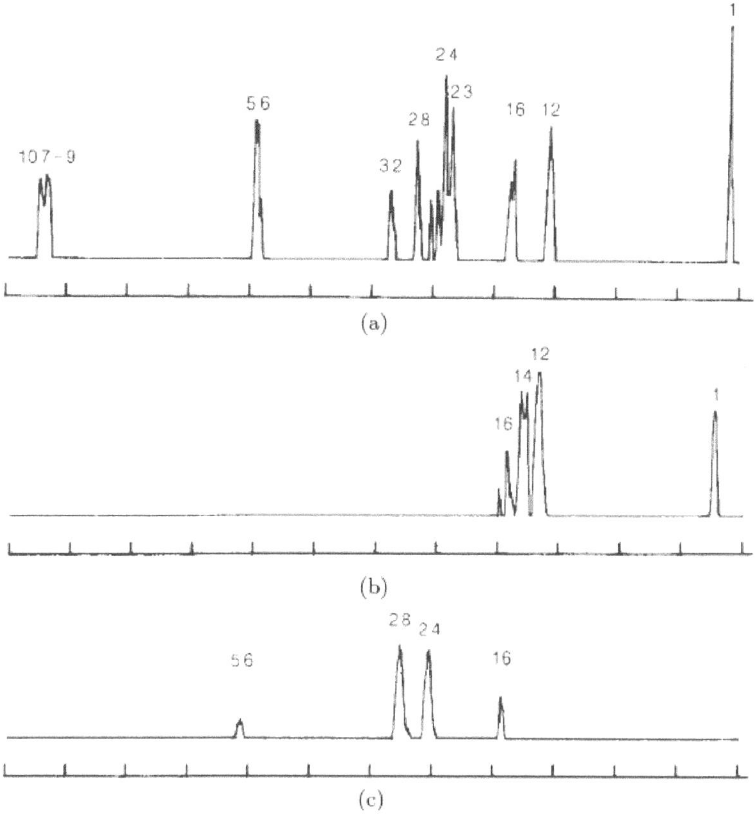

Fig. 2.14. Time-of-flight atomic mass spectra of three grains in comet Halley recorded by the PIA spectrometer on board Giotto. Panels *a*, *b* and *c* show mixed, CHON and silicate grains, respectively. The intensity scales are logarithmic, and the 107–109 mass peak for grain *a* is caused by silver ions from the target plate. Adapted by permission from MacMillan Publishers Ltd: Kissel, J. *et al.*, *Nature* **321**, 336–337 (1986), © 1986 Nature Publishing Group.

the analysis of returned Wild 2 material yielded only sparse and biased information (Sandford *et al.* 2006). However, the presence of nitrogen-rich polycyclic aromatic hydrocarbons (PAHs) was established along with a component that is poor in aromatics. Amino acids including glycine were also reported as found in Stardust-returned foil samples that captured gaseous material in the comet coma (Elsila *et al.* 2009). Overall, the organic material captured in the Stardust aerogel was found to be more akin to that of interplanetary dust particles of cometary origin than to the organic residues derived from carbonaceous chondrite meteorites (see below).

Some more evidence came from spectral imaging of the coma in Halley's comet. It was found that the light emission from some radicals that contribute to the visible gas coma (like C_2, C_3 and CN) was anisotropically distributed around the nucleus. There was a good correlation between the maximum radical brightness and the dust jets seen in continuum images. This shows that the parent molecules of those radicals are not only emitted from the nucleus but also released from dust grains in the coma. However, the specific parent molecules and their abundances cannot be established with certainty.

Near-infrared spectroscopy of the 67P nucleus was performed within the Rosetta mission by the VIRTIS instrument (Capaccioni *et al.* 2015). From the albedo and spectral slopes seen across more or less the whole surface, it can be concluded that a pristine, organic refractory material is seen, which is different from tholins or other products of radiative processing. According to Quirico *et al.* (2016), it is a dark refractory, polyaromatic, carbonaceous material mixed with opaque minerals like iron sulfides and iron-nickel alloys. As illustrated by Fig. 2.15, between 2.9 and 3.6 μm wavelength there is an evident

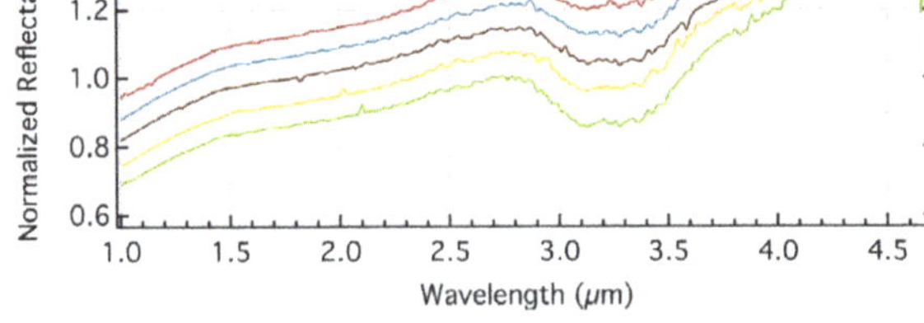

Fig. 2.15. Vertically shifted, near-infrared mean spectra of five regions on the 67P nucleus obtained by the VIRTIS instrument on board Rosetta. Reprinted from Quirico, E. *et al.*, *Icarus* **272**, 32–47 (2016), with permission from Elsevier.

absorption band, centered at 3.2 μm, which previously has not been seen on comet nuclei.

The paper by Quirico *et al.* provides an in-depth discussion of the likely carriers of this absorption. Like in the Stardust analyses, this is based on comparison with analog materials of known composition. These include both substances made in the laboratory and cosmic samples. Among the latter, there are the so-called *insoluble organic matter* (IOM) samples derived from carbonaceous chondrites, and the *interplanetary dust particles* (IDP) collected by stratospheric aircraft and *Antarctic micrometeorites* (AMM), which certainly contain an important fraction of cometary material. In the case of the 3.2 μm band, none of these analogs provides a satisfactory match, and preference is given to a semi-volatile component, plausibly containing carboxylic acids and the NH_4^+ ion. Interestingly, no hydrated minerals have been identified, and no genetic link with the CI, CR and CM chondrite types has been established, consistent with the above-mentioned Stardust results. There is no evident ice absorption in the spectra obtained early in the mission, but mapping of the 3.2 μm band across the nucleus surface indicated that water ice was a significant contributor in the neck region though more or less absent elsewhere.

The semi-volatile component of cometary organics is apparently of great importance. Its decomposition in the solar heat may release radicals like CN, C_2 and C_3, which tend to appear in comets within a few AU of the Sun. The case for this is the above-mentioned association of those radicals with dust features in the coma. If the semi-volatiles thus contain unsaturated carbon-chain molecules hosting C_2 and C_3, one may suspect the presence of a preserved interstellar component of comet material (Mumma and Charnley 2011). Moreover, below the surface of the comet nucleus, the organic semi-volatiles may act as a sintering agent, being decomposed at the very surface and providing material that to some extent diffuses downward through the porous matrix before condensing at grain interfaces.

One puzzling result came from a photometric narrow-band survey of production rates of the mentioned radicals and others in many comets by A'Hearn *et al.* (1995), and largely confirmed by the spectroscopic survey of Fink (2009). These indicated that comets belong to either of two different categories. In one case, the production rates $Q(C_2)$ and $Q(C_3)$ vary proportionally to $Q(CN)$ in a typical way, whereas in the other case, these two radicals are clearly depleted relative to CN. It was found that essentially all long-period and Halley Type comets behave in the typical way, but there are two kinds of Jupiter Family comets: typical and depleted. This is illustrated by Fig. 2.16. All the radicals are largely produced from organic grains, but in some comets these grains are depleted in carbon chain molecules. Those comets strongly prefer the Jupiter Family in the sample analyzed by A'Hearn *et al.* (1995) but less so in a more recent sample (A'Hearn, private communicaton).

2.6.4. *Cometary material: The interstellar model*

There are two competing scenarios for the origin of the material that comet nuclei are made of. Let us first deal with the one

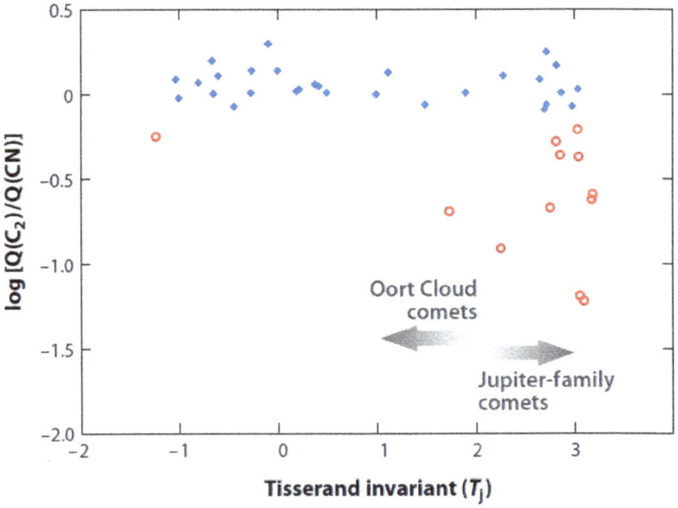

Fig. 2.16. Relative production rates of C_2 and CN in comets observed by A'Hearn *et al.* (1995), plotted versus the jovian Tisserand parameter. Carbon chain depleted comets are marked by red open circles. Credit: M. J. Mumma. Reproduced with permission of *Annual Review of Astronomy and Astrophysics*, Vol. 49. © by Annual Reviews.

that was favored in the wake of the Halley exploration, mainly because of the similarity between the compositions of comets and the prestellar material observed in star-forming regions. Its popularity has decreased, but it is still a strong contender to offer at least part of the explanation. We may call it the *interstellar model.*

The story begins in the extended, cool atmospheres of late-type giant or supergiant stars. This is an environment, where tiny grains of silicates or graphite may condense, and these get trapped in the outflow (stellar wind) that observations have revealed. Hence, they reach into interstellar space as members of the general population of interstellar grains. Such grains are known to exist since almost one century, because they cause an extinction of the light from distant stars, and this extinction is wavelength dependent, so that the stars get reddened. From the amount of interstellar reddening, a typical size of $\sim 0.3\,\mu\text{m}$ has been estimated for the grains.

The properties of the interstellar gas vary from region to region, and this concerns in particular the density. But the grain/gas mass ratio remains about the same at $\sim 1\%$. The most important regions for grain evolution are the *dark molecular clouds.* These are the largest and densest of all, and from a vantage point inside such a cloud the grain population is effectively like an opaque wall preventing all visual and ultraviolet light of stars from entering. This means that there is very little heating of the gas, which takes an equilibrium temperature not far from absolute zero (specifically, $\sim 20\,\text{K}$).

With a relatively high gas density, an extremely low temperature and an abundance of grain surfaces, the conditions are excellent for forming molecules — in the first place, simple molecules like H_2 or OH. Due to the absence of UV radiation, such molecules may survive for a considerable time and are thus able to move around in the gas phase and meet other molecules. Thus, chemical reactions may occur, and new molecules may be formed. The high gas density implies that such reactions can proceed at an important rate, thus changing the chemical composition of the gas.

In fact, hundreds of molecules have been observed in interstellar gas, in particular in protostellar regions within giant molecular clouds. Thus, more and more has been learned about the relevant chemistry, which is very different from laboratory chemistry. The latter proceeds in equilibrium due to the furious rate of interactions at high densities, but the molecular clouds present conditions that are truly vacuum-like. This means that extremely reactive radicals can exist for some time in the absence of collisions with other species, so these can participate in the chemical reaction network in a way that is very different from equilibrium chemistry. Another consequence is the survival of species like CO, which is found in large abundance in spite of the ubiquitous presence of H_2, which in equilibrium would transform this into CH_4 and H_2O.

The gas molecules are bound to collide with the grains. They often stick to the grain surfaces and thus change the composition of the grains by forming *icy mantles* on the pre-existing *refractory cores*. Due to the extremely low temperature, the rate of sublimation from the grain mantles may be very low even for very volatile species. Hence, sticking of new molecules may proceed faster than sublimation, and the mantles grow in thickness. Spectroscopic observations of proto-stellar objects have revealed the presence of several molecules in both gaseous and solid phases.

Thus, over long enough time, interstellar grains may become incorporated into molecular clouds over and over, and this provides for an important evolution. During a visit into a molecular cloud an icy mantle forms on the grain as seen, and this will include extremely reactive radicals. At the extremely low cloud temperatures these radicals stay practically immobile and thus do not react. However, they retain a large chemical potential, should the grain be heated. Now, the molecular clouds do not live forever. To some extent, they are consumed by star formation, and they tend to be dispersed by the shock waves around supernovae, which arise quite rapidly due to the short lifetimes of the most massive stars. When a molecular cloud is thus dispersed, lots of ice-mantled grains find themselves in a new environment with much less extinction, and the temperature increases.

As a consequence, the chemical potential of the radicals is released explosively, since the reactions are exothermic and cause further heating. The result of such explosive chemistry (which has been reproduced in the laboratory) is a sort of carbon-rich, refractory substance that is sometimes dubbed "yellow stuff" for lack of detailed knowledge about the chemical composition. Hence, after each dive into a molecular cloud a grain will end up without any ice but with its core surrounded by a mantle of *refractory organics*. This cycle of evolution was mainly explored by the American astrophysicist J. Mayo Greenberg.

The story ends in a particular molecular cloud — the one in which the Sun was formed. A certain part of it collapsed to form the solar system, and we may call this the *presolar cloud*. As shown in Fig. 2.17, the grains residing in the presolar cloud would have a structure involving both a silicate core, an organic refractory mantle, and an outer, icy mantle. Thus, each grain might have the composition of a typical comet, and comets may be aggregates of such grains (Greenberg and Hage 1990).

2.6.5. *Cometary material: The solar nebula model*

In the other scenario we start from the accretion disk around the nascent Sun, which is usually called the *solar nebula*. This formed

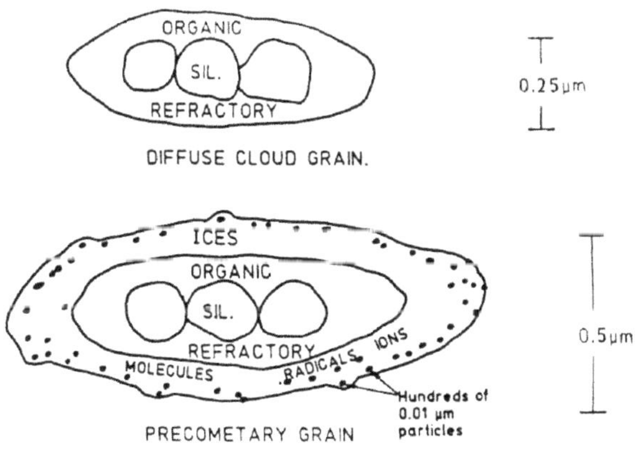

Fig. 2.17. Schematic illustrations of typical interstellar grains in two environments. The top picture shows a grain situated in a usual atomic hydrogen cloud ("diffuse cloud"), which has undergone one or more passages through molecular clouds and acquired a refractory organic mantle. The bottom picture shows such a grain during a subsequent stay in a molecular cloud, where an extra, icy mantle has been formed. Reproduced from Greenberg and Hage (1990) © AAS. Reproduced with permission.

out of the presolar cloud and should at least originally have contained the same type of grains as we described above. However, it is clear that the formation of the solar nebula involved substantial heating. For instance, meteorites often contain inclusions made of extremely refractory minerals, showing that there was a place and a time, where the temperature was high enough to keep only such minerals in the solid state. In the case of the CAIs, the condensation temperature is in excess of 1500 K.

The question is, what temperatures were reached in the part of the solar nebula, where comets grew? Only one thing seems certain, namely, that the organics were not vaporized, since it would be very unlikely for the complex, hydrogen-poor molecules to arise again after having been UV-dissociated in the gas phase. But there is no telling what happened to the ice, so it is worth considering the consequences, if the ice vanished by sublimation from the presolar grains and had to recondense in the solar nebula.

One constraint is given by the high abundance of oxidized species like CO and CO_2 in comets. When these molecules were set free to move in the solar nebula, which was dominated by H_2, they would fall victim to reduction into CH_4 and H_2O unless the density was low enough for these reactions to be kinetically inhibited (i.e., to have too long time scales). The fact that the gas-phase carbon in comets is oxidized rather than reduced thus sets a lower limit to the heliocentric distance of the comet-forming region.

If comets contain less water than the presolar grains did, as suggested by the Rosetta results, this should have some implications for our understanding of the way comets formed. H_2O molecules were available in such quantities that, when recondensation occurred, water ice could easily dominate over the refractory components. There would hence have to be some factor inhibiting this recondensation, at least partially. One possibility is that the time scale for grain growth and sedimentation was not very long compared to the time scale for the growth of icy mantles by condensation. This would more likely occur at smaller heliocentric distances, but the turn-over distance remains to be estimated by detailed calculations.

However, there is a different mechanism that seems established by observations and which may lead to a similar result. This is the above-mentioned radial mixing in the solar nebula. The high abundance of silicate minerals with high formation temperatures in the dust of comet 81P/Wild 2 can thus be explained, but if those minerals account for a large fraction of the entire silicate component, it also represents an important extra supply of refractories into the comet material.

2.6.6. *Cometary ice: Crystalline or amorphous?*

There are lots of solid phases of H_2O, but most of these are stable only at high pressures and are thus of no concern for comets. Here we deal with low temperatures and very low pressures, and then the options are few. Usual ice (I_h) is crystalline with a hexagonal crystal structure, and at low temperatures it may transform into cubic ice (I_c), which has a different crystal structure as the name implies. However, there is also an amorphous form of ice, which can be made in the laboratory by vapor deposition onto a cold plate at temperatures $T \sim 20$–$40\,K$ or less.

The ice in comets is certainly not pure H_2O ice. Other molecules must also be present in the ice phase, since they too contribute to the outgassing. Different possibilities exist. The first to be proposed was *clathrate hydrates*. The clathrates are a class of compounds, in which "guest" atoms or molecules are imprisoned into the cavities offered by the lattice geometry of the crystalline ice. In hexagonal ice, one example often mentioned in the literature is methane clathrate ($CH_4 \cdot {}^7H_2O$), since equilibrium models of solid formation in the solar nebula — when extended to very low temperatures — show that methane molecules would creep into the ice lattice at $T \sim 100\,K$.

In the actual low-pressure conditions of the outer parts of the solar nebula, this is unlikely to proceed, but clathrates have nonetheless been popular. The reason is that the guest molecules are freed, as the ice sublimates, and one can thus explain why the usually observed coma species like C_2 and CN appear in comets approaching the Sun roughly at the distance of about 2.5 AU, where the H_2O ice starts to sublimate. If the parent molecules in question would form their own ice, these would start sublimating at different distances. Note, however, that modern results show such coma species to originate largely from a semi-volatile component of comet material rather than volatiles frozen into ice.

Despite the popularity that clathrates once enjoyed, these are no longer a prime subject, when the chemistry of comets is discussed. It has been recognized since a long time that some comets outgas very large quantities of CO — much more than a clathrate could accommodate. Moreover, the fact that the outgassing patterns in comet Hale-Bopp shown in Fig. 2.5 show the minor species in general to follow the production curve of CO rather than that of H_2O is clearly contrary to the clathrate prediction.

In the framework of the interstellar model it is more tempting to think of the ice as amorphous, because the icy mantles on the cometary grains would have been deposited at temperatures similar to those of the laboratory experiments. Another attractive feature is the fact that amorphous ice has been found to be extremely rich in micropores, which form good trapping sites for guest molecules even in huge quantities (Bar-Nun and Kleinfeld 1989). From the work of Akiva Bar-Nun and his colleagues using a mixture of H_2O and other molecules for deposition, as illustrated in Fig. 2.18, we learn that the trapping efficiencies are very sensitive functions of the plate temperature (exponential fall-off), and below 40 K the amount of trapped gas even exceeds the amount of H_2O ice.

Amorphous ice is not absolutely stable. Upon slow heating, some release of trapped gas occurs already at very low temperatures due to annealing, but the main effect is due to crystallization, which proceeds at an exponentially increasing rate (Schmitt *et al.* 1989). Thus, at temperatures well below 100 K the crystallization time is millions of years or more, while at 130–140 K it is just a few hours. In addition, the settling of the H_2O molecules into a crystalline lattice is exothermic, so the ice gets heated and crystallizes even faster. This has led to the idea of episodic, explosive crystallization in comets occurring when the local ice temperature reaches the above-mentioned range, as described by Dina Prialnik and colleagues (e.g., Prialnik and Bar-Nun 1987).

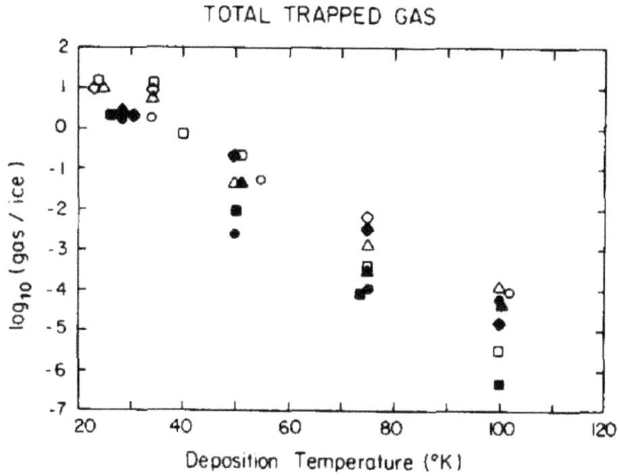

Fig. 2.18. Trapping efficiency of CH_4, CO, N_2 and Ar into amorphous ice, plotted with different symbols, versus the deposition temperature. Filled and open symbols denote different compositions of the flowed vapor. The plotted quantity is the log of the ratio between the amounts of trapped gas and ice. Reprinted from Bar-Nun, A. and Kleinfeld, I., *Icarus* **80**, 243–253 (1989), with permission from Elsevier.

Upon crystallization, most of the once trapped gases escape, as shown in Fig. 2.19. They are then free to diffuse through the pore system in either direction: toward the interior where they may recondense at lower temperatures, or toward the surface where they may flow into space — possibly dragging grains along. Thus, explosive crystallization offers a means to explain outbursts of activity sometimes seen in comets (see Sec. 4.4.3). It may also cause an increased activity of new comets from the Oort Cloud on their inward orbital branch (see Sec. 4.5.1). However, even though the amorphous-crystalline phase transition is certainly exothermic, evidently the released energy is not fully available for heating of the ice. It is possible that, depending on the amount of trapped gases, most or all of this energy may be consumed by unbinding the guest molecules from their trapping sites. In addition, most of the available energy will go into heating the dominant, refractory component of the comet material.

Still, there are reasons to be skeptic about cometary ice being amorphous. These have to do with the origin of the ice and the

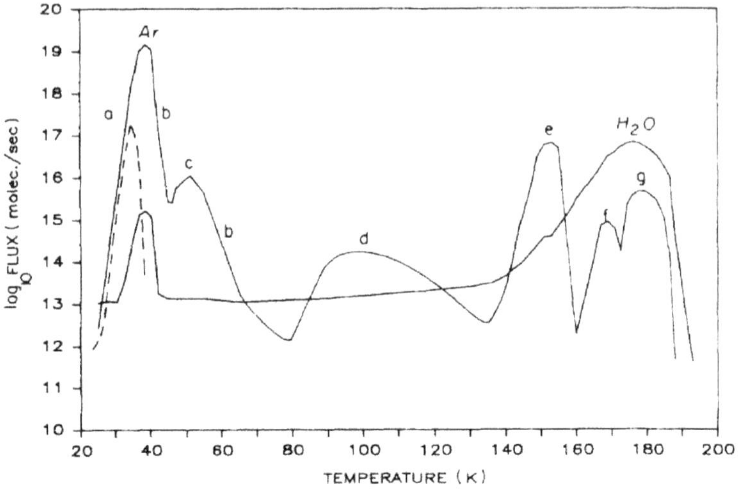

Fig. 2.19. Fluxes of argon and water from amorphous ice with trapped Ar, deposited at 20 K, versus the temperature during slow heating. Near 30 K frozen Ar sublimes from the ice surface, dragging along some H_2O. The next major peak in the Ar flux starts near 140 K and is due to crystallization of the ice. Near 180 K the ice sublimes. Reprinted figure with permission from A. Bar-Nun *et al.*, *APS Phys. Rev. B* **35**, 2427–2435 (1987). © (1987) by the American Physical Society.

thermal history of the cometary grains in the solar nebula. Although the formation of icy mantles on the primordial grains in the presolar cloud occurred at the same temperature as the cold plate, where amorphous ice was formed in the laboratory, there is an enormous difference in the ambient gas density, and thus the time scales of deposition are very different. Thus, while in the laboratory the deposited molecules have too little time to find their place in a crystal lattice, in the presolar cloud they might not have encountered this problem.

Second, if the presolar ice was indeed amorphous, it could possibly have crystallized due to shock heating, as the grains and their surrounding gas fell into the solar nebula at important velocities. It has even been proposed that the ice could have been vaporized, so that the extant cometary ice must have formed by condensation in the solar nebula under the ambient conditions. According to calculations by Kouchi *et al.* (1994), this would rather result in crystalline ice throughout the region of comet formation.

In the presence of such uncertainties, one has to ask if there is any direct evidence of amorphous ice being responsible for cometary outgassing. It appears that such evidence may exist, but it is not unquestionable. One case in point is comet 29P/Schwassmann-Wachmann 1. With a perihelion distance of 5.7 AU and a discovery made in 1927, this comet is arguably the first discovered Chiron type comet. It has an orbit of very low eccentricity but is extremely variable in its activity. Most of the time it stays quiescent, but it undergoes sporadic, major outbursts with a very large increase in brightness (see Sec. 4.4.3).

Explosive ice crystallization has been proposed as the mechanism behind the outbursts of comet 29P, but there is no detailed, convincing argument in favor, so the issue is open. However, the low-level activity in the quiescent phase has been studied by radio astronomical observations with interesting results. These have revealed a persistent, anisotropic outflow of CO molecules from the nucleus plus a symmetric production of CO from grains surrounding the nucleus in a wide cloud (Gunnarsson *et al.* 2002). The nuclear source, accounting for about 1/4 of the molecules, is of particular interest. As seen in Fig. 2.20, this is a jet-like source emanating from the nucleus region facing the observer. Since the comet is always near opposition, this means a nuclear source near the subsolar point at all times.

It is obvious that there cannot be frozen CO on the nuclear surface, since this would disappear at once by sublimation. The source of the CO is therefore to be found below the surface. If it were located deeper than several rotational skin depths from the surface, the local temperature would stay almost constant, and the production of CO would occur on both day and night sides of the nucleus, which is not observed. Due to the low thermal inertia of comet material (Sec. 1.5), this skin depth is so small that the CO source must be very close to the surface. It is therefore natural to seek the explanation for the CO release among mechanisms that are triggered at the temperature of the 29P subsurface layer. This can be

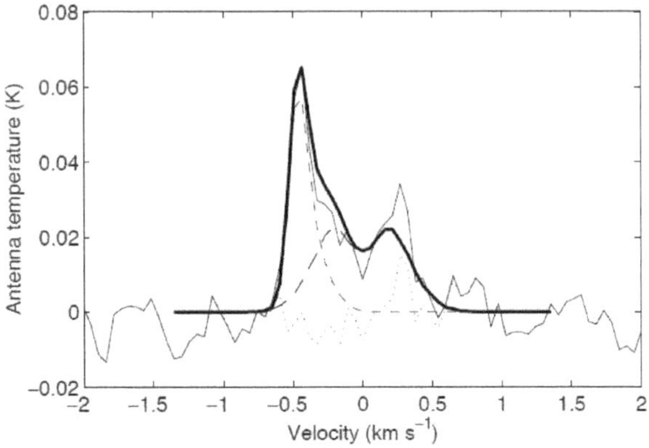

Fig. 2.20. Spectrum of comet 29P, observed in 1998 using the ESO/SEST telescope (thin, solid line). The emission is due to the 230 GHz CO ($J = 2 - 1$) transition. The thick line shows a best model fit, using a nuclear outgassing source (short dashes) and a distributed coma source (long dashes). The dots show the residual between model and observations. Reprinted from Gunnarsson, M. *et al.*, *Icarus* **157**, 309–322 (2002), with permission from Elsevier.

estimated at ~130 K in good agreement with the H_2O crystallization temperature.

Hence, at any given time, the subsolar latitude of the nucleus would have a thin surface layer containing crystalline ice. Below this there is a crystallization front producing the CO outflow, which recedes at the same average rate as the surface is eroded by the release of icy grains. The crystallization front must be prevented from explosive recession, and thus very little of the latent heat of crystallization can be available for heating the ice. The grains seen in the 29P coma would consist of this material and might produce H_2O (though still unseen) by sublimation due to the low albedo of the grains. This seems like a good working model to study the matter further, but there is no proof that it represents the truth.

Chapter 3

Comet Dynamics

As we shall see, comet dynamics is a rather heterogeneous subject, and different features are important for different groups of comets. There is no common denominator, but in nearly all the situations, comets move on heliocentric orbits that are relatively stable for significant amounts of time. The two-body problem (Sun–comet) is then a zero-order approximation, to which one may add small perturbations — typically, due to the gravity of the planets.

These perturbations induce changes of the heliocentric orbital elements. Usually, when the comet is not very close to a planet, the changes are slow. However, close approaches may often occur, and these are typically of short duration but may change the heliocentric orbit dramatically. In its general formulation, the N-body problem of the Sun, several planets and the comet is not fully integrable, and this holds even for the 3-body problem, where only one planet is involved. However, under certain conditions, the latter problem allows us to derive an analytic formulation of the comet's energy of motion, which is independent of time and thus serves to constrain the way the orbit may evolve.

3.1. Circular Restricted Three-body Problem

We consider the theoretical problem of the motion of three bodies, where one of these is so small that it has no influence at all on the motions of the other two. This is the *restricted 3-body problem*, and it deals with the motion of the small body, because the other two move on unperturbed orbits around their center of mass. These orbits are typically assumed to be either elliptic or circular. In the elliptic problem, there is no closed expression constraining the small body's motion. However, the *circular, restricted 3-body problem* allows the formulation of an energy integral. This is of great use in comet dynamics, where as a first approximation one considers the comet as massless and the Sun and Jupiter to represent the rest of the Universe with Jupiter's orbit around the Sun being circular. In reality, this is quite a good approximation, so the energy integral serves as a good hint on the orbital evolution of the comet.

Briefly, the derivation of the energy integral goes as follows. We place ourselves in a rotating coordinate system, where the two massive bodies are at rest. The acceleration of the massless body is now due to the gravities of the massive bodies and the centrifugal force. All these components are conservative, and the massless body moves in a fixed potential. Choosing the units of mass, distance and time such that the total mass of the system is unity, the distance between the massive bodies is unity, and the angular velocity of the circular motion is also unity, we can at once write:

$$C = -2E = (x^2 + y^2) + 2\left\{\frac{1-m}{\rho_1} + \frac{m}{\rho_2}\right\} - v^2, \qquad (3.1)$$

where we have taken a rotating (x, y, z) frame with the z-axis parallel to the angular momentum of the 2-body system, m is the smaller of the two masses, ρ_1 and ρ_2 are the distances of the massless body from the other two bodies, and v is the velocity of the massless body in the rotating system. Note that with the units chosen, the gravitational constant also has the value unity. The quantity E is the constant energy of motion per unit mass of the small body, and C goes under the name *Jacobi integral*, having been first derived by C. G. J. Jacobi in 1836.

Allowing for the factor -2, the term x^2+y^2 in Eq. (3.1) represents the centrifugal energy, the terms in the curly brackets represent the potential energy, and the last term represents the kinetic energy. In 1889, F.F. Tissérand used the Jacobi integral to derive a criterion to look for the possible identity of comet pairs, where one was observed before and the other after an encounter with Jupiter, which led to a large change of the orbital elements (the so-called *Tisserand criterion*). This is a simple expression in terms of the orbital elements, which would be quasi-constant in the circular problem as long as the comet moves far from Jupiter, and it reads

$$T = \frac{a_J}{a} + 2\sqrt{\frac{a}{a_J}(1-e^2)}\cos i. \tag{3.2}$$

Here a_J is the semi-major axis of Jupiter's orbit, and a, e and i are the semi-major axis, eccentricity and inclination of the comet orbit. The latter should in principle be measured with respect to Jupiter's orbital plane, but in practice the ecliptic inclination can be used as well. With the units chosen for the Jacobi integral, we of course have $a_J = 1$.

The quantity T is called the *Tisserand parameter*, and it means the same as C. It is rarely used for the purpose it was conceived for, but as seen in Sec. 1.4, it plays an important role when classifying comets in terms of their orbits. It is only approximately constant, because the ellipticity of Jupiter's orbit may cause changes — in particular during close encounters — and the perturbations due to the other planets have no reason to leave T unaffected. In principle, each planet can in theory be equipped with its own Tisserand parameter, which would be useful for comets that are under that planet's control. To allow for this possibility, one sometimes writes T_J instead of T to avoid ambiguity. This was done in Sec. 1.4, when the Tisserand parameter was used for orbital classification of comets.

However, in practice T is used mainly for short-period comets, and these are typically under Jupiter's control and rarely affected significantly by any other planet. Together with the relative smallness of Jupiter's eccentricity (0.048), this means that the orbital evolutions experienced by short-period comets are in fairly good agreement with the predictions of Eq. (3.2) using constant T. In Table 3.1 we list several large orbital transformations undergone by Jupiter Family comets in connection with close encounters with Jupiter. It is seen

Table 3.1. Examples of major orbital transformations experienced by numbered periodic comets. Abbreviations used are: 'Ch-G' for Churyumov-Gerasimenko, 'S-Ch' for Smirnova-Chernykh, and 'W-K-I' for West-Kohoutek-Ikemura. The year of closest approach to Jupiter and the minimum jovicentric distance ($\Delta_{\min}$) attained are listed. For comets 74P and 82P the encounters were long-lasting and involved two perijove passages. The perihelion (q) and aphelion (Q) distances and the Tisserand parameter (T) are given before and after the respective encounters, separated by a hyphen.

Comet	Year	$D_{\min}$ (AU)	q (AU)	Q (AU)	T
16P/Brooks 2	1886	0.001	5.47–1.95	14.48–5.43	2.98–2.89
39P/Oterma	1937	0.165	5.79–3.39	8.06–4.56	3.02–3.03
39P/Oterma	1963	0.095	3.39–5.45	4.56–8.99	3.03–3.00
67P/Ch-G	1959	0.052	2.76–1.29	5.90–5.73	2.76–2.75
74P/S-Ch	1955–1963	0.245–0.467	5.65–3.55	11.92–4.80	3.01–3.01
76P/W-K-I	1972	0.012	4.93–1.40	13.45–5.29	2.73–2.68
81P/Wild 2	1974	0.006	4.97–1.48	21.53–5.23	2.77–2.88
82P/Gehrels 3	1970–1973	0.001–0.041	5.68–3.43	8.19–4.67	3.02–3.03

that T was not much perturbed in spite of dramatic changes of the standard orbital elements. The changes that did occur were typical of those that Jupiter's orbital eccentricity induces during very close or long-lasting encounters.

The comets of the Jupiter Family are characterized by low inclinations (see Fig. 1.4), and they generally have $\cos i \approx 1$. A co-planar orbit ($i = 0$) will remain co-planar when perturbed by Jupiter, and in general, low-inclination orbits also stay with low inclinations. It is therefore of interest to use Eq. (3.2) to draw evolutionary curves in the (a, e) plane for constant T with $i \equiv 0$. In Fig. 3.1 we see a rendering of such a diagram using the aphelion and perihelion distances (Q and q) on the axes.

We see that the observed Jupiter Family comets are likely to have evolved into their current orbits from a source population with larger perihelion distances and larger aphelion distances as well. As the development in discovery techniques has allowed comets with larger perihelion distances to be discovered, our records thus start to probe deeper into this source population. However, the diagram concentrates on the vicinity of the observed Jupiter Family and does

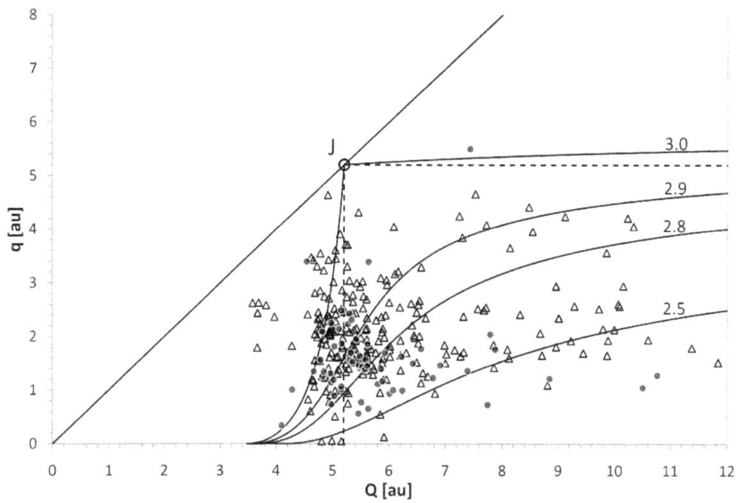

Fig. 3.1. Level curves of constant T for zero inclination, plotted in a diagram of perihelion versus aphelion distance. The symbols refer to the orbits of the numbered periodic comets. Filled circles denote comets discovered before 1950, and open triangles denote more recent discoveries. Courtesy T. Wiśniowski.

not allow to gain insight into the ultimate source of the comets. This issue will be dealt with in Sec. 5.5.

There are some interesting differences between the evolutionary curves in Fig. 3.1. In the circular 3-body problem, only comets with $q < a_J$ or $Q > a_J$ can have very close encounters with Jupiter. But for $T \approx 3$ the curves lead into orbits that do not satisfy this requirement, so the question arises if such evolutions are possible. In practice, they are possible due to the ellipticity of Jupiter's orbit together with the fact that slow encounters may bring the comets close to Jupiter even if the comet motions upon approach aim relatively far from the planet (see Sec. 3.2.2).

In fact, as shown by Ernst Öpik (1951), the value of T is directly related to the speed of approach. Suppose the comet to be situated close to Jupiter's orbit. If we consider Eq. (3.2) with $a_J = 1$, according to the vis-viva integral, the term $1/a$ equals $2 - V^2$, where V is the speed of the comet in a fixed frame. The second term can be identified with twice the component of the comet's angular momentum perpendicular to Jupiter's orbital plane, which can also be written $L_z = V_{tr}$, where V_{tr} is the comet's velocity component parallel to Jupiter's velocity in the fixed frame upon encounter. In the normalized units, the Tisserand parameter can therefore be expressed as

$$T = 2 - V^2 + 2V_{tr}. \tag{3.3}$$

The relative encounter velocity U between the comet and Jupiter can be written

$$U^2 = (\mathbf{V} - \mathbf{C})^2 = V^2 + 1 - 2\mathbf{V} \cdot \mathbf{C} = V^2 + 1 - 2V_{tr}, \tag{3.4}$$

where $\mathbf{C}$ is Jupiter's velocity vector ($C = 1$).

From Eqs. (3.3) and (3.4), we obtain

$$U^2 = 3 - T. \tag{3.5}$$

Here we see that the speed of approach (U) can have a real value only if $T \leq 3$, and for $T = 3$, the speed vanishes. This is consistent with what was said above, namely, that close encounters are formally impossible in the circular problem for orbits with $T > 3$. We also see that comets that encounter Jupiter with $T \approx 3$ will do so with a very small velocity, so that Jupiter's gravity may attract them even from large distances.

Finally, let us note that the definition of Jupiter Family comets in Sec. 1.4 essentially means that these comets would approach Jupiter with speeds less than Jupiter's own orbital velocity, while the opposite is the case for Halley-type comets.

3.2. Close Encounters

Table 3.1 showed several cases, where the orbits of Jupiter Family comets were changed more or less dramatically as a result of close encounters with Jupiter. But the recent dynamical history of the entire Jupiter Family is full of such encounters and their resulting jumps in orbital element space. Only a minority of comets have been spared of these events. Thus, for the Jupiter Family it is clear that close encounters play a dominant role in the orbital evolution. Moreover, the dynamics of long-period and Halley-type comets features orbital jumps at close encounters as an important phenomenon as well. Let us therefore take a closer look at the workings of the close encounters.

The fact that the heliocentric orbit may change dramatically means that the approximation of elliptic motion breaks down during the close encounter. Hence, the planet is surrounded by a *sphere of influence*, inside which it replaces the Sun as the main arbiter of the comet's motion. A rough estimate of the size of the sphere of influence can be obtained by comparing the magnitudes of the forces exerted on the comet by the Sun and the planet. In the heliocentric frame the central force is approximately $GM_\odot/a_p^2$, if the comet is situated in the vicinity of the planet (a_p is the semi-major axis of the planetary orbit). The perturbing force by the planet is GM_p/Δ^2, where Δ is the distance between the comet and the planet. Equality of the two forces occurs for

$$\Delta = R_\mathrm{h} = a_p m_p^{1/2}, \tag{3.6}$$

where $m_p = M_p/M_\odot$. As long as $\Delta > R_h$, the central force is larger than the perturbing force, and the heliocentric motion is relatively stable, but when the opposite inequality holds, the heliocentric ellipse cannot be expected to describe the comet's motion even approximately.

If we instead place ourselves in the planetocentric frame, we can ask the same question for the two-body motion of the comet under the gravity of the planet. In this case, the central force is GM_p/Δ^2, while the perturbing force due to the Sun is a tidal force, which can be expressed as $GM_\odot\Delta/a_p^3$. Equality of the two forces now occurs for

$$\Delta = R_\mathrm{p} = a_p m_p^{1/3}. \tag{3.7}$$

Hence, the planetocentric motion can be approximately described as a conic section as long as $\Delta < R_\mathrm{p}$ but not for larger Δ values.

Both these spheres around the planet, with radii R_h and R_p, can be called spheres of influence. Obviously, $R_\mathrm{h} < R_\mathrm{p}$, so there is a range of Δ between the two radii, where both the heliocentric and planetocentric two-body orbits are relatively useful to describe the motion. If we take Jupiter as the planet, we have $R_{\rm h} \simeq 0.16\,\text{AU}$ and $R_{\rm p} \simeq 0.52\,\text{AU}$.

3.2.1. *Hyperbolic deflection*

Let us first assume that the comet approaches the planet with a non-negligible speed. To be precise, we take this speed to be large enough for the planetocentric orbit to be hyperbolic. This is the usual situation, and the peculiar case where it does not apply will be treated afterwards. Of course, an accurate computation of the outcome of a close encounter cannot be based on the two-body approximation, but the model of a close encounter as a *hyperbolic deflection* is still useful.

In practice, the hyperbolic deflection can be realized as follows. The comet moves along an osculating elliptic orbit around the Sun at the moment, when it enters into the planet's sphere of influence. Its planetocentric velocity vector at that time can be written $\mathbf{U} = UV_p\hat{\mathbf{U}}_{\rm in}$, where U was defined in Sec. 3.1, V_p is the orbital speed of the planet in arbitrary units, and $\hat{\mathbf{U}}_{\rm in}$ is the unit vector defining the direction of $\mathbf{U}$. At the given entry point at planetocentric distance $R_{\rm p}$, $\mathbf{U}$ defines a hyperbolic orbit around the planet. The comet is assumed to follow this orbit until it reaches the surface of the sphere of influence again on the outward branch of the hyperbola. At this moment, the planetocentric velocity vector $\mathbf{U}' = U'V_p\hat{\mathbf{U}}_{\rm out}$ is used to compute a new, osculating heliocentric orbit. It is not strictly necessary to use $R_{\rm p}$ for the radius of the sphere of influence — any value down to $R_{\rm h}$ is equally admissible.

From the properties of hyperbolic motion, we realize that $U' = U$. Now, consider the approximation that this motion takes no time at all. This means that the perturbation of the heliocentric orbit is an instantaneous change of the velocity vector $\mathbf{V}$, which is triggered by an instantaneous rotation of the planetocentric direction of motion $\hat{\mathbf{U}}$ — i.e., the hyperbolic deflection. We note that this perturbation will leave T unaffected in accordance with Eq. (3.5).

To determine the amount of deflection, let us consider the so-called *b-plane* (Valsecchi *et al.* 2003), which contains the center of the planet and is perpendicular to $\hat{\mathbf{U}}_{\rm in}$. We approximate the actual deflection, which occurs over a finite part of the hyperbola, by the full deflection between the two asymptotes, so that $\hat{\mathbf{U}}_{\rm in}$ defines the direction of the incoming asymptote. We also take UV_p to be the hyperbolic velocity at infinity, the so-called unperturbed encounter velocity. The point in the b-plane, to which the comet aims, is situated at distance b from the center of the planet, and for this we use the term *impact parameter*.

The plane of the hyperbola is perpendicular to the b-plane. Denoting the angle between its asymptotes by $\pi - \gamma$, we have

$$\tan\frac{\gamma}{2} = \frac{GM_p}{b(UV_p)^2}. \tag{3.8}$$

This shows that the deflection angle (if small) is directly proportional to the mass of the planet and inversely proportional to the impact parameter and the square of the unperturbed encounter velocity.

The instantaneous change of $\mathbf{V}$ implies changes in the orbital energy and angular momentum of the comet. To describe these changes, it is useful to consider the angle θ between the vectors $\mathbf{U}$ and $\mathbf{V}_p$, as illustrated in Fig. 3.2. When describing the orientation of $\mathbf{U}$ with respect to $\mathbf{V}_p$, θ is the polar angle, and there is also an azimuthal angle ψ, which is not shown in the figure. The angle between the in- and outgoing vectors $\mathbf{U}$ and $\mathbf{U}'$ is γ, and both vectors

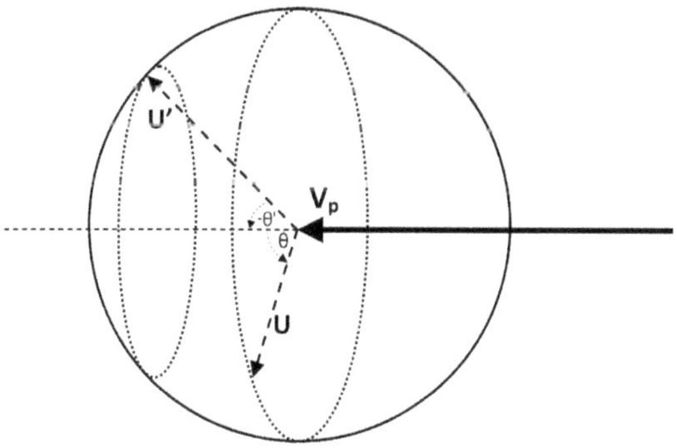

Fig. 3.2. The geometry of velocity space at a close encounter, using vectors that are defined in the main text. The dotted ellipses represent latitude circles at different co-latitudes θ and θ'. Courtesy T. Wiśniowski.

are situated in the *scattering plane*. However, in general, $\mathbf{V}_p$ is not situated in this plane. Hence, $\psi' \neq \psi$, and $\gamma \neq \theta - \theta'$.

Referring the comet's velocity vector approximately to the position of the planet, the orbital energy is given by the vis-viva integral as

$$E = -\frac{GM_\odot}{2a} = \frac{1}{2}V^2 - \frac{GM_\odot}{r_p}, \tag{3.9}$$

where r_p is the orbital radius of the planet. Using the cosine theorem, we can write

$$E = \frac{1}{2}V_p^2(1 + U^2 + 2U\cos\theta) - \frac{GM_\odot}{r_p}, \tag{3.10}$$

before the encounter and the same expression with θ replaced by θ' after the encounter. We thus have:

$$\Delta E = UV_p^2(\cos\theta' - \cos\theta). \tag{3.11}$$

For the perturbation of the angular momentum, we have

$$\Delta L = r_p \cdot UV_p(\cos\theta' - \cos\theta). \tag{3.12}$$

Using Eqs. (3.11) and (3.12), the perturbations of inverse semi-major axis and eccentricity can easily be calculated from the values of U, θ and θ'. It is easy to realize that the expression $\cos\theta' - \cos\theta$ takes its largest value for a given γ, if $\mathbf{V}_p$ is situated in the scattering plane and $\theta' = \theta - \gamma$. In this case, for the inverse semi-major axis we can derive:

$$\Delta\left(\frac{r_p}{a}\right) = 2U\{\cos\theta - \cos\theta'\} = 4U\sin\frac{\gamma}{2}\sin\left(\theta + \frac{\gamma}{2}\right). \tag{3.13}$$

Hence we see that, for a given impact parameter, encounters with very low velocity ($U << 1$) yield very small energy perturbations. From this, we may expect that the speed of transfer along the evolutionary curves in Fig. 3.1 caused by close encounters with Jupiter is reduced for Tisserand parameters approaching the critical value of 3. However, this is only valid as long as the encounter speeds are large enough to warrant the approximation of instantaneous, hyperbolic deflection. Thus, let us also consider the other extreme

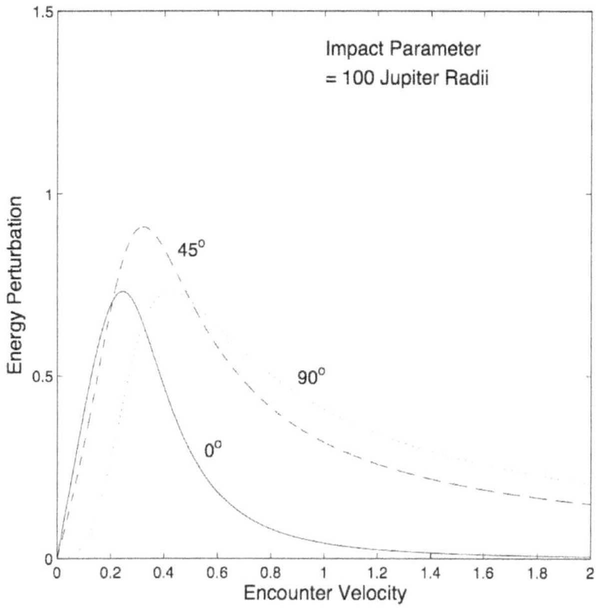

Fig. 3.3. Perturbations of a_J/a, computed with Eqs. (3.8) and (3.13), for a common impact parameter $b = 100R_J$ and three values of the angle θ characterizing the approach direction. Credit: H. Rickman, "Cometary Dynamics," in Lecture Notes in Physics, Vol. 790 (2010), pp. 341–399, © Springer-Verlag Berlin Heidelberg 2010. With permission of Springer Nature.

case of extremely large velocity ($U \to \infty$), for which Eq. (3.8) tells us that $\sin\frac{\gamma}{2} \to 0$ proportional to U^{-2}. Consequently, the energy perturbation again tends to zero.

This behavior is illustrated in Fig. 3.3, which uses three different approach directions with a common impact parameter of 100 jovian radii for encounters with Jupiter. The largest energy perturbations are generally obtained with encounter velocities $U \simeq 0.3$–0.4, corresponding to Tisserand parameters $T \simeq 2.8$–2.9. This is the range that is most frequently populated by observed Jupiter Family comets.

The method of calculating orbital perturbations induced by close encounters with a planet, which was just described, was introduced by Öpik (1951). It has been found to be very useful as a tool for statistical investigations into the dynamical evolutions of objects that are subject to such encounters, like comets and early solar system planetesimals. In addition, it serves well to provide an insight into how the encounters work, which is not available when tracing the motions by numerical integrations. It is not necessary for the encounters to be very deep or very fast — the only scenario where the method fails miserably is the one of very slow encounters. Otherwise, it is not very accurate but at the same time not deceptive.

3.2.2. *Slow encounters*

Let us now face the problem of the very slow encounters. These can be discussed with the aid of a concept called *zero-velocity surfaces*, which relates to the Jacobi integral. In Eq. (3.1), let us put $v^2 = 0$, which means that the comet is at rest in the rotating frame. For any particular value of C, the equation then defines a three-dimensional surface in (x, y, z) space, called the zero-velocity surface. This can be approached by the comet, but only at a vanishing speed, and it can never be crossed.

If C surpasses the critical value of 3 by a significant amount, the comet may move in three domains: far outside the planetary orbit, within a large ovoid around the Sun, or inside a small ovoid enclosing the planet. These domains are disconnected. However, if we let C decrease toward 3, the zero-velocity surfaces change shape and new situations may occur. At $C = 3.0388$, the two ovoids meet at the inner *Lagrangian point* on the line between the planet and the Sun, usually denoted L_1. For Jupiter, this is situated 0.35 AU from the planet. For slightly smaller values of C, there are two allowed domains of motion: the exterior one, and a common one formed by the merger of the heliocentric and planetocentric ovoids. At $C = 3.0375$, this common domain meets the exterior one at the outer Lagrangian point L_2, which for Jupiter is situated 0.36 AU from the planet. For slightly smaller values of C, comets are free to move between the inner and outer heliocentric regions by passing at small velocity through the ovoid-like zone between the Lagrangian points.

In Fig. 3.4 we show a cut of two zero-velocity surfaces with Jupiter's orbital plane, yielding the dashed and full-drawn curves. The dashed, quasi-elliptic curve represents the largest jovicentric region, from which there is no escape in the circular restricted 3-body

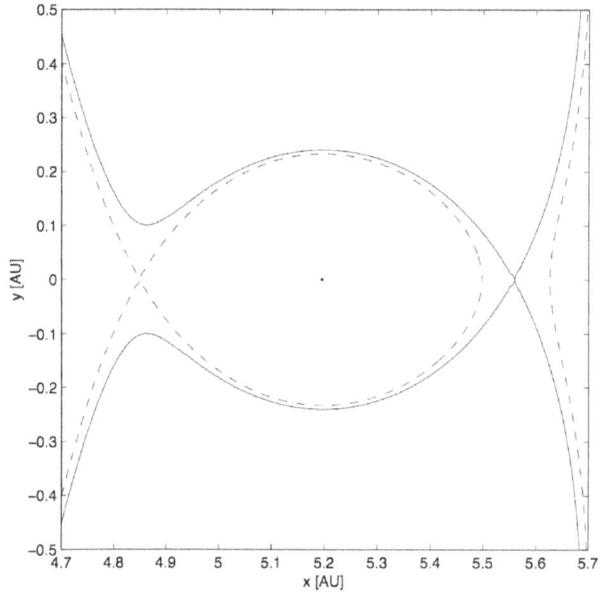

Fig. 3.4. Zero-velocity curves in Jupiter's orbital plane for two values of the Jacobi constant, passing through the inner (dashed) and outer (full-drawn) Lagrangian points. The Sun is at the origin, and Jupiter is at $x = 5.2$ AU. Credit: H. Rickman, "Cometary Dynamics," in Lecture Notes in Physics, Vol. 790 (2010), pp. 341–399, © Springer-Verlag Berlin Heidelberg 2010. With permission of Springer Nature.

problem. Such a region exists around any planet and is usually referred to as the *Hill sphere* due to its roughly spherical shape. Its radius is the distance from the planet to the L_1 point and can be shown to be

$$r_H = r_p \left(\frac{m_p}{3} \right)^{1/3}, \tag{3.14}$$

where m_p is the mass of the planet expressed in solar masses.

The surface of the Hill sphere can be seen as a stability limit for planetocentric satellite motion. Within the circular restricted 3-body problem, as mentioned, there is no escape for objects moving inside this limit, but there is also no access into the Hill sphere for objects moving outside. The fact that reality in the solar system differs from this idealized problem has important consequences in comet dynamics. Imagine a comet that approaches Jupiter from the solar or anti-solar direction with a C value close to the critical range as just described. If C is such that there is a gap at the Lagrangian point in question, this gap would remain forever in the idealized problem, but in reality (due mostly to the eccentricity of Jupiter's orbit and the perturbations by Saturn) the gap may temporarily close — only to open up again at a later time. This allows for the possibility of *temporary satellite captures* around Jupiter, when a comet may spend an extended period of time orbiting within Jupiter's Hill sphere and may in fact be gravitationally bound to the planet. Such captures in the near past or future are known for several Jupiter Family comets, and an example from orbital integrations by Tancredi *et al.* (1990) is shown in Fig. 3.5.

Clearly, a comet may enter into Jupiter's Hill sphere from one side and exit on the other independent of the details of the intervening motion. In such a case, the heliocentric orbits before and afterwards are very different from each other. Due to the requirement of a

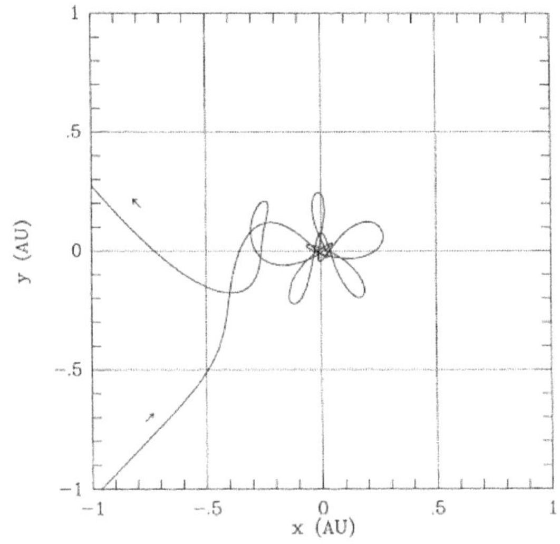

Fig. 3.5. Trajectory of comet 111P/Helin-Roman-Crockett with respect to Jupiter (plotted at the origin) during a temporary satellite capture predicted to occur around the year 2075. The frame rotates so that the Sun is always on the negative x-axis. Both entry and exit occur close to the L_1 point. Credit: G. Tancredi *et al.*, *A&A* **239**, 375–380 (1990), reproduced with permission © ESO.

very slow motion relative to Jupiter, the heliocentric radial velocity must be small, and the comet is either close to perihelion or close to aphelion. Thus, as explained by Carusi and Valsecchi (1979), there is a route between comet orbits that are tangent to Jupiter's at perihelion or aphelion via low-velocity encounters, possibly long-lasting and involving temporary satellite captures.

When comets exit from Jupiter's vicinity via the L_1 gap after low-velocity encounters, they enter into a special kind of heliocentric orbits, as explained by Tancredi *et al.* (1990). These orbits have aphelia somewhat inside the L_1 point and periods close to 2/3 that of Jupiter. Since this resonance is also shared by the Hilda group of asteroids, the comets in question are often called *quasi-Hildas*.[1] Contrary to the asteroids, these have no protection from close encounters with Jupiter, and numerical tracing of the motions of real quasi-Hildas reveals that they are just temporary visitors into the Jupiter Family.

Comet 39P/Oterma is a case in point. This comet had a long-lasting encounter with Jupiter with closest approach in 1937, leading from an outer orbit with $q \simeq 5.8\,\mathrm{AU}$ and $Q \simeq 8.1\,\mathrm{AU}$ into an inner orbit with $q \simeq 3.4\,\mathrm{AU}$ and $Q \simeq 4.6\,\mathrm{AU}$. It was discovered in 1943 by Finnish astronomer Liisi Oterma and kept under observation in the 1940s and 1950s while orbiting three times around the Sun. Then a new slow encounter with Jupiter with closest approach in 1963 transferred the comet back into an outer, Chiron-type orbit with $q \simeq 5.5\,\mathrm{AU}$ and $Q \simeq 9.0\,\mathrm{AU}$, and observations ceased until 2001, when the comet was rediscovered by a team led by American astronomer Yanga Fernández. The two orbital transformations were listed in Table 3.1.

3.3. Lidov–Kozai Cycles

While the observable comets are almost always susceptible to close encounters with Jupiter in the long run, it is worth considering the likelihood of a close encounter for a particular set of orbital elements.

[1]These are marked with sky-blue dots in Fig. 1.5.

Generally speaking, this depends on two factors (Rickman *et al.* 2014). First, one has to determine the smallest distance between the elliptical orbits of the comet and Jupiter to see if a close encounter is at all possible. This involves a calculation of the *MOID* (minimum orbit intersection distance; see Bowell and Muinonen 1994), for which a numerical method has been described and made public by Wiśniowski and Rickman (2013). Second, if the MOID is small enough to allow a close encounter, one needs to evaluate the probability of both objects passing their MOID points close enough in time for a close encounter to occur. Such a probability was analyzed for the case of an actual collision, assuming random elements, by Rickman *et al.* (2014), but the method is easily applicable to close encounters with Jupiter or any other planet.

The stochastic approach to the timing problem has obvious drawbacks for Jupiter Family comets because of the influence of mean motion resonances, and to some extent, the same is true for Halley Type comets too. We will now discuss *secular perturbations* in the absence of close encounters, focusing primarily on long-period comets. These are favored by their much higher average inclination, which tends to keep the MOID values large for most of the time.

Secular perturbations mean long-term variations of the heliocentric orbital elements under the influence of a perturbing planet. By long-term variations we mean that the orbital position of the comet does not matter, and the differential equations governing these variations are averaged over the orbital periods of the comet and the planet, which then acts like an elliptical arc of matter spread along its actual orbit. A first approximation is offered by a linear solution, where the semi-major axis of the comet remains constant. The two orienting angles ϖ (longitude of perihelion) and Ω (longitude of the ascending node) are then linear functions of time characterized by their secular frequencies. The eccentricity oscillates with the frequency of ϖ, and the sine of the inclination oscillates with the frequency of Ω.

These secular frequencies are the rates of circulation of the apsidal and nodal lines, respectively. *Secular resonances* occur, when either of these rates coincides with the rate of the perturbing planet.

The linear approximation then breaks down, and the eccentricity or inclination may undergo much larger variations. This phenomenon is of importance in the asteroid belt and, at least potentially, for Jupiter Family or Halley Type comets, but not for long-period comets. Halley Type and long-period comets may experience a different type of resonance, which can be visualized as follows.

Over a long enough period of time, the planet's apsidal and nodal lines circulate so that it acts approximately like a circular annulus of matter in the invariable plane,[2] centered on its constant semi-major axis. In this situation, neither ϖ nor Ω is important for the secular evolution of the comet orbit. In fact, the comet experiences the attraction of a circular annulus of matter in the fixed, invariable plane. This perturbing force cannot change the comet's orbital energy, and since it is confined by circular symmetry to the meridional plane spanned by the normal vector of the invariable plane and the comet's heliocentric radius vector, the associated torque does not change the comet's angular momentum component perpendicular to the invariable plane. Thus, the expression $L_z = \sqrt{a(1-e^2)}\cos i$ remains constant, as does a.

However, the perturbing force and its torque do perturb the comet orbit, and this happens by changing the total angular momentum vector $\mathbf{L}$. Due to the constancy of L_z and a, there are coupled variations of e and i, which are associated with the absolute value and direction of $\mathbf{L}$, respectively. These variations relate to the variation of the only angular element that matters, namely, the argument of perihelion $\omega = \varpi - \Omega$. Depending on ω, the nodes of the comet orbit may have different positions with respect to the planetary annulus, and this governs the variation of $\mathbf{L}$.

The ω variation may be a circulation, in which case ϖ and Ω circulate at different rates. Alternatively, ω may librate around $\pm\pi/2$, in which case there is a 1:1 resonance between the rates of ϖ and Ω. In the (L, ω) parametric plane, trajectories near the separatrix between these two modes exhibit large variations of L

[2]This is a plane perpendicular to the constant, total angular momentum of all the planetary orbits.

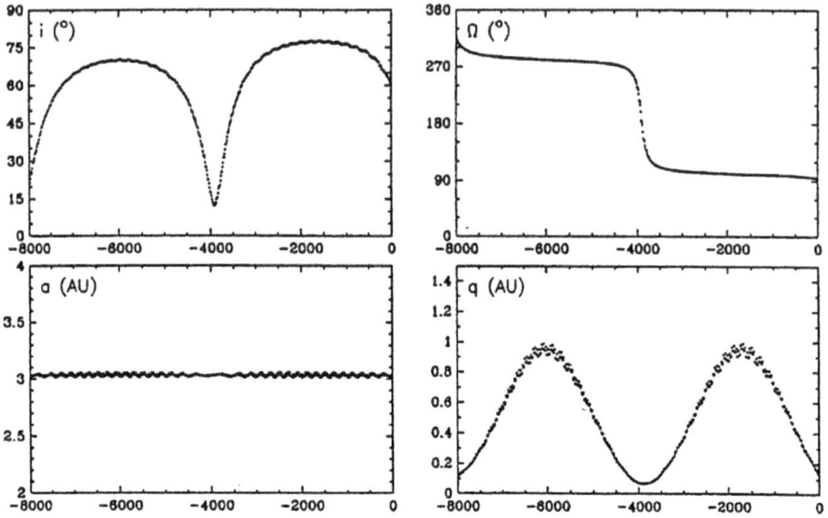

Fig. 3.6. Long term variations of the inclination (i), longitude of the ascending node (Ω), semi-major axis (a) and perihelion distance (q) of comet 96P/Machholz 1. The time scale is in years and goes backward from a zero-point in 1986. Reproduced from Gonczi *et al.* (1992) with permission.

and thereby of e and i, especially when L_z is small. As required by the constancy of L_z, these variations are in anti-phase, so that i has its minimum when e is at maximum and the converse. Alternatively, the variations of perihelion distance and inclination are in phase. Figure 3.6 shows a rare example of such variations among short-period comets. Comet 96P was discovered by American amateur astronomer Donald Machholz in 1986. With a Tisserand parameter of only 1.94 in spite of a short orbital period, L_z is indeed small, and the perihelion distance spans a wide range. The comet is close to the borderline between the Jupiter Family and Halley Type comets, and hence its dynamical origin is not clear.

The mechanism behind such substantial variations of e and i was first explained in the framework of asteroid dynamics by Japanese astronomer Yoshihide Kozai (1962). The terms *Kozai resonance* for the 1:1 resonance of $\dot{\varpi}$ and $\dot{\Omega}$ and *Kozai cycle* for the coupled oscillation of e and i have become generally accepted, like the term *Kozai mechanism*. However, it has recently been recognized that the

Russian physicist Mikhail Lidov discovered the mechanism somewhat before Kozai in the different framework of satellite dynamics. Hence, the name Kozai is sometimes replaced by Lidov–Kozai.

The Lidov–Kozai mechanism is very important for comet evolution, as will be seen several times in the following chapters. Examples are the physical destruction of comets in sungrazing or Jupiter-impacting orbits attainable by this mechanism. We will now instead treat an issue of great importance for the arrival of the new comets from their distant source.

3.3.1. *Galactic tides*

Consider comets belonging to the Oort Cloud (see Sec. 1.4) at typical distances of more than 10 000 AU from the Sun. These are subject to a tidal force due to the whole Galaxy, which perturbs their orbital motions around the Sun. According to a simple axisymmetric model of the gravitational field of the Galaxy, this force has two components: one radial with respect to the Galactic center, and one perpendicular to the Galactic plane. Due to the offsets of classical Oort Cloud comets from the Sun in the radial or normal directions, this tidal force can be noticeable.

The tidal acceleration can be expressed in terms of the local density of the Galactic disk (ρ_o) and the kinematic parameters of Galactic differential rotation — the so-called Oort constants A and B. Using cartesian coordinates (x', y', z) centered on the Sun such that the unit vectors $\hat{\mathbf{x}}'$ and $\hat{\mathbf{y}}'$ point toward the Galactic anticenter and transversely along the local circular velocity in the Galactic plane, and $\hat{\mathbf{z}}$ is perpendicular to this plane, the equation of motion can be written

$$\ddot{\mathbf{r}} = -\nabla U_\odot + (A - B)(3A + B)x'\hat{\mathbf{x}}' - (A - B)^2 y'\hat{\mathbf{y}}' \\ - [4\pi G\rho_o - 2(B^2 - A^2)]z\hat{\mathbf{z}}, \tag{3.15}$$

where $U_\odot = -GM_\odot/r$ is the solar potential.

With modern estimates of $A = +13$ km/s/kpc and $B = -13$ km/s/kpc (Gunn *et al.* 1979) and $\rho_o = 0.1\, M_\odot/\mathrm{pc}^3$ (Holmberg and Flynn 2000), we realize that only the term involving ρ_o is non-vanishing in the z component, and this term is almost ten times larger than the coefficients of the the x' and y' components. Thus the disk tide is much stronger than the radial tide.

By neglecting the latter, we hence get a first approximation to the long-term dynamical behavior of Oort Cloud comets by using the equation

$$\ddot{\mathbf{r}} = -\nabla\left\{-\frac{GM_\odot}{r} + 2\pi G\rho_o z^2\right\} \tag{3.16}$$

(Heisler and Tremaine 1986). Let us now note that the perturbing acceleration in Eq. (3.16) is equivalent to that of an infinite disk of matter in the Galactic plane with density ρ_o. This disk can be seen as an infinite superposition of annuli similar to the one representing Jupiter in the Kozai mechanism. Therefore, the result is similar, and the vertical Galactic tide of Eq. (3.16) causes perturbations that we may interpret as due to a Kozai cycle. Like in the case of the long-term jovian perturbations, we deal with an integrable dynamical system, and the solution can be described by analytical formulae.

Heisler and Tremaine (1986) developed such a theory. They averaged the comet's Hamiltonian function over the orbital period, so that the mean anomaly disappeared from the Hamiltonian equations of motion, thereby securing an energy integral. The conservation of L_z then made the system fully integrable, and the angular variable is the Galactic argument of perihelion ω_G. The resulting Lidov–Kozai cycle for the eccentricity is illustrated for two representative cases in Fig. 3.7. The plotted quantity $1 - e$ and the Galactic inclination i_G both vary with ω_G, in phase with each other.

The integrals characterizing the curves are the energy, or Hamiltonian:

$$C = 1 - e^2 + 5e^2 \sin^2 i_G \sin^2 \omega_G, \tag{3.17}$$

and the normalized, vertical angular momentum component:

$$L_z = \sqrt{1 - e^2} \cos i_G. \tag{3.18}$$

The separatrix between libration and circulation of ω_G corresponds to $C = 1$. As will be seen in Sec. 5.2, it is of interest to derive

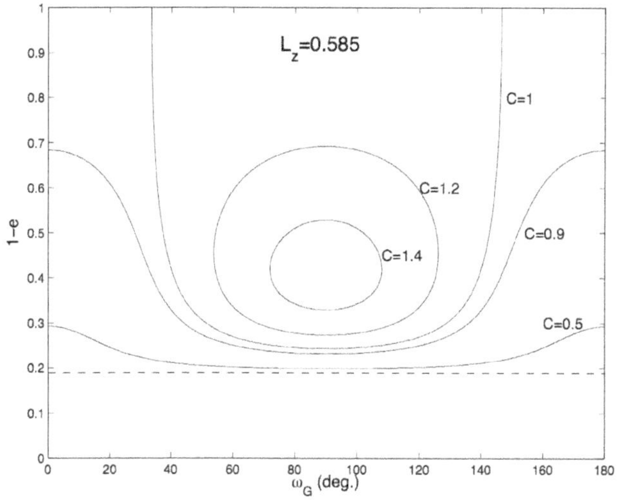

(a)

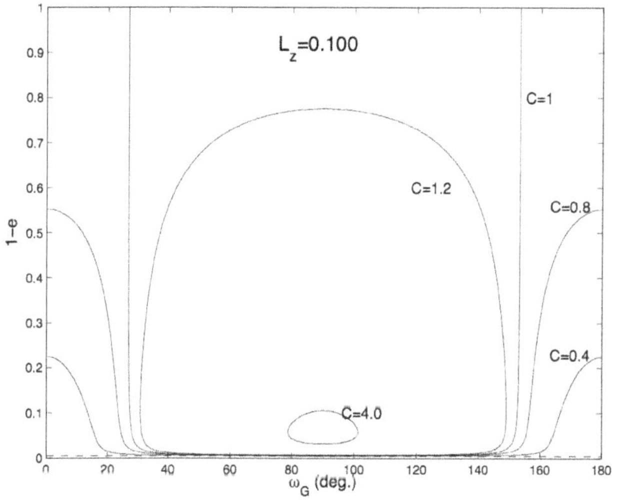

(b)

Fig. 3.7. Variations of $1 - e$ with ω_G according to the Galactic disk tide theory by Heisler and Tremaine (1986). Panels (a) and (b) exhibit two values of the normalized, vertical angular momentum component $L_z = \sqrt{1 - e^2} \cos i_G$, and the curves show projections of the phase space trajectories for selected values of the Hamiltonian. The dashed lines correspond to $L = L_{\rm min} = L_z$. Modified from H. Rickman, "Cometary Dynamics," in Lecture Notes in Physics, Vol. 790 (2010), pp. 341–399, © Springer-Verlag Berlin Heidelberg 2010. With permission of Springer Nature.

an expression for the variation of the perihelion distance caused by the Galactic tide over one orbital revolution. For comets that are either observable or nearly so, the eccentricity is close to unity, which implies the approximation $L \propto \sqrt{q}$ for the absolute value of the angular momentum. Hence we have

$$\frac{dq}{dt} \propto \sqrt{q}\frac{dL}{dt}, \tag{3.19}$$

and the expression for the averaged Hamiltonian yields a formula for the rate of change of the angular momentum:

$$\left|\frac{dL}{dt}\right| \simeq 5\pi G\rho_o a^2 |\sin 2\beta_G|, \tag{3.20}$$

where β_G is the Galactic latitude of perihelion. For the change of perihelion distance during one orbital revolution, i.e., a time interval proportional to $a^{3/2}$, we get

$$|\Delta q| \propto \sqrt{q} a^{7/2} |\sin 2\beta_G|, \tag{3.21}$$

showing that the maximum changes occur for orbits with Galactic perihelion latitude equal to $\pm\pi/4$.

Let us finally note that the integrability of the equations of motion was obtained by the orbital averaging of the Hamiltonian and therefore rests upon the correctness of this procedure. A verification of this may be obtained by showing that the orbital period is much shorter than the period of ω_G libration. This is indeed the case for orbits with semi-major axis $a \sim 10\,000$ AU or less, but the libration period varies inversely with the orbital period, and for $a \gtrsim 40\,000$ AU the integrability breaks down. In addition, the radial component of the tidal acceleration plays an increasingly important role and may drive comets out of the Sun's Hill sphere in the Galaxy, whereby they are lost from the solar system.

3.4. Stellar Perturbations

Let us recall the meaning of the Galactic tide. This is the result of comets moving relatively far from the Sun in the gravitational field of the Galaxy. Here one considers the smooth potential of the entire stellar system similar to what is done when tracing stellar orbits. But the analogy between cometary dynamics and Galactic dynamics reaches further. The smooth potential situation is a very good approximation most of the time, when the star is far from the potential wells of other stars, but encounters do occur so that the stellar orbits are deflected. The accumulated effect of such deflections over a very long time is called *relaxation.*

In the case of Oort Cloud comets, we also have to deal with stellar encounters. In principle, there are other objects with much larger mass, which the solar system may also encounter during its history. Giant molecular clouds (so-called GMCs) and stellar clusters are examples. But these are extremely rare, and in general they are neglected in analyses of Oort Cloud dynamics. Galactic field stars, on the other hand, are always present in the solar neighborhood, and they even penetrate into the Oort Cloud relatively frequently.

In terms of kinematics, the Sun is a normal member of the thin disk. Its excursions from a circular orbit in the Galactic mid-plane are small — both vertically and radially. This property is shared by a great majority of the nearby stars, so a local centroid of stellar motions can be defined as the average of their velocity vectors in an arbitrary reference frame. Relative to the circular motion, this centroid is somewhat lagging behind, since more stars are in the outer than the inner part of their Galactic orbits. Each star has a peculiar velocity with respect to the centroid, and so has the Sun as well. This amounts to about 20 km/s and is directed toward the *apex*, which is situated in the constellation of Hercules.

A crude estimate of the time scale for close encounters between the Sun and its neighbors can be derived using the mean stellar velocity (V_*) and the local number density of stars (n_*):

$$T_* = (\pi R^2 V_* n_*)^{-1}, \tag{3.22}$$

where encounters within a distance R are considered, and T_* is the average time between successive encounters. It is not enough to substitute the apex velocity for V_*, because the dispersion of the peculiar velocities of other stars is also very important.

Figure 3.8 shows the result of using Eq. (3.22) with $V_* = 30\,\mathrm{km/s}$ and $n_* = 0.1\,\mathrm{pc}^{-3}$. Let us estimate that the Oort Cloud extends to a

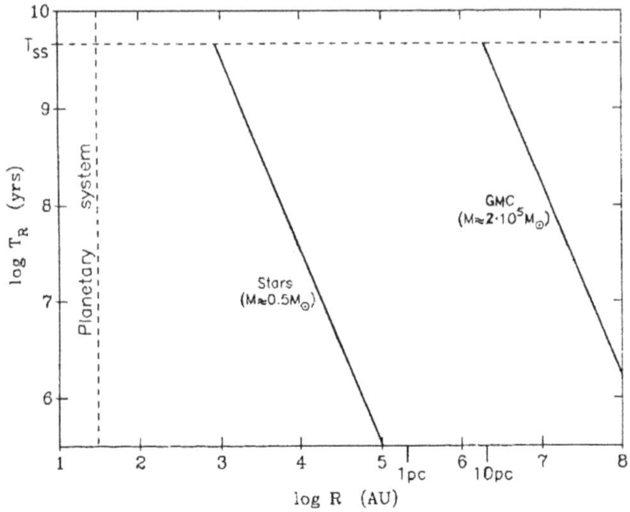

Fig. 3.8. The log of the average time interval between consecutive encounters between the Sun and stars of the current solar neighborhood (left line) or Giant Molecular Clouds (right line) versus the log of the miss distance. T_{SS} is the age of the solar system, and the vertical dashed line indicates the radius of the planetary system. Credit: H. Rickman, "Cometary Dynamics," in Lecture Notes in Physics, Vol. 790 (2010), pp. 341–399, © Springer-Verlag Berlin Heidelberg 2010. With permission of Springer Nature.

radius of about 50 000 AU. It then follows that the average interval between penetrating stellar encounters is about 1.7 million years. We easily realize that this is similar to — or even shorter than — the orbital periods of many Oort Cloud comets. Even though a typical stellar passage has a large effect only on a minority of these comets, it is clear that such effects have to be considered when discussing Oort Cloud dynamics.

Another piece of information contained in Fig. 3.8 is that the closest stellar encounter to be expected during the age of the solar system has a distance $R \sim 1\,000$ AU. This should not be taken literally, because the Sun has likely experienced neighborhoods quite different from the present one, when it was a young star, and this allows for a possibility of even closer encounters (see Sec. 6.2). On the other hand, no encounter with a GMC even close to the Oort Cloud radius is likely to have occurred.

We now turn to the dynamics of the stellar encounters. To simplify the problem, let us only consider the interactions between the Sun, one passing star and the comet. Since the comet is practically massless, this is a restricted three-body problem. This is usually a good approximation, since in most cases the star-star interactions are weak, and the effects of two stars on the Sun and the comet are thus additive, even if these stars should happen to encounter the Sun nearly simultaneously. However, we should be aware that stars often form binary or multiple systems. When the Sun encounters a binary star, the situation is benign in two limiting cases. Many binaries are very tight, in which case they may be approximated by a single point mass at their center of mass. Wide binaries form another frequent category, and then the two components can be treated as individual stars following two different but parallel trajectories. Between these two extremes there are in principle cases that require a more realistic treatment, but this is rarely done in practice.

The weakness of the interactions can be illustrated by using Eq. (3.8). Assume that a star approaches the Sun with a velocity of 30 km/s and an impact parameter of 50 000 AU. One can then derive a hyperbolic deflection angle of only eight arcseconds. Even if the star aims at only 10 000 AU from the Sun, the deflection amounts to less than one arcminute. It is thus common practice to approximate the stellar trajectories by straight lines when treating encounters with Galactic field stars.

It is thus possible to treat the problem in a heliocentric frame, where the star travels with constant speed along a straight line. Concerning the comet, the simplest treatment is to assume that it stays at rest during the passage of the star. The argument behind this assumption is that the speed of an Oort Cloud comet is limited by the speed of escape, which at a distance of 10 000 AU from the Sun amounts to 0.4 km/s. Since the speed of the star is ~ 100 times larger, it is natural to neglect the motion of the comet as a first approximation.

If one integrates the acceleration imparted to the Sun or the comet over time during the whole passage of the star, one obtains the resulting impulse or velocity change caused by the star. This is a vector situated in the plane spanned by the accelerated object and the stellar trajectory. This has two components — one directed along the trajectory and the other perpendicular to it. The latter points from the object to the closest point on the trajectory. It is easily seen from the symmetry around the closest point that the accelerations caused by the star along its trajectory before and after crossing this point cancel out. Thus, the impulse vector reduces to the perpendicular component, which is readily shown to be

$$\mathbf{I} = \frac{2GM_*}{V_*}\frac{\hat{\mathbf{d}}}{d}, \tag{3.23}$$

where M_* is the mass of the star, V_* is its velocity, d is the distance from the object to the closest point on the trajectory, and $\hat{\mathbf{d}}$ is the unit vector along this direction.

The impulse imparted to the comet in the heliocentric frame is obtained as the difference between the cometary and solar impulses:

$$\Delta\mathbf{V}_c = \frac{2GM_*}{V_*}\left\{\frac{\hat{\mathbf{d}}_c}{d_c} - \frac{\hat{\mathbf{d}}_\odot}{d_\odot}\right\}. \tag{3.24}$$

By placing the comet at the desired place in its orbit and applying the impulse derived from Eq. (3.24) to the heliocentric velocity at this point, one can compute new orbital elements from the new velocity. This way of accounting for stellar perturbations is called the *Classical Impulse Approximation* (CIA), and it dates back to the earliest works where this problem was considered. The relevant geometry is illustrated in Fig. 3.9.

Nonetheless, it is worth considering the shortcomings of the CIA and ways to improve it. If, for instance, the comet moves in the direction of the vector **d** during the lapse of the stellar encounter, there will be a cometary impulse component in the direction of the stellar motion, which translates directly into the heliocentric impulse $\Delta\mathbf{V}_c$. Other motions of the comet will also have effects on the calculated impulse. Generally speaking, those effects grow in importance for comets in smaller orbits. For the inner core of the Oort Cloud (see Chap. 5), the CIA is not very accurate.

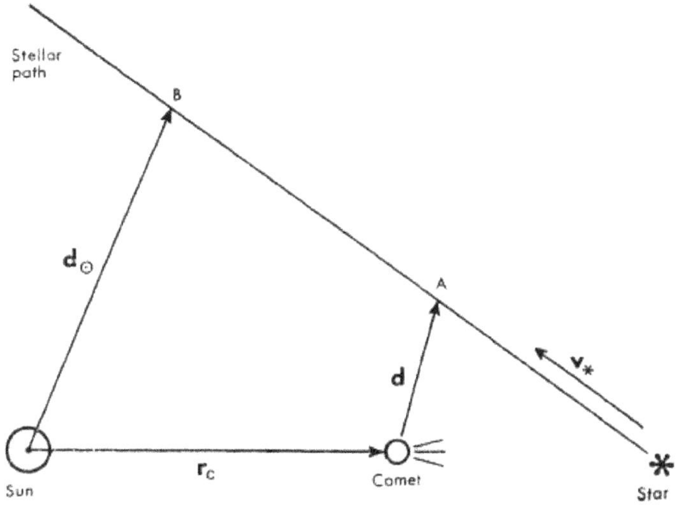

Fig. 3.9. The geometry of a stellar encounter as considered in the classical impulse approximation. The vectors **d** and $\mathbf{d}_\odot$ make right angles to the stellar trajectory at points A and B, respectively. © Publishing House of the Czechoslovak Academy of Sciences (1976).

Due to the approximations inherent in the treatment of stellar encounters as single, isolated events, there is no point in performing accurate numerical integrations. Improvements of the CIA have involved accounting for the actual, hyperbolic shape of the stellar trajectory (Dybczyński 1994) and splitting the stellar passage into finite segments and adding up the impulses received during each of these (Eggers and Woolfson 1996; Rickman *et al.* 2005). This sequential impulse approximation has been found to yield good results in terms of accuracy versus computing time for almost any kind of cometary orbit.

An undeniable advantage of the CIA is its ability to yield analytical results that allow us to estimate the statistical influence of stellar perturbations on comets at large. Since most stellar passages through the Oort Cloud do not involve particularly close encounters with any given comet or the Sun, they cannot be treated by neglecting any of the terms in Eq. (3.24). The only simplification that is generally valid to some extent is achieved by assuming the star to pass relatively far from the Sun-comet pair, as illustrated by Fig. 3.10.

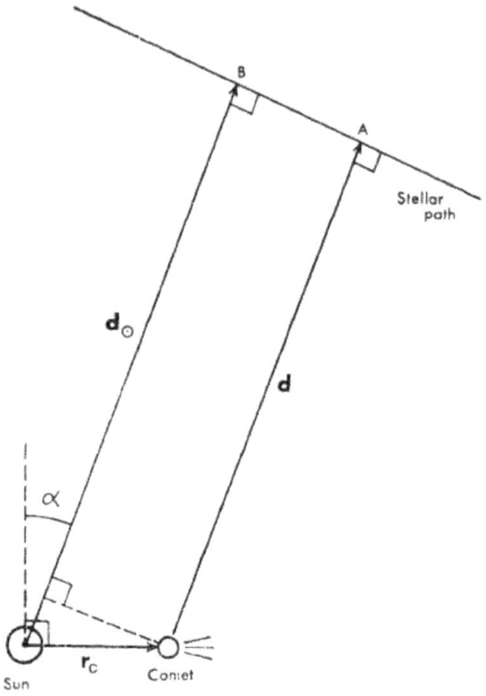

Fig. 3.10. The geometry of a distant stellar encounter leading to a tidal version of the classical impulse approximation. In this case, the differential impulse on the comet relative to the Sun is estimated. © Publishing House of the Czechoslovak Academy of Sciences (1976).

A tidal approximation to the heliocentric impulse then follows as:

$$\Delta \mathbf{V}_c \approx \frac{2GM_* r_c \sin \alpha}{V_* d_\odot^2} \cdot \hat{\mathbf{d}}_\odot, \tag{3.25}$$

where the symbols are as shown in the figure. Since the perturbation of the angular momentum is $\Delta \mathbf{L}_c = \mathbf{r}_c \times \Delta \mathbf{V}_c$, its absolute value is seen to be proportional to r_c^2. Thus, exposing comets with different semi-major axes a to the same set of stellar perturbations, we obtain $|\Delta L| \propto a^2$. Since the perturbation of the perihelion distance is approximately $\Delta q \propto \sqrt{q}\Delta L$, the expectance of $|\Delta q|$ is

$$\mathcal{E}(|\Delta q|) \propto \sqrt{q} \cdot a^2. \tag{3.26}$$

Now, consider a time interval short enough that no more than one stellar encounter can be expected. Its probability of occurrence is then proportional to the length of the interval. This is a fair approximation for the orbital period of an Oort Cloud comet, which is $P \propto a^{3/2}$. We hence get the expectance of the change in perihelion distance during one revolution as

$$\mathcal{E}(|\Delta q|_P) \propto \sqrt{q} \cdot a^{7/2}. \tag{3.27}$$

3.5. Energy Diffusion

In Sec. 3.3, we treated the secular evolution of comet orbits (in the absence of close encounters) under the assumption that the perturbing planet acts like a uniform, circular annulus of matter, thus causing no changes in the orbital energy of the comet. This is of course a highly idealized situation, which is useful to understand the secular changes in the angular momentum of the comets but which does not hold true in reality. In particular, the orbital energy of the comets does change, and we will now account for these variations.

The histogram in Fig. 3.11 from Rickman (2010) shows the distribution of changes of inverse semi-major axis experienced by observed long-period comets during their passages through the planetary system. The initial and final orbits in question are barycentric, i.e., referred to the joint center of mass of the Sun and the planets. The underlying comet sample is not biased in a dynamical sense, since the only criterion is that they were discovered and well enough observed to establish their orbits with good accuracy.

The distribution exhibits a narrow peak surrounded by wide tails. The latter are due to the few comets that had fairly close encounters with Jupiter and will not be discussed here. We estimate the half width at half maximum of the peak to be about 0.0006 AU^{-1}. There is no obvious departure from symmetry around zero — positive and negative perturbations are about equally common.

For a theoretical background, we can write the equation of motion of a comet in the heliocentric frame as

$$\ddot{\mathbf{r}} = -\nabla U_{\odot} - \nabla R_p, \tag{3.28}$$

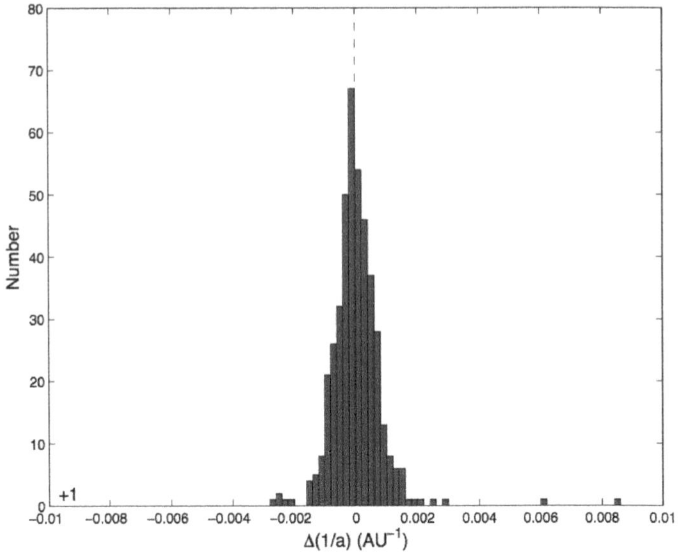

Fig. 3.11. Distribution of planetary perturbations of inverse, barycentric semi-major axis for 418 observed long-period comets, derived from data listed by Marsden and Williams (2005). Credit: H. Rickman, "Cometary Dynamics," in Lecture Notes in Physics, Vol. 790 (2010), pp. 341–399, © Springer-Verlag Berlin Heidelberg 2010. With permission of Springer Nature.

where $U_\odot = -GM_\odot/r$ is the solar gravitational potential and

$$R_p = -GM_p\left\{\frac{1}{|\mathbf{r}-\mathbf{r}_p|} - \frac{\mathbf{r}_p\cdot\mathbf{r}}{r_p^3}\right\} \tag{3.29}$$

is the so-called perturbing function, which looks like a potential in Eq. (3.28) but varies with time due to the motions of the comet and the planet (radius vectors $\mathbf{r}$ and $\mathbf{r}_p$), as seen from Eq. (3.29). It has to be accurately monitored when solving the equation of motion. Of course, when several planets are involved, R_p is replaced by the sum of all such expressions for the different planets.

We are now interested in the rate of change of the inverse semi-major axis of the comet. For this purpose, we note that the orbital energy is given by

$$E = \frac{1}{2}\dot{\mathbf{r}}^2 + U_\odot = -\frac{GM_\odot}{2a}, \tag{3.30}$$

where a is the semi-major axis. By taking the time derivative of Eq. (3.30) we get

$$GM_{\odot}\frac{d}{dt}\left(\frac{1}{2a}\right) = -\dot{\mathbf{r}}\cdot\ddot{\mathbf{r}} - \dot{\mathbf{r}}\cdot\nabla U_{\odot}, \tag{3.31}$$

and by taking the scalar product of Eq. (3.28) with $\dot{\mathbf{r}}$, we recognize that

$$GM_{\odot}\frac{d}{dt}\left(\frac{1}{2a}\right) = -\dot{\mathbf{r}}\cdot GM_p\nabla\left\{\frac{1}{|\mathbf{r}-\mathbf{r}_p|} - \frac{\mathbf{r}_p\cdot\mathbf{r}}{r_p^3}\right\}. \tag{3.32}$$

Introducing $m_p = M_p/M_{\odot}$ and the dimensionless position vectors $\mathbf{s} = \mathbf{r}/a_p$ and $\mathbf{s}_p = \mathbf{r}_p/a_p$ with the planetary semi-major axis a_p as unit of length, we finally get

$$\frac{d}{dt}\left(\frac{1}{a}\right) = \frac{2m_p}{a_p}\dot{\mathbf{s}}\cdot\left\{\frac{(\mathbf{s}_p-\mathbf{s})}{|\mathbf{s}_p-\mathbf{s}|^3} + \frac{\mathbf{s}_p}{s_p^3}\right\}. \tag{3.33}$$

The contribution by any planet to the time derivative of $1/a$ is thus proportional to m_p/a_p times an expression depending on the geometrical configuration and the velocity of the comet. In the common situation, when the comet is quite far from every planet, it is clear that the jovian contribution dominates over those of the other planets. Thus, the width of the central peak in the perturbation distribution of Fig. 3.11 is essentially set by the jovian perturbations.

A very accurate determination of such a perturbation requires a numerical integration of the equations of motion, yielding the functions $\mathbf{s}(t)$ for the comet and $\mathbf{s}_p(t)$ for all the planets. However, an approximate solution can be obtained by considering the unperturbed, Keplerian orbits and the corresponding, analytically derived position vectors. This method must be reserved for perturbations that change the comet orbit only insignificantly so that the action of each planet is independent of those of other planets.

Now, consider for instance the typical jovian perturbations that build up the central peak of the distribution. These can be positive or negative, and larger or smaller, the outcome depending on the time sequence of geometrical configurations exhibited by the comet and Jupiter. Of course, the different comets that contribute to Fig. 3.11 are independent of each other, and their respective geometrical patterns are uncorrelated. But is this the case also for the consecutive passages of the same comet, as it returns to perihelion on successive orbits? If so, the sequence of perturbations would be like a set of random drawings from a relevant parent distribution. Alternatively, will the comet retain a memory of its previous perihelion passage, as it passes perihelion again, so that the geometries — and thus the perturbations — are in fact correlated? If so, there would be some regularity or predictability in the sequence of perturbations.

If we consider comets with very long periods — say, tens of thousands of years or more — there is no regularity in the sequence, and the dynamical evolution appears like a random walk along the energy axis. The reason why such comets behave this way is that the geometrical pattern followed by a comet during a given perihelion passage is extremely sensitive to the way the comet was perturbed during the previous perihelion passage. Even an insignificant change in this perturbation may change the timing of the following perihelion passage by several years, and this is enough to cause a major change in the perturbation exerted by Jupiter.

The resulting unpredictability is the signum of *dynamical chaos*, which means that nearby orbits diverge exponentially on a time scale called the *Lyapunov time*. The orbital evolution of long-period comets is hence chaotic, and the random walk they experience along the energy axis is an expression of this chaos. As long as the steps are small, representing the central peak of the probability distribution, a relevant mathematical description is offered by diffusion theory. This has been frequently used in early works but suffers from important limitations due to the existence of the wide tails surrounding the central peak. These tails do not extend to infinity, and the very large perturbations are extremely rare, but given enough time they may dominate the total effect.

When comets with shorter periods are considered, a limit may be reached, beyond which the random walk picture is no longer relevant. A small difference in the perturbation at one perihelion passage will no longer lead to drastic changes in the following perturbation. Nearby orbits will still diverge rapidly, but the chaos is gone. In fact, resonant behavior starts to appear so that the energy may oscillate, as the comet orbit librates around a commensurability with Jupiter's mean motion. This is known to occur for some Halley Type comets (Carusi *et al.* 1987). The mentioned limit is not sharp but is usually considered to correspond to a semi-major axis of ~ 100 AU.

Let us finally mention a somewhat different type of chaos, which occurs for Jupiter Family comets and the related Centaurs. This has its roots in close encounters — in particular, with Jupiter. From the above description of close encounter dynamics, it is clear that the outcome of a hyperbolic deflection is sensitive to the initial conditions. Hence, the difference between two neighboring heliocentric orbits gets blown up considerably during a close encounter, and if a new encounter occurs only a few decades or a century later, the perturbations may differ drastically. Over longer times, a practical identity of two initial orbits may become totally unrecognizable. An example of this is shown in Fig. 3.12.

The difference between the chaos expressed by the random walk of long-period comet energies and the semi-major axis evolutions plotted in Fig. 3.12 for 95P/Chiron is that the latter chaos is intermittent, arising from short episodes (i.e., close encounters) between which the evolution appears quite regular. Nevertheless, chaos is pervasive in the long-term orbital evolutions of Jupiter Family comets and Centaurs in general, and the reason is the close encounters. Between those, the objects in question appear stable or show regular oscillations, including librations around mean motion resonance with Jupiter.

In the latter case, the commensurability of the orbital periods can stabilize the comet orbit by offering protection against close encounters. Consider for instance the 2:1 resonance between a comet and Jupiter. Typically, the aphelion of the comet orbit is close to Jupiter's orbit, and the perihelion is much closer to the Sun. If Jupiter is 90° away from the comet's aphelion direction, when aphelion passage occurs, there is no close encounter. In exact resonance, Jupiter has moved around half its orbit at the time of the next

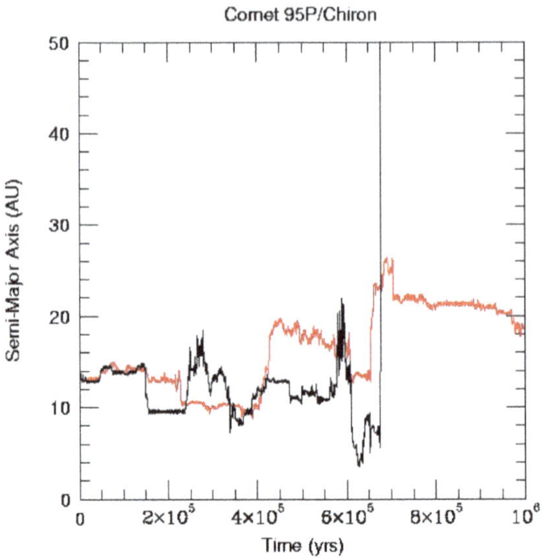

Fig. 3.12. Orbital evolution of the Centaur 95P/Chiron, obtained from accurate numerical integrations. The semi-major axis is shown as a function of time from the start of the integration. The red and black curves represent two variants with a minute difference in the initial conditions. Due to sequences of close encounters with giant planets, the two orbits diverge considerably within $\sim 10^5$ years. Hence, the two objects lose all memory of their initial vicinity, and the dynamics is clearly chaotic. Courtesy H. F. Levison. Credit: H. Rickman, "Cometary Dynamics," in Lecture Notes in Physics, Vol. 790 (2010), pp. 341–399, © Springer-Verlag Berlin Heidelberg 2010. With permission of Springer Nature.

aphelion passage and is therefore 90° away again. The same situation is bound to repeat over and over. However, if the resonance is not exact, the angle between the radius vectors of the comet and Jupiter at the comet's aphelion passage will change, so that one of the two consecutive aphelion passages gets more and more dangerous as the angle decreases. It can be shown that Jupiter's gravity tends to counteract this evolution, so that the decrease can be halted and the angle starts increasing again. Hence the angle may oscillate or "librate" around 90°, and close encounters can be avoided as long as this libration persists. However, in reality there are always small interferences by other planets, and a librating comet is bound to have a large libration amplitude to begin with. Thus, the librations are always short-lived, being broken by close encounters.

3.6. General Road Map

We have described the elements of comet dynamics in terms of the dynamical processes that govern the changes of their orbits. The perturbing agents are the planets, the passing stars, and the Galaxy as a whole. This has demonstrated that comet orbits are generally unstable and subject to major changes over long enough time intervals. Let us now sketch the patterns of orbital evolution that arise from the perturbations suffered by the comets.

Observable comets are ephemeral. Being Jupiter-crossing, their orbits are unstable and in addition, the observable comets may end their lives by physical effects, as will be discussed in Chap. 4. Due to the transfer routes provided by comet dynamics, the observable population is replenished and may be kept in a steady state. The source for these transfers is present in distant parts of the solar system in the form of giant reservoirs of icy bodies, which turn into comet nuclei as they enter into observable orbits.

Our description of the transfer routes had better start from the distant sources. Let us first consider the most distant of all — the Oort Cloud. This includes comets with a wide range of semi-major axes. While they all are subject to external perturbations due to Galactic tides and stellar encounters, the effects are the strongest for the most distant comets. The orbital energy is not immune to these perturbations, but the most profound changes are suffered by the angular momentum. There are hence large-scale changes of the perihelion distance, which may bring comets all the way between nearly circular orbits and nearly parabolic ones with perihelia in the observable region. This is the origin of the so-called *new comets*, which have the largest orbital energies (i.e., negative but closest to zero) among all observed comets.

Oort Cloud comets with smaller semi-major axes experience the same kind of external perturbations but on longer time scales. Their routes into observable orbits, to be described in Chap. 5, differ from those of the more distant comets but nonetheless end up in the same category of observed, new comets. In the current Galactic environment there is a limit to the semi-major axis, below which the perturbation time scale grows larger than the age of the solar system. Comets in those orbits would be isolated from the rest of the solar system and would never become observable. If the Oort Cloud does include such fossilized comets, these must have found their way into their current abode at a time, when the solar system was placed in a different environment. We shall return to such possibilities in Chap. 6, but for now there is no reason to discuss comets situated beyond the current transfer routes.

Let us instead return to the new comets. The constituent comets of the Oort Cloud often have semi-major axes in excess of 10 000 AU, and thus the new comets typically have inverse semi-major axes between zero and 0.0001 AU^{-1}. Since the typical perturbations of $1/a$ according to Fig. 3.11 are several times larger than this, it follows that the new comets are very often expelled from the solar system on hyperbolic orbits. The subsequent Galactic and stellar perturbations will rarely protect them from expulsion, and thus the solar system feeds comets into interstellar space as long as the flux of comets from the Oort Cloud persists.

However, at least as many new comets experience the opposite perturbations, thus becoming more tightly bound in orbits with smaller semi-major axes. A certain fraction will be lucky enough to be captured more tightly than the average, and these stand a very good chance of experiencing many more perturbations on subsequent returns. The resulting evolution is a random walk in orbital energy as described above. A salient feature of this evolution is the presence of an absorbing wall at zero energy, because there is no return from expulsion from the solar system. Increasing the number of perihelion passages (N), more and more comets are expelled, and the number of survivors decreases as $1/\sqrt{N}$. But these include an increasing fraction of comets that arrive into orbits, where the periods are too short for the random walk picture to apply. Even Halley Type comets might result from such an evolution.

In Fig. 3.13 we see a comprehensive outline of the cometary transfer routes in the solar system. The ones just described are included together with some that remain to be discussed. The latter involve a different category of comets. In fact, we are dealing with the two classes defined in Chap. 1.4: nearly isotropic comets and ecliptic

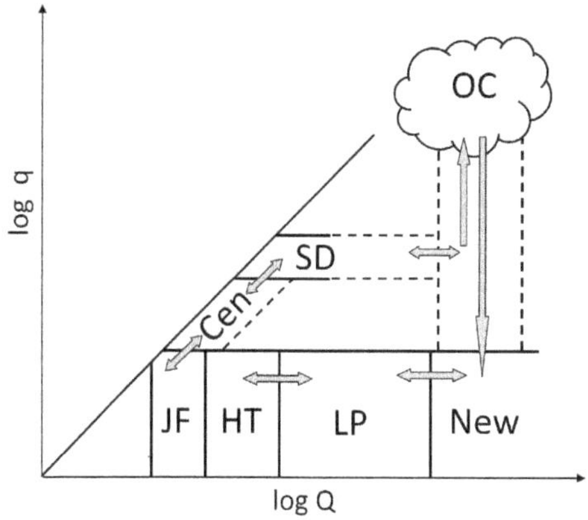

Fig. 3.13. Illustration of the main transfer routes of comets in the solar system using a generic diagram with the logs of the aphelion and perihelion distances on the axes. The routes are shown by shaded arrows. OC is the Oort Cloud; SD is the Scattered Disk; Cen means Centaurs; New stands for new comets; LP means long-period comets; HT means Halley Type comets; and JF is the Jupiter Family. Courtesy T. Wiśniowski.

comets. The Oort Cloud is the source of the nearly isotropic comets, while the ecliptic comets come from the flattened trans-neptunian populations — in particular, the *Scattered Disk*, which will be the subject of extensive discussions in Chaps. 5, 6 and 7. The observable component of the ecliptic comet population is the Jupiter Family. An indication of how comets are transferred into the Jupiter Family was obtained in Sec. 3.1 from Fig. 3.1, showing evolutionary curves for constant Tisserand parameter.

These curves connect the Jupiter Family with a part of the Centaur population, where some Chiron-type comets are found. In fact, as will be seen in Chap. 5, the Centaurs have a similar connection to the Scattered Disk and may be seen as a station on the way from the Scattered Disk into the Jupiter Family.

Finally, is there also a connection between the Scattered Disk and the Oort Cloud? The answer is yes. It is also easy to see the main direction of this transfer, namely, from the Scattered Disk into the Oort Cloud. Comets must have formed much closer to the Sun than both the Oort Cloud and the outer part of the Scattered Disk. It is then natural for the new-born comets to move on low-inclination orbits, forming an initially very flat population. The Scattered Disk has some resemblance to this, even though the difference is substantial. As we shall see in Sec. 6.3.1, this disk is thought to have been created out of a primordial reservoir of comets. It then acted as a source, from which the Oort Cloud was built.

The dynamics of this process has some similarity to the above-described energy diffusion of long-period comets. However, the scattered disk objects have their perihelia near or slightly beyond the orbit of Neptune, and they experience gravitational scattering due to Neptune rather than Jupiter. Close encounters are once again rare, but in the long run they dominate the scattering. The time scale is measured in hundreds of millions of years. As a result, more and more disk objects are placed into orbits that reach out far enough for external perturbations to perturb their angular momenta, thus lifting the perihelia far away from all planets. Due to those perturbations and the following stellar encounters during billions of years, the orbits get fully randomized, losing the memory of an ecliptic origin. As we shall see in Sec. 6.3.2, this holds at least for the outer part of the Oort Cloud, while the inner parts still maintain a preference for low inclinations.

Of course, there is nothing to prevent Oort Cloud comets on their way toward observable orbits to have this process interrupted by close encounters with Neptune or any other giant planet. If they hence get decoupled from the external perturbations, they may get captured into scattered disk or Centaur orbits and eventually penetrate into Jupiter Family or Halley Type orbits. It is important to realize that all dynamical processes in a dissipation-free system are time reversible, but the question is which processes dominate in the real solar system. This question is often controversial, and sometimes the answers are only preliminary, as we shall see in later chapters.

Chapter 4

Physical Evolution in Observable Comets

Comet nuclei are characterized by high porosities and very low bulk strengths. This makes them vulnerable to physical destruction. They also evolve in terms of their gas-producing activity, which is their defining property as comets and determines our ability to detect and study them. As a consequence, in addition to their ever changing orbits, their observability is variable and their lifetimes are finite: they may actually cease to exist or become temporarily undetectable.

To understand the role of comets in solar system formation and evolution, we have to characterize the essentially unseen, distant populations that the observed comets derive from. Therefore, the issue of comet lifetimes is of critical importance. Let us now take a closer look at both observational evidence and theoretical models as relevant to those lifetimes.

4.1. Mass Loss and Erosion

The gas-producing activity of comets implies the escape of material from their nuclei. Since this activity is a defining property of comets, mass loss is also a fundamental property. The amount of mass loss from the nucleus per orbital revolution (ΔM) is a convenient way to express the long-term erosion that the nucleus undergoes. Unfortunately, this quantity is poorly known for practically all observed comets, but we still have to evaluate the current knowledge based on available observations.

It may be naive, but let us first see if there is a reasonable sample of comets, for which ΔM can be estimated with some confidence and accuracy using observations. In the first place, we concentrate on the loss of H_2O molecules, which is the easiest to estimate. Since the relevant observational techniques have not been accessible for very long, only relatively recent comets can be considered. One nice example is comet C/1995 O1 (Hale-Bopp), which was bright and famous, and hence very well observed. Production curves for various molecules in this comet are shown in Fig. 2.5 (Biver *et al.* 2002).

The H_2O production curve can be taken as the OH curve from radio observations with the Nançay telescope — an internally consistent data set. At heliocentric distances less than 3 AU before perihelion and 2 AU after perihelion, this curve can be approximated by straight lines in the log–log plot, corresponding to a power-law slope of -2. Thus, the H_2O production rate varied as the inverse square of the heliocentric distance in this part of the orbit. At larger distances, a more rapid fall-off is seen, which makes it reasonable to take the total H_2O mass loss as the integral over the near-perihelion part. If we denote the production rate by Q, the calculation of the number N of molecules lost is simply

$$N = \int Q\,dt = \int Q_o r^{-2} \frac{df}{\dot{f}} = \frac{Q_o \Delta f}{h}, \qquad (4.1)$$

where f is the true anomaly and h is the orbital angular momentum.

Estimating the production rate at $r = 1$ AU (Q_o) to be 6×10^{30} and 8×10^{30} molecules per second before and after perihelion, respectively, we easily evaluate the total integral as 9.3×10^{37} H_2O molecules. The corresponding mass is $\Delta M(H_2O) = 2.8 \times 10^{12}$ kg. This is a crude estimate due to the approximations employed. Of course, the integral can be evaluated more precisely, and covering the apparition of the comet more completely. However, for several reasons, it is not certain that such accuracy is warranted. First, one should consider also other data on the H_2O production rate, which are sometimes but not always consistent with the curve that we used. Second, there are obviously other molecules that contribute importantly to the total mass loss, which makes the use of $\Delta M(H_2O)$

less obvious. It is possible to integrate the other curves in Fig. 2.5 and thus estimate those contributions, but for instance, CO_2 is a molecule that is likely important but is missing from the plot for lack of observations. A third reason is that the absolute mass loss is less interesting than the fractional mass loss, measured in percent of the mass of the nucleus. Unfortunately, for comet Hale-Bopp this mass is rather poorly known.

Perhaps the most serious concern remains to mention. As discussed in Sec. 2.2.3, the mass loss in comets is likely dominated by refractories, forming large grain aggregates or chunks that are difficult to trace by remote sensing. While Earth-based observations have actually shown these to exist, the quantitative assessment due to close-up Rosetta measurements still came as a surprise. A natural question to ask is, if comet 67P/Churyumov-Gerasimenko may be a peculiar case, so that its large DGR cannot be extrapolated to other comets like Hale-Bopp, but there is no firm answer to this question. Even restricting attention to other Jupiter Family comets, it is difficult to be confident about an overall generalization of the 67P DGR. Therefore, an important uncertainty over the total ΔM remains, no matter how well we determine the gaseous mass loss.

As another example we take the apparition of comet 1P/Halley in 1986, for which the water production data are summarized in Fig. 4.1. This shows that large discrepancies of $\Delta M(H_2O)$ may result from using different data sets to define the production curve (Rickman 1989). Basing the estimate on the OH data only, a much lower value is obtained than if attention is focused on the observations of the hydrogen coma or the near-IR spectra showing H_2O emission bands.

Faced with this situation, Rickman (1989) chose to define two values of $\Delta M(H_2O)$: a lower one of 2.5×10^{11} kg that mainly uses the OH data, and a higher one of 4×10^{11} kg based on the maximum curve in Fig. 4.1. Even today, in retrospect, it does not seem possible to do better than this. Rickman estimated the contribution by other volatiles to be ~15–20% and took the DGR to be ~0.5–1.0. The latter was larger than would have been estimated before the close-up observations of 1P/Halley but is clearly much less than the Rosetta observations have suggested.

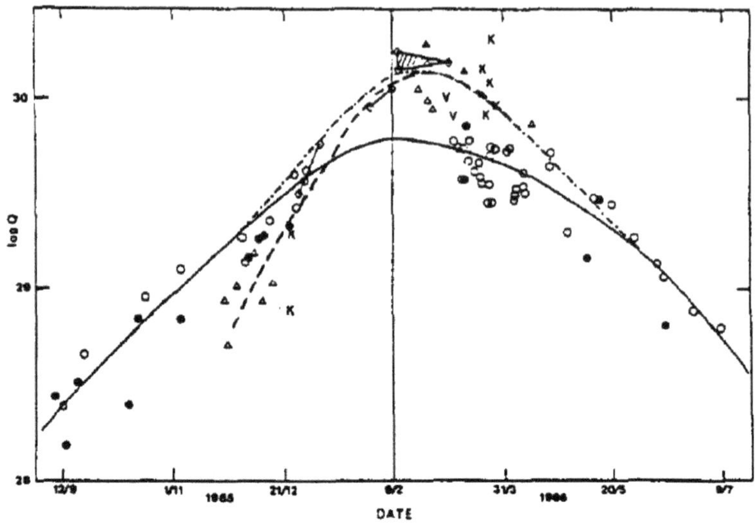

Fig. 4.1. Observed H_2O production rates (Q) in comet 1P/Halley, Sept. 1985–July 1986. The symbols are explained in the paper by Rickman (1989). The solid curve is a fit to the OH data, and the dashed curve is based on near-perihelion atomic hydrogen observations. The dash-dotted curve shows the basis for a maximum estimate of the integrated H_2O mass loss. Reprinted from Rickman, H., *Adv. Space. Res.* **9**, 59–71 (1989), with permission from Elsevier.

Comets in general present even larger problems. Good coverage of the H_2O production curve by direct observations is rare — especially for Jupiter Family comets, which usually stay rather faint. Hence the situation is fundamentally bad. The option to use visual light curves, as discussed in Sec. 2.3, has not yet been pursued and also has its problems, since it is tedious if at all possible to define the true, total magnitude from a heterogeneous set of observations. Moreover, the correlation of heliocentric magnitude with H_2O production rate in Eq. (2.8) is only a statistical relation and might be deceptive when used for individual comets at varying heliocentric distance.

In practice, mass loss by surface erosion has been included in a crude manner into simulations of comet evolution on the basis of thermal model calculations of the H_2O sublimation rate (Weissman 1980, 2011). A more comprehensive approach was introduced by Rickman *et al.* (2017a) in the following way. A thermal model including surface sublimation of H_2O is used to set up an interpolation table of the H_2O escape flux versus r. Sequences of orbital elements for consecutive revolutions of simulated comets are obtained by numerical integration, and for each of these a cumulative amount of H_2O mass loss per unit area is obtained by stepping through the orbit and consulting the table. This is augmented by escaping dust using an assumed, constant DGR, and the erosion depth per revolution is calculated by assuming also a density for the surface material. Finally, the free sublimation fluxes listed in the table are reduced by applying a constant activity f according to Sec. 2.4.1.

This approach is very crude, since the actual production curves — being likely affected by a variable f — are not realistically simulated, and the complexities of surface evolution revealed by Rosetta — likely including a variable DGR — are not at all modeled. Strictly speaking, the only realism may be the account of the variation of erosion depth with perihelion distance, and the calibration to observed gas production rates through the choice of the parameter f.

One way to illustrate the effect of surface erosion on the Jupiter Family as a whole is to consider the age distribution of its members. As sketched in Sec. 3.6, this population may be kept in a steady state, where a gain rate through dynamical captures from the source is balanced by the combination of physical losses and dynamical ejections. We may consider the physical losses to occur by shrinkage of the nuclei below a lower limit to the radius, as caused by surface erosion. The age of a Jupiter Family comet may be defined as the time that has passed since the first time it got a perihelion distance below 2.5 AU — this has often been taken as a *visibility limit* for comets.

Clearly, if we compare simulation results for the Jupiter Family with and without surface erosion, the former must involve younger ages. But the size of this effect depends on the assumed size distribution of the newly captured comets. In Sec. 2.1 the cumulative size distribution was discussed, and its power law index was found to be most likely in the range from 1.5 to 2.5. For the latter value, the distribution is steeper, and more comets are small, making them vulnerable to shrinkage or destruction by erosion. Figure 4.2 shows

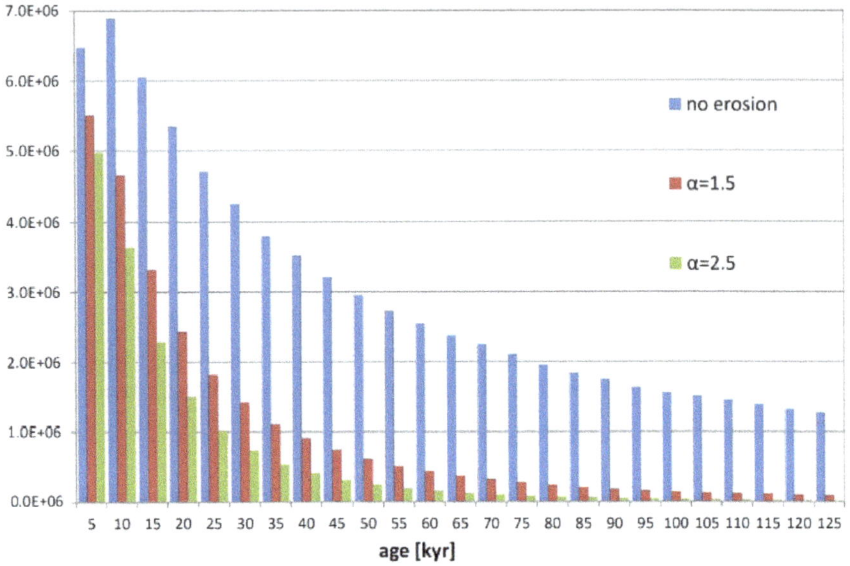

Fig. 4.2. Age distribution for a steady-state Jupiter Family according to dynamical simulations. The comets in question have perihelion distances $q < 1.5$ AU and nuclear radii $R > 1$ km. One case without erosion and two cases with erosion for different size distribution index α are shown. The integrals of the distributions are proportional to the total population of the Jupiter Family. Credit: H. Rickman *et al.*, *A&A* **598**, A110 (2017), reproduced with permission © ESO.

simulation results obtained in the way just described by Rickman *et al.* (2017a) for the steady-state age distribution of Jupiter Family comets. It exhibits the expected limitation of ages due to erosion and shows that this is more pronounced for a steeper size distribution.

4.2. Dust Mantling

It has been a long time since comet nuclei could be imagined as fresh snowballs or drifts of newly fallen snow. They have been known to consist of a mixture of chemical and mineralogical species with very different condensation temperatures (also referred to as *volatility*). Water ice is neither the most, nor the least volatile of the constituents. As H_2O sublimates near the surface, the surface layer naturally gets enriched in refractories. Since these are thought to have a composition similar to that of the dust grains in the coma and tail, the term *dust mantling* has been in common use since more than 35 years for this process of enrichment.

However, the physical nature of the mantling process cannot be specified until we know the nature of the nucleus material. Advances in this knowledge will lead to updates in our picture of mantling. This has already happened in the past and is now happening again. It therefore seems convenient to take a historical approach when describing this subject.

By the time the term dust mantling was coined, comet material was thought to be dominated by ice, which in turn was dominated by H_2O. The overall dust/ice ratio was considered to be much less than one, and the common picture was an icy matrix with refractory inclusions of various sizes. This way the fresh surface was essentially icy, and it was generally modeled as flat and free-sublimating into a surrounding vacuum. With the recession of the surface, the refractory inclusions (dust grains) would be freed, finding themselves in an outward flux of gas. Balancing the force applied by the gas drag with the gravity of the nucleus, one could find the appropriate radius of a grain, assumed to be spherical and obeying this balance — the so-called *critical radius*. This would separate the smaller grains, which would be carried away by the gas flow, from the larger ones, which would remain on the surface. The concept dates back to the pioneering paper by Fred Whipple (1950), where the physics of a cometary nucleus was first introduced.

From (Rickman *et al.* 1990), we take an expression for the critical radius (a_c) in the case of a non-rotating nucleus with radius R_n and density ρ_n as

$$a_c = \frac{9C_D u_0 m Z}{32\pi G R_n \rho_n \rho_d}, \tag{4.2}$$

where C_D is the drag coefficient (a dimensionless number of order unity), m is the mass of a gas molecule, G is the gravitational constant, and ρ_d is the density of the dust grains. The gas flux is denoted by Z, and $u_0 = \sqrt{2kT_s/\pi m}$ is the mean vertical gas speed at the surface (T_s is the surface temperature, and k is Boltzmann's constant). An example of the variation of a_c with heliocentric

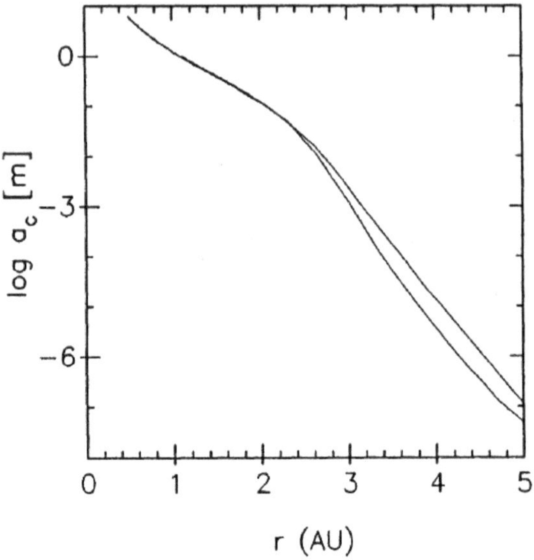

Fig. 4.3. The log of the critical radius versus heliocentric distance, computed for a rotating nucleus wth radius 1 km and density $0.2\,\mathrm{g/cm^3}$. Beyond $r = 2$ AU, the lower curve corresponds to the pre-perihelion branch and the upper curve to the post-perihelion branch of the orbit. Credit: H. Rickman *et al.*, *A&A* **237**, 524–535 (1990), reproduced with permission © ESO.

distance r, where Z was derived as sketched in Sec. 2.4.1, is shown in Fig. 4.3.

According to this picture, even micron-sized grains would stay on the surface near the aphelion of a Jupiter Family comet. However, the surface recession is also vanishingly slow. When the comet approaches the Sun, the gas flux increases rapidly, and at $r = 3$ AU even mm-sized grains get carried away. At perihelion, the surface is purged of all grains with $a < a_c(q)$. Unless the perihelion distance decreases, the remaining large grains will assemble on the surface in growing numbers, until they form a coherent layer that prevents the escape of smaller grains. At this point, we may speak of a dust mantle.

Clearly, for a given perihelion distance, larger nuclei would get mantled quicker than smaller ones, and the mantles would likely be thinner. For a given nuclear radius, assuming a steep grain size distribution, comets with larger perihelion distances would develop thinner mantles more quickly, compared to comets with smaller perihelion distances. The mantles are predicted to be relatively porous, allowing gas to diffuse from the icy surface at the bottom of the mantle into space. A mantled surface would thus produce some gas, but the flow would be quenched.

Due to this quenching, a substantial gas pressure may build up at the icy interface. This might not threaten the stability of the mantle in case the orbit does not change, or the perihelion distance increases. However, a sudden, substantial decrease of the perihelion distance is accompanied by a strong increase of the gas pressure at the interface and is thus likely to cause the mantle to be blown off. This way, a fresh icy surface is once more exposed, and the grain collection process starts anew.

There is no reason to describe this model any further, because in retrospect it has to be deemed unrealistic. The first problem is the assumed structure of the material. A very strange comet formation scenario would have to be invoked in order to argue that, first, refractory grains would accrete to form a population with an extremely wide size distribution, and then, ices would condense to form km-sized objects that would trap dust aggregates of all sizes. In fact, there was no such theory, and the structure was just an ad-hoc assumption. The situation changed with the exploration of Halley's comet (see Sec. 1.5.1).

From the observed grain compositions, the large porosity, and the very pristine elemental composition it appeared that comet material was essentially an aggregate of the primordial, interstellar grains of the pre-solar cloud — a picture advocated by Mayo Greenberg (see Sec. 2.6.4). In principle, each constituent sub-micron grain would capture the bulk cometary composition, being made up of a silicate core, an organic cover, and an icy mantle. This picture became very popular, and when dust mantling models of the kind described above were constructed at the end of the 1980's, these already appeared slightly suspect, since the structure assumed for the comet material was so different from the expected reality.

The composition of a Greenberg grain is roughly 1/3 each of silicates, organics and ice. This is quite far from that of the presumed icy matrix with embedded dust grains, and in particular, the assumption of an icy matrix appears questionable. Moreover, the observations made in comet Halley indicated that the refractory component might dominate more than the Greenberg model predicted. As mentioned in Sec. 2.2.3, the concept of the nucleus as a dirty snowball started to weaken and subside in favor of a more dust-rich object with a significant but not dominant icy component.

In line with this development, in 1994, Ekkehard Kührt and Uwe Keller introduced a very different mantle model. They did not treat its formation but assumed the comet material to be a refractory matrix with internal cohesion, within which the ice would exist as inclusions. Sublimation of the ice in the layer next to the surface and escape through the pores of the refractory matrix would naturally create an evacuated surface crust, whose cohesion would stabilize it against blow-off due to gas pressure (Kührt and Keller 1994). Due to thermal insulation and mechanical shielding, the gas flow from the icy interface would be severely quenched. Thus, the persistent and significant activity observed in many comets could not be explained, unless the comet material is heterogeneous on a macroscopic scale with some surface regions being ice-dominated. An alternative suggestion is that the crust may be punctuated or broken up by processes like meteoritic impacts or thermal cracking, thus allowing local gas-producing activity to proceed.

But the situation has changed once more with the Rosetta exploration of 67P/Churyumov-Gerasimenko. The gas producing activity of this comet went from local to global on the way to perihelion and then back to local again. The layer next to the surface was uniformly icy although ice on the very surface was never seen to be particularly common. The ejected material was found to contain about six times as much refractories as water ice by mass over the whole pre-perihelion branch (Fulle *et al.* 2016). One has to ask in earnest, if dust mantling is a proper way to explain the reduced activity of comets in the light of these observations. We shall return to relevant findings by Rosetta below.

If, including other volatiles than H_2O, there is only $\sim$15% ice in the comet material, a relevant picture may be the sub-surface sublimation of the ice at a rate depending on the local temperature, and the diffusion of the gas through the pores into space. The gas flux will then necessarily be much lower than from a free-sublimating icy surface. If the surface material responds to the thermal and mechanical strains by breaking up into pieces that are lifted off by the gas flow, the growth of an evacuated crust may be avoided, and the activity may proceed at a reduced, perhaps variable but finite level.

4.2.1. *Evolution of comet 67P*

Let us start with a few remarks about the orbital evolution. We saw in Sec. 1.5.5 that comet 67P was recently captured into its current orbit and previously had a much larger perihelion distance. But a more fundamental question concerns the evolution on a longer time scale and this one cannot be answered due to the chaotic nature of this evolution (Guzzo and Lega 2015). The comet may have had similar or smaller perihelion distances for long times in the past, or it may be making one of its first approaches to the Sun. Which is the case, we cannot say.

This begs the question if the appearance of the nucleus or its state of activity may give any clues. Would a very old Jupiter Family comet look any different from a fresh one? Based on what was described so far in this chapter, one would certainly think so. Comets would age somewhat like people do due to a monotonous mantling process and it would be as easy to distinguish an old comet from a young one as it is to tell an old man from a youngster. However, in Sec. 4.5.2 we shall see that reality may be quite different as far as comets are concerned. They certainly decrease in size, but the size itself does not say how old a comet is, and during this process the surface is likely to change back and forth between looking fresh and looking worn-out.

A related issue is the explanation for the bilobate shape of the 67P nucleus. In popular wording, the two main ideas may be described as the apple-core and snowman theories. The latter means that the two lobes are separate precursors that came gently together during the formation process, a bit like when the head is placed carefully onto a snowman. The former involves reshaping by erosion of an originally monolithic nucleus, whereby the neck was preferentially eroded and thus carved out. Here the answer appears clear. The comprehensive thermal model by Keller *et al.* (2015) showed that, on average over a whole orbit, the neck is actually the least heated region of the entire nucleus and thus should be the least eroded. Hence, the snowman scenario has to be preferred, and the overall shape of the 67P nucleus does not tell us how old the comet is in terms of physical evolution.

The way erosion works on this comet is largely conditioned by the axial orientation. The obliquity is as high as 52°, which leads to a strong seasonal variation of the insolation. In addition, the orientation is such that the difference between the hemispheres is nearly maximized. All orientations should in principle be equally possible, so the one of 67P may be coincidental and thus quite peculiar. The erosional behavior of 67P may then be non-typical compared to Jupiter Family comets at large. As seen in Sec. 2.4.3, thermal modeling shows that the sublimation-caused erosion of 67P is strongly concentrated to the southern hemisphere. This does not only imply gaseous mass loss but solid particles as well. Overall, about 3/4 of the recent mass loss in comet 67P is estimated to have occurred during southern summer, thus mainly affecting the southern hemisphere (Keller *et al.* 2015).

While microscopic dust particles are easily carried into space by the expanding coma gas, there is a larger size range where the terminal velocities are of the order of the escape speed from the surface. In this situation, it is not certain that they will actually escape. Fall-back onto less active regions on the nucleus may occur — possibly far from the place where the particles left the surface. Inhomogeneities in the gas flow near the nucleus also contribute to such lateral transport, since particles can be lifted in a dense gas jet and then travel into the low-density environment, whereupon they are likely to fall back.

The importance of such phenomena was first pointed out by Nick Thomas and co-workers (2015b). The authors started from morphologic evidence for lateral transport of particles along the

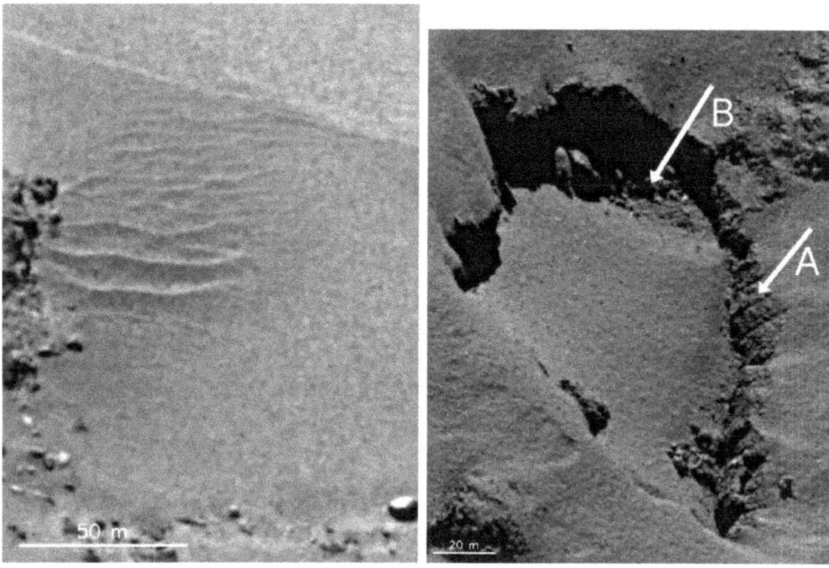

Fig. 4.4. Left panel: Aeolian ripples in the Hapi region. Right panel: Smooth-floored pit in the Ash region with arrows indicating the fractured wall (A) and a rough talus deposit (B). Credit: N. Thomas *et al.*, *A&A* **583**, A17 (2015), reproduced with permission © ESO.

surface of the nucleus, like ripples and dunes, or deposition into locally depressed areas. Examples are shown in Fig. 4.4. They also introduced the commonly used term *airfall*, which is borrowed from volcanology. In fact, the northern hemisphere exhibits a lot of smooth terrain likely caused by airfall originating from the south, while the southern regions are rough, apparently exposing consolidated material.

The boulders strewn along the Hapi valley may be another example of airfall. The ragged appearance of local landscapes on the 67P nucleus indicates that surface erosion sometimes occurs by the separation of large chunks of solid material — something we may call *fragmentation*. Note that such large pieces are not hard to lift in a low-gravity environment, especially near the equator where most of the gravity is cancelled out by the centrifugal force. They have been observed orbiting the 67P nucleus rather close to the surface, probably after having been lifted during the preceding perihelion passage (Rotundi *et al.* 2015), and similar fragmentation/lifting events may have occurred during the time of Rosetta observations as well (Fulle *et al.* 2016). Those boulders that return to the nucleus are likely to gather in places like the Hapi valley, where the gravitational potential is very low.

In fact, the shuffling of material from the southern to the northern hemisphere is an important aspect of surface evolution in comet 67P, but as noted above, we cannot be sure that this phenomenon is common among comets in general. According to Keller *et al.* (2016), the material has to be icy, meaning it contains water ice but probably no frozen super-volatiles. This implies that the smooth surfaces of the northern hemisphere become covered by icy material in connection with the high near-perihelion activity, and on the following inward orbital branch activity starts with H_2O sublimation from the northern regions as observed by Rosetta.

We finally note that secular effects of surface evolution are extremely difficult to characterize even with Rosetta accompanying the comet from long before to long after perihelion. The reason is the surface dichotomy caused by the particular pattern of erosion and the fact that the nucleus turns different sides to the Sun at similar distances before and after perihelion. Ideally, one would have liked to detect a change in the level of activity or the appearance of the surface from the beginning till the end of the mission, but the changes seen are more likely due to observing different parts of the nucleus.

4.3. Dormancy and Rejuvenation

No theory is yet available to predict how the long term evolution of comet activity should work. Complications are obvious from the detailed monitoring of surface phenomena by the Rosetta/OSIRIS camera as described above, but we also need to confront any such, forthcoming theory with the evidence on physical aging, dormancy and activity variations in comets at large. Let us now turn to this issue.

4.3.1. *Missed apparitions*

The observational records of short-period comets are quite diverse. While many comets show great regularity having practically no missed returns since their discovery, others have behaved in the opposite way. In Table 4.1 we list several comets, which have been missed during many consecutive returns to perihelion and then were observed again. The possible reasons for this behavior will be discussed by reference to the individual cases.

Table 4.1. Basic data concerning the history of observations for a selection of short-period comets that have been missed for long intervals. N_{obs} is the number of observed apparitions before the temporary loss of the comet and N_{mis} is the number of consecutive, missed returns. The years of the discovery apparition, first and last missed returns, rediscovery apparition and last observed apparition are also listed.

Comet	Discovered	N_{obs}	Missed (N_{mis})	Redisc.	Last obs.
9P/Tempel 1	1867	3	1886–1961 (12)	1967	2016
17P/Holmes	1892	3	1913–1957 (7)	1964	2014
18D/Perrine-Mrkos	1896	3	1916–1949 (6)	1955	1968
41P/Tuttle-Giacobini-Kresák	1858	1	1863–1901 (8)	1907	
	1907	1	1912–1945 (7)	1951	2006
54P/de Vico-Swift-NEAT	1844	1	1850–1888 (8)	1894	
	1894	1	1901–1959 (10)	1965	
	1965	1	1973–1995 (4)	2002	2017
64P/Swift-Gehrels	1889	1	1899–1963 (8)	1972	2009
72P/Denning-Fujikawa	1881	1	1890–1969 (10)	1978	
	1978	1	1987–2005 (3)	2014	2014
80P/Peters-Hartley	1846	1	1854–1974 (16)	1982	2014
97P/Metcalf-Brewington	1906	1	1914–1983 (10)	1991	2011
113P/Spitaler	1890	1	1897–1979 (13)	1986	2015
139P/Väisälä-Oterma	1939	1	1949–1988 (5)	1998	2008
205P/Giacobini	1896	1	1903–2002 (16)	2008	2015
206P/Barnard-Boattini	1892	1	1899–2003 (19)	2008	2008
226P/Pigott-LINEAR-Kowalski	1783	1	1790–1996 (32)	2003	2009
274P/Tombaugh-Tenagra	1931	1	1940–2004 (8)	2013	2013
289P/Blanpain	1819	1	1825–1998 (34)	2003	2014

In most cases, the comet was discovered long ago and was lost after the first apparition. As a consequence, the ephemerides were too poor for search efforts to relocate the comet during most of the missed returns. In such a case, assuming the absolute brightness to remain the same, there is nothing strange unless it can be shown that the comet should have been independently rediscovered at one or more of the missed returns, considering the observational circumstances and the observers and instruments available. For other comets the situation may be different, because good ephemerides should have been published, so that focused searches were indeed possible. This concerns the first two comets, 9P/Tempel 1 and 18D/Perrine-Mrkos, and also 41P/Tuttle-Giacobini-Kresák, 54P/de Vico-Swift-NEAT and 72P/Denning-Fujikawa, for which the second or third suite of missed returns followed two apparitions that were already linked. One may then conclude that the comet must have stayed fainter than some limiting magnitude during the missed returns to explain the lack of recoveries. Of course, this only holds for those returns, when the comet was actually looked for.

Let us start with comet 9P/Tempel 1. As mentioned in Sec. 1.5.3, the lack of observations of this comet from 1886 to 1961 may be due to changes in the orbital parameters. In particular, the perihelion distance rapidly increased from 1.5 AU to 2.0 AU, and this must have made the comet fainter. Let us see by how much. We consider the best observing geometry, when the comet is at perihelion in opposition. It is then at its brightest at the same time as it culminates at midnight and is most easily observed. We now use some results mentioned in Chap. 4. The H_2O production rate $\mathcal{Q}$ at perihelion can be assumed to be proportional to q^{-2}, where q is the perihelion distance, because this is the behavior of the H_2O sublimation flux, and we assume that neither the radius nor the activity of the nucleus changes. For $q' = 4q/3$, as in the present case, we get $\mathcal{Q}'/\mathcal{Q} = 9/16$. Using Eq. (2.8) in Sec. 2.2.2, the heliocentric magnitude at perihelion is seen to be 1.0 units fainter with the more distant orbit. Using the preceding Eq. (2.7) with $\Delta'/\Delta = 2$, the apparent magnitude at perihelion is 2.5 units fainter.

Thus, the brightness of comet 9P/Tempel 1 under favorable conditions is expected to have dropped by 2.5 magnitudes as a result of the orbital change, in case nothing happened to the nucleus. In the beginning, with no development of instrumentation, this was serious and the problem could have worsened, if the first apparitions were not geometrically favorable. Over nearly a century, the observing techniques got much better, but the ephemerides got worse. Finally, when the brightness could get back to normal levels, it was still largely due to the great effort by Brian Marsden (see Sec. 1.5.3) that the comet was actually recovered. This example shows that care is needed before stating that a comet must have become inactive for the reason that it was not recovered for many returns.

Another example is comet 41P/Tuttle-Giacobini-Kresák. This one did not undergo large orbital changes and yet went unobserved for first eight, then seven returns. Here, two facts need to be kept in mind. The comet is intrinsically very faint under normal conditions, like during the three discovery apparitions. It was thus close to the limit of observability, and therefore it could very well go unobserved in between. The other fact will be discussed in Sec. 4.4.3, namely, that its brightness has behaved irregularly during later returns. Comet 41P is among the comets that have produced the most extensive outbursts. Thus, we know that some comets do experience short intervals, when they outperform themselves in terms of brightness by producing gas and dust at a much enhanced rate. As a consequence, some comets may have gone unobserved simply by failing to repeat the outburst during which they were discovered.

By checking the detailed information about the comets in Table 4.1, one can identify several cases, where this may have happened. Comet 54P/de Vico-Swift-NEAT presents one of these. At its first discovery it was an easy object with the telescopes used to look for comets in the mid-19th century. It was not recovered at the favorable return in 1855, which indicates that something had already happened to the comet. At Swift's rediscovery in 1894, it was very faint and continued to fade. In recent times the comet has been even fainter. Overall, the comet has certainly dropped in activity in addition to a gradual increase of perihelion distance. Whether the first discovery was aided by an outburst cannot be stated with certainty, but this appears likely.

Comet 226P/Pigott-LINEAR-Kowalski must obviously have been fairly bright at its first discovery in 1783, but at its rediscovery in 2003 it was of 19th magnitude and did not get much brighter. This is a considerable fading that occurred without any decrease of the water flux due to increasing perihelion distance. A similar situation holds for comet 289P/Blanpain, which must also have been fairly bright at the first discovery in 1819, especially since there was an independent discovery by Pons. When it was rediscovered by LONEOS in 2003, it was of 18th magnitude and mistaken for an asteroid. The intervening increase in perihelion distance is by far too small to account for this change in behavior. Both these comets have undergone a major decrease of activity, which is either gradual or stepwise — in the latter case, they were likely in outburst when first discovered.

Table 4.1 also includes several comets that show evidence neither for decreasing activity nor for outbursts or other irregular behavior. These are simply faint comets, which sometimes happened to be discovered long ago but have lingered near the limit of observability. In this group we may count 64P/Swift-Gehrels, 80P/Peters-Hartley, 113P/Spitaler, 139P/Väisälä-Oterma and 206P/Barnard-Boattini.

The preceding discussion has introduced us to different kinds of evolution of brightness or activity among comets. In Sec. 2.4.2 we saw that different comets have widely different values of the activity parameter f, in particular considering the small nuclei with radius ~1 km or less. Now we see that the f value of the same comet can also change considerably as a function of time. Generally, there is no accurate information about the actual values of f, but in relative terms f has decreased conspicuously in some comets. This may be a settling to a normal, low value of f after an outburst, or a gradual fading away of the activity so that the comet enters a "dormant" mode. On the other hand, for some comets f seems to have stayed fairly constant at the level where the comet is just about active enough to be observed. This level must be different for different comets with different nuclear radii. If the nucleus is very large like for 28P/Neujmin 1 or 49P/Arend-Rigaux, the activity can be observed even if the level is very low (Sec. 2.4.2).

Clearly, if a marginally observable comet with a sub-km nucleus undergoes a further drop of f, it may be altogether lost, depending on the perihelion distance and the size of the nucleus. The above-mentioned comet 54P may be near such a fate, since a failure to detect it in the thermal IR indicates a nuclear radius less than 0.6 km (Fernández *et al.* 2013). Has this evolution already occurred in other comets? Possibly yes, and a case in point may be comet 18D/Perrine-Mrkos, which has been lost for almost 50 years. However, there is no proof of a major decrease in activity and the comet was previously lost for almost 40 years and then rediscovered, so another future rediscovery cannot be excluded.

This leads thoughts to comet 5D/Brorsen, which was much more active than 18D during its observed apparitions in 1846–1879 but never seen afterwards. Here the question is if there is anything left of the comet, or if its nucleus was completely disintegrated. The nucleus may have been quite small, and the comet may have had a large f as is the case for some of the currently observed Jupiter Family comets. But it was not seen to split like its famous colleague, 3D/Biela (Sec. 4.4.2), and a complete disintegration without any previous warning seems unlikely. Therefore, the loss of comet Brorsen may rather be a question of a major drop in activity that left an object behind, which is too small to be easily rediscovered.

From this discussion, we may characterize "dormant" comets by very low or possibly no activity. These may be marginally observable or not observable at all as comets. In addition to the outbursts, the more or less rapid transition of Jupiter Family comets into the dormant state seems rather common. The opposite transition was studied by L'ubor Kresák (1987). He carefully determined the absolute magnitude of the comets and used this to extrapolate their expected apparent magnitude at unobserved apparitions, mostly before the comets were discovered. From this he identified time intervals, when it seems hard to believe that the comets were not observed, if they did not have a significantly fainter absolute magnitude. For instance, comet 3D/Biela was discovered in 1772 and should definitely have been recovered at its first return in 1778, which however was not the case. Comet 5D/Brorsen should have been discovered in 1835, two revolutions before the actual discovery — this adds credibility to the above argument that the comet went dormant after 1879 and that the observed, high activity was a temporary affair.

For comet 72P/Denning-Fujikawa, which is listed in Table 4.1, Kresák found that the absolute magnitude in 1881 when it was first discovered must have been extraordinary. Otherwise it would likely have been discovered at its 1829 apparition, or at least a few months earlier in 1881. At rediscovery in 1978, the absolute magnitude was 4.5 units fainter, and this is likely to have been the case for most of the unobserved interval. We thus see that comets can be woken up, or *rejuvenated*, from dormancy — sometimes just for one apparition and sometimes for several.

4.3.2. *Comets in asteroidal disguise*

Since the existence of dormant comets is established, it is natural to ask if comet nuclei in a state of deep dormancy[1] can be observed and permanently classified as asteroids. There is no doubt that this has happened and we have several independent lines of evidence. First, some asteroids have been found to move in typical Jupiter Family orbits, which are quite distinct from any typically asteroidal orbits. This started to happen in the 1980s, and since then many discoveries have been made in the framework of the surveys for Near Earth Asteroids. Table 4.2 lists a selection of asteroids in Jupiter Family orbits.

All the asteroids in Table 4.2 have $T_J < 3$, as appropriate for comet candidates, but many more asteroids fulfill this criterion. As many as 30% of all near-Earth objects (NEOs) may have $T_J < 3$ (Stuart and Binzel 2004). Our list has been trimmed by selecting

[1]Such nuclei are often called "extinct", but the distinction between dormant and extinct seems rather academic as long as we cannot tell, whether a dormant comet will wake up again.

Table 4.2. Asteroids from the IAU/MPC database of August 2016 with perihelion distance $q < 1.5$ AU, aphelion distance $Q > 4.9$ AU, and $H \leq 16.8$ corresponding to $R \geq 1.5$ km, assuming an albedo of 0.04. Listed are orbital parameters including the inclination i and the jovian Tisserand parameter T_J, and the estimated average radius. Numbered asteroids are listed in chronological order of discovery.

Designation	Discovery	q (AU)	Q (AU)	i (°)	T_J	Radius (km)
(3552) Don Quixote	1983 SA	1.232	7.268	31.2	2.312	9.2
(5370) Taranis	1986 RA	1.212	5.444	19.1	2.730	3.0
(8373) Stephengould	1992 AB	1.460	5.099	40.8	2.587	5.0
(9767) Midsomer Norton	1992 EB_1	1.455	5.309	21.5	2.771	1.7
(20086)	1994 LW	1.201	5.169	21.8	2.770	1.5
(26166)	1995 QN_3	1.173	5.429	14.8	2.753	1.5
(85490)	1997 SE_5	1.191	6.215	2.4	2.644	3.7
(301964)	2000 EJ_{37}	1.368	7.896	10.1	2.441	6.6
(322713)	2000 KD_{41}	1.408	5.366	5.5	2.840	1.5
(401857)	2000 PG_3	0.407	5.238	22.0	2.549	2.0
(139359)	2001 ME_1	0.353	4.944	5.8	2.672	1.6
(385268)	2001 RC_{12}	1.159	5.281	27.4	2.689	2.4
(405058)	2001 TX_{16}	1.438	5.721	8.1	2.770	4.8
(37384)	2001 WU_1	1.372	4.942	20.5	2.852	3.6
(196256)	2003 EH_1	1.190	5.055	70.9	2.064	1.9
(115916)	2003 WB_8	1.441	5.039	20.3	2.837	4.8
(144870)	2004 MA_8	1.480	5.058	13.2	2.884	4.2
(248083)	2004 QU_{24}	1.284	5.363	23.3	2.724	2.0
(248590)	2006 CS	0.878	4.944	52.4	2.441	1.7
(428223)	2006 WW	0.354	5.004	32.7	2.542	2.0
(355256)	2007 KN_4	1.234	5.448	12.5	2.771	1.5
(365246)	2009 NE	0.360	4.984	35.2	2.535	2.2
(414586)	2009 UV_{18}	1.162	5.181	8.3	2.835	2.0

only relatively large, numbered objects with large aphelion distances. This way, the orbits are secure and the likelihood of cometary origin is enhanced. Nonetheless, according to DeMeo and Binzel (2008), only 54% of the NEOs with $T_J < 3$ show evidence of being extinct comets based on spectra and albedos.

This means that there is a dynamical route for Main Belt asteroids to reach into NEO orbits with $T_J < 3$, and Table 4.2 may contain a significant fraction of such objects in addition to the extinct comets. To distinguish the comets from the asteroids, one needs observations of physical properties. Unfortunately, there are not many of those. The most important concern the first object, (3552) Don Quixote, which has proved to be a comet by showing both a coma and a tail when observed with the Spitzer Space Telescope (Mommert *et al.* 2014). At the time of writing, it has not yet received a dual classification with a periodic comet number, but this may happen soon. Don Quixote was, already before the Spitzer observations, strongly suspected of cometary nature, since its spectral type[2] is typical of objects beyond the asteroid Main Belt.

Comet-like spectral behavior has also been established for 1997 SE_5, 2000 PG_3 and 2001 ME_1, while 2001 RC_{12} is not a comet candidate (DeMeo and Binzel 2008). However, the last mentioned has not been observed spectrally and was excluded from the comet candidates only because its albedo is marginally larger than the typical cometary range (Fernández *et al.* 2005). Some of the asteroids of Table 4.2 have orbits that are affected by the 2:1 resonance with Jupiter, so that one may ask if these could belong to the small group of stable asteroids at this resonance. In this regard, the long-term motions of (5370) Taranis, (8373) Stephengould and (9767) Midsomer Norton were studied by Roig *et al.* (2002) with the result that all three belong to what they called "strongly unstable asteroids" — a property that may suggest a cometary origin.

Comparing the orbital elements of the listed asteroids with those of Jupiter Family comets, we find that the asteroids have somewhat smaller Tisserand parameters and, above all, higher inclinations.

---

[2]The so-called D type is characterized by low albedo and a red color.

Such behavior is expected of dynamically old comets, as discussed in Sec. 5.6.

There are asteroids in other kinds of cometary orbits too. Those in Halley Type orbits are called *Damocloids* after their namesake, (5335) Damocles, and some asteroids have been discovered in long-period orbits. These will play a role in the discussion of comet aging in Sec. 4.5.1. Finally, a special case needs to be mentioned. Comet 2P/Encke is one of the few comets, whose orbit defies the usual criteria for cometary orbits — being decoupled from Jupiter's influence by a very small aphelion distance (see Sec. 1.4). It is not a typical asteroidal orbit, but in 1978 an asteroid was discovered whose orbit is very similar to the one of comet Encke. This is (2212) Hephaistos. The orbital similarity and the fact that a whole complex of small bodies seems related to comet Encke (see below) means that a genetic relationship between Encke and Hephaistos is often assumed. Hephaistos is hence assumed to be an extinct comet.

There is yet another kind of evidence that extinct comet nuclei have been observed as asteroids. This has to do with the fact that the production of large grains and boulders from active comets has the consequence that *dust trails* are formed, which evolve into streams of particles moving more or less along the comet orbit. When the Earth crosses such a stream, we observe a *meteor shower*, and the streams are referred to as *meteor streams*. Under normal circumstances, an asteroid would not produce such a stream, and hence, when it does, we tend to think it is an extinct comet.

The most famous association between an asteroid and a meteor stream was suggested by Fred Whipple in 1983. Asteroid 1983 TB had been discovered using the IRAS satellite, and Whipple recognized that its orbit practically coincides with the orbit of the well-known Geminid meteor shower, which is observed in December each year. The asteroid has become known as (3200) Phaethon, a name that points to its very small perihelion distance of 0.14 AU. It is beyond doubt that Phaethon is the source of the Geminid meteors, but the orbit is not cometary at all with an aphelion distance as small as 2.4 AU. However, this does not mean that Phaethon cannot have a cometary origin. Whether it was once a main belt asteroid or a comet, a special dynamical mechanism is required to explain the present orbit.

Several other well observed Earth-crossing asteroids have also been proposed as sources of meteor streams, but for now, let us only pay attention to the *Taurid complex.* This is a broad meteor stream with low inclination and small semi-major axis, which the Earth interacts with for a good fraction of its orbit, involving several meteor displays (Porubčan *et al.* 2006). The two night-time showers are called the Northern and Southern Taurids. It is known since more than 80 years (Denning 1928) that the stream has a complex structure, and the suggestion of a relationship with comet Encke was made by Whipple (1940).

Recent in-depth studies of the Taurid complex have revealed more than a dozen filaments. Several of those appear to be associated with recently discovered, small asteroids, which in turn move on orbits similar to comet Encke. Thus, there is a whole group of asteroids including Hephaistos that more or less share their orbits with comet Encke and contribute to the Taurid complex — thereby showing their cometary nature. However, only comet Encke has so far shown activity, while its companions seem dormant or extinct.

4.3.3. *The case of comet Encke*

There is one comet that by itself carries a lot of weight in the discussion of the physical evolution of comets, namely, comet 2P/Encke. This is largely due to two special circumstances. One is that comet Encke has by far the shortest orbital period of all comets and was discovered very long ago, so that it has performed many more orbital revolutions while being under observation than any other comet. The other is that this particular orbit keeps it safe from encounters with Jupiter, unlike the typical Jupiter Family comets, so that the orbit is quite stable on the time scale of centuries, while over much longer times it is subject to secular evolution and may thus change dramatically.

The long term dynamics of comet Encke illustrates an important issue about simulating comet motions. Levison and Duncan (1994) showed by long term integrations that the comet is practically certain to disappear by hitting the Sun on a time scale of 10^5–10^6 years due to eccentricity pumping by the secular ν_6 resonance.[3] Now, the dynamics under consideration is conservative, i.e., Hamiltonian, and this means that the evolutions are time reversible. So, does this result mean that comet Encke is likely to have been ejected from the Sun far back in time, because backward integrations would lead to that result? No, this obviously cannot be the case, and the lesson to learn is that the fate awaiting objects in the future may be statistically evaluated by Monte Carlo sampling, but such statistics is rather useless when it comes to evaluating the dynamical origin of the objects. Here one has to look for the true route rather than the statistically likely one, though by backward integrations it might be like finding the needle in the haystack.

A related issue deals with the origin of comets in general. Forward integration of the motion of observed comets, when extended long enough into the future, will likely lead to ejection into interstellar space through gravitational perturbations by Jupiter. For the same reason as mentioned above, one might thus conjecture that all comets come from interstellar space and have been captured into the solar system by Jupiter. But the real sources of comets are found inside the solar system, such as the Oort Cloud. For those comets that come directly from the Oort Cloud, this can be demonstrated by accurate backward integration (see Sec. 5.1), while for other comets, whose connection to their source is more remote, any backward integration is likely to be led astray by the small but unavoidable errors. Hence the actual route and the true source would almost certainly be missed.

The question remains, what is the dynamical origin of comet Encke? The best explanation so far was provided by Levison *et al.* (2006). In principle, the problem is that the secular mechanisms at work do not change the semi-major axis of the orbit, so they cannot change the orbit into its present aphelion distance of 4.1 AU from an

[3]This means that the apsidal line of comet Encke's orbit precesses at nearly the same rate as Saturn's apsidal line.

orbit with aphelion close to Jupiter's orbit. Moreover, for the same reason, close encounters obviously cannot do the job. Two options remain: nongravitational effects or close encounters with terrestrial planets. The first has the problem that a consistent reduction of the orbital period over a vast number of orbits would be required, and the observations of nongravitational effects found in different linkages of Jupiter Family comets do not indicate such consistency. Moreover, the nongravitational effect of comet Encke itself has been observed since two centuries, and it has not been stable. In the beginning, there was a notable decrease of orbital period, but this has by now vanished and turned into an increase.

Instead, Levison *et al.* could show by dynamical simulations based on forward integrations that, by including the interactions with the terrestrial planets, the capture of comets into the Jupiter Family might explain the presence of not just one comet in an Encke-like orbit but actually as many as 12 with nucleus diameters larger than 1 km. However, this result was reached assuming that the physical lifetimes were infinite. This is of course unlikely, but the most difficult problem was to explain how even one of the 12 comets could be active, like comet Encke is. This would require an active lifetime that is unreasonably long when compared to the expectations for Jupiter Family comets (see Sec. 5.6).

As a way out of this problem, Levison *et al.* made the following suggestion. In a typical case, the evolution into an Encke-like orbit would take ~1 Myears, but for almost all this time the orbit would undergo very small perturbations after an initial decoupling from Jupiter by an encounter with a terrestrial planet. The comet may then be dormant and thus survive intact, until finally another such encounter brings it into the ν_6 resonance and the eccentricity starts to increase. Then the perihelion distance decreases, and as found by Rickman *et al.* (1991), this may reactivate the comet (see Sec. 4.5.2). In this scenario, comet Encke would be dynamically very old, but it spent almost the whole time in a dormant state. It is impossible to judge from orbit integrations, how long ago the reactivation took place, but obviously, this must not have been too long ago.

Comet Encke has now been under observation for more than two centuries and more than 60 revolutions around the Sun. Has it been fading? Although claims of fading have been made, it appears that these are not reliable. In a study of the long term brightness behavior of 2P/Encke, Lars Kamél (1991) found that there has been a gradual shift in the light curve with its peak approaching the perihelion passage. This evolution may have caused the gradual decrease of the nongravitational effect experienced by the comet. But it also made it increasingly difficult to observe the comet at its brightest, which according to Kamél would have resulted in an apparent fading.

Concerning the long term physical evolution, it should be noted that 2P/Encke is a comet with very low activity. This was estimated at about 1% by Tancredi *et al.* (2006), who found the nuclear radius to be 2.1 km. With the resulting amount of relative mass loss per revolution, the comet may survive for very long, and it may also have been reactivated very long ago without having been exceedingly large to begin with.

4.4. Splits and Outbursts

Comet nuclei are unique among the macroscopic objects in the solar system in having been both suspected and seen to split apart without undergoing any collision. This phenomenon may be of great importance for their evolution. On the one hand, we know of at least one case, where a comet has disappeared just after it was observed to split. On the other hand, it is clear that the mass loss of comets in the long run must be accelerated, if from time to time major parts of the nuclei get detached and leave the comets. If these are large enough, they may survive for long, and thus comets may multiply by division, increasing their population size. However, even though there is evidence that this has happened, under normal circumstances the effect is probably small.

The light curves of comets are often somewhat irregular, and sudden increases of brightness are common. But real outbursts with an increase of a hundredfold or more are very rare. Nonetheless, these may represent important mass loss episodes and should therefore play a role in the long term evolution of the comets in question. In this respect, they play a role similar to the splitting events. We now pass to a brief description of observed splits and outbursts.

4.4.1. *Tidal splits*

In Sec. 2.5.2 we discussed the physics of tidal splitting in relation to two cases, where comets were disrupted by Jupiter's tidal force: D/1993 F2 (Shoemaker-Levy 9) and 16P/Brooks 2. However, the Sun can obviously act in a similar way on comets with very small perihelion distances. In turn, such orbits with essentially zero angular momentum can be produced by the Lidov–Kozai effect (see Sec. 3.3) or by other secular resonances. Now, consider a comet that experiences such an evolution. Its actual perihelion distance decreases with each new revolution, until a minimum is reached and an increase follows. If the smallest value of this perihelion distance is larger than the radius of the Sun but so much smaller than the radius of the solar Roche sphere that fragments may be shed from the nucleus, there will be a tidal split without the loss of the fragments by solar impact.

Of course, there is no guarantee that all the fragments will survive until the dynamical evolution has brought them back into more normal values of the perihelion distance. The rate of insolation experienced around perihelion by sungrazing comets is enormous. Even if the time is too short for total destruction by surface sublimation of ice, other effects like, e.g., thermal cracking may add to the vulnerability of sungrazers. A recent example is comet C/2012 S1 (ISON), apparently a new comet from the Oort Cloud, which did not survive its perihelion passage in November 2013 at $q = 0.01244$ AU although its early observations had not indicated a particularly small nucleus. It is thus unclear, which role is played in the real solar system by the process in question.

If the dynamical evolution involves a perihelion passage inside the solar photosphere, this means the demise of the comet and its fragments from a possible, preceding tidal split. We may be witnessing this phenomenon in the case of the *Kreutz group* of

Fig. 4.5. Contemporary drawing of the Great March Comet of 1843.

sungrazing comets. Several of the brightest and most impressive comets ever observed were members of this group like the Great March Comet of 1843, which is illustrated in Fig. 4.5. Unfortunately, it has not been possible to determine their orbital periods accurately, but these range around 500–1000 years. Extrapolating backward in time, the great comet of 1106 is suspected to be the direct precursor of some, though not all, of the Kreutz group members through a tidal split at its perihelion passage. The presence of other members, which do not share the same heritage, means that a further generation is needed, and speculations have been made about comets of the antique that might be a Kreutz group grandparent.

As mentioned in Sec. 1.4, most comets on record are actually Kreutz group members, since a vast number of tiny sungrazers belonging to this group have been discovered by satellite-based, sun-staring telescopes like SOHO. In general, these have not survived their perihelion passages — sometimes because of collision with the Sun, but mostly because they were too small to withstand the fierce erosion due to solar heating. It is not possible to say with certainty, if any Kreutz group comet (including the largest ones) will eventually survive.

With the aid of the mentioned coronagraphic observations, a few more groups of sungrazers have been identified. These are the Kracht, Marsden and Meyer groups. The Kracht and Marsden groups appear related to comet 96P/Machholz 1 with an orbital period of only about five years, while the Meyer group seems to have orbits of substantially longer periods. Comet Machholz 1 was briefly mentioned in Sec. 3.3 as an example of the Lidov–Kozai mechanism operating on a short-period comet. The large-scale, secular variations of its orbit combined with a gradual disintegration process — possibly related to tidal splits — leads to the formation of a whole complex of interplanetary objects including meteoroids in quite different orbits, forming different streams and observed as different showers.[4] Apparently, the complex also involves larger objects seen as sungrazing mini-comets.

From long-term integrations of the motion of comet Machholz 1, Bailey *et al.* (1992) found that the Lidov–Kozai cycle of the comet evolves in such a way that a collision with the Sun may take place in the future, more than 14 000 years from now. Due to the uncertainty of the actual, current orbit, this prediction is not safe, but the point is that a future tidal split with ensuing solar impact is indeed a possibility for one of the observed short-period comets. As mentioned, it is also possible that the complex of objects related to Machholz 1 has its origin in a past, similar episode.

4.4.2. *Non-tidal splits*

Most comet splits occur far from any gravitating body and so do not depend on tidal forces. The physical mechanism behind these splits (or more than one mechanism!) is not known, although some candidates are typically favored over the others. We shall not make further comments on these — see, however, the discussion in Sec. 2.5.2.

[4]The well-known Quadrantid shower seen in January belongs to these.

In many cases the observed splits have been quite modest phenomena that required careful monitoring to be seen (Chen and Jewitt 1994). Only minor fragments left the nuclei with no direct, dramatic consequences. However, there have also been some very spectacular events, which deserve specific attention. Historically, the most important event was the split of Biela's comet, which was noted in the winter of 1846 soon after this periodic comet had been recovered. It was first seen in 1772 and then again in 1805 without a possibility to prove the identity of these two comets. Its name comes from Wilhelm von Biela, who observed the comet in 1826 and was able to establish the short period of its orbit.

Biela's comet was again observed in 1832 and then missed in 1839 due to very unfavorable observing conditions. The 1846 return was well observed, especially after observers had noted an unexpected duplicity of the comet. Two components were discerned about one arc minute apart, and some evidence suggests that lots of smaller fragments were also present between the two main ones. A contemporary drawing is shown in Fig. 4.6. The time when the comet split is somewhat uncertain, but it might have been close to the preceding aphelion passage, when the comet was more than 6 AU from the Sun.

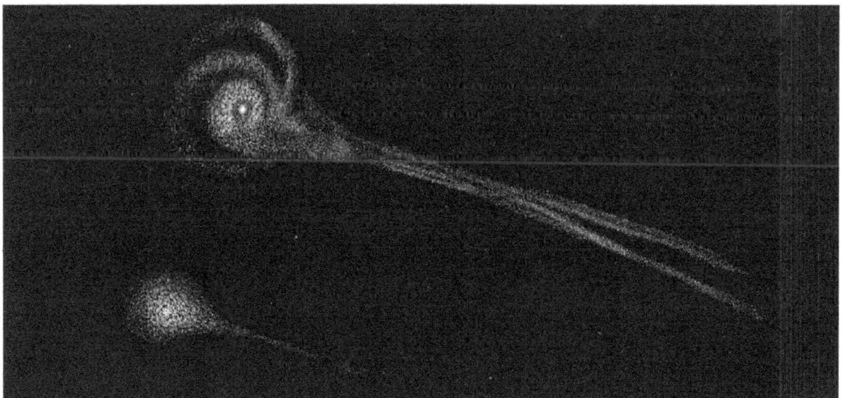

Fig. 4.6. Contemporary drawing of comet 3D/Biela from February 1846.

In 1852, both components came back to perihelion as observable comets. Observations continued as foreseen, but after the comets faded out of sight, they were never seen again. Instead, when the Earth crossed the vicinity of the comet orbit in 1872, a very spectacular meteor shower[5] was observed. This was obviously caused by debris from the split comet, and the persistent impression is that both components were completely disintegrated. However, some skepticism is warranted concerning this conclusion, and there have been numerous attempts to reobserve the main fragments and to identify them with other observed comets. None of these have been successful, and it is indeed likely that not much is left of this, once rather impressive comet. It is nowadays designated as 3D/Biela.

There is also a recent case, where a comet has split and apparently disintegrated completely before the eyes of the observers. This was comet C/1999 S4 (LINEAR), a long-period comet that seemed to have a very small nucleus. Its perihelion passage should have occurred on 26 July 2000, and the perihelion distance was 0.765 AU. However, after a minor split on 5 July 2000 a burst in brightness that is thought to have marked the complete disintegration of the comet was observed on 23 July 2000.

It is thus rather clear that a comet can disappear after being split, but there are many more cases where comets have split without disappearing and there are also several comets that have disappeared without any observed splitting episode. A recent example of a comet that has undergone repeated splits without having as yet disappeared is 73P/Schwassmann-Wachmann 3.[6] Seemingly unprovoked disappearances are on record for several recent long-period comets including, e.g., C/1996 Q1 (Tabur) and C/2002 O6 (SWAN). The reasons for these disappearances are not clear.

We hence need to estimate the role of splits in the evolution and mass loss of comets using statistical data on the frequency of observed

[5]This shower bears the name *Andromedids* but has more or less disappeared due to orbital perturbations from Jupiter, which have dispersed the stream.
[6]At the time of writing, the jury is out, because the next perihelion passage is imminent and no recovery has yet been reported.

splits and the average amount of mass loss involved in all these splits. According to Boehnhardt (2004), estimates of the splitting frequency tend to result in about one event per comet and century, mainly based on observations of short-period comets. Long-period comets and, in particular, new comets may have a higher splitting frequency, but the observed sample is quite small. It should be noted that the observed frequency may be a lower limit, since many splits may have gone unnoticed. However, it is difficult to imagine that the missed splits should contribute a larger total mass loss than those observed.

There is a wide range of fragment masses, and the estimated average has a large statistical uncertainty. At face value, comparing with the time spent by Jupiter Family comets in their observable orbits, it seems that the splits may give an important contribution to the total mass loss for initially km-sized nuclei. However, as seen above, such comets may be dormant for a large fraction of the time, and it is not known if dormant comets can also split like the active comets do. Thus, the above-mentioned estimate should be taken with a grain of salt.

4.4.3. *Outbursts*

The three comets most worthy of notice here are periodic comets 17P/Holmes, 29P/Schwassmann-Wachmann 1 and 41P/Tuttle-Giacobini-Kresák (here listed in numerical order). In terms of strength or magnitude of the outbursts, these comets are not absolutely outstanding when compared to other comets with irregular brightness behavior, but still they can be singled out as the most extreme. The fact that a small group of short-period comets can thus be singled out, and that their outbursts are irregularly recurrent, may provide some clues as to the underlying mechanism. Before returning to this issue, let us briefly describe them one by one.

Comet 17P/Holmes was included in Table 4.1 because after three observed apparitions, it was missed during the following seven returns. It is normally very faint, but it could be discovered in 1892 apparently due to an outburst that even rendered it a naked eye object for a short time, before it faded away. After its recovery at a faint magnitude in 1964, it continued to be a very unspectacular object until 23 October 2007. Over a couple of days, it then brightened by nearly 15 magnitudes, i.e., nearly a million times. This is the most extreme outburst ever observed for a comet. It became easily visible to the naked eye and for a time counted as the third brightest object in the constellation Perseus. Then its coma swelled to a radius of nearly a million km, larger than the Sun. Afterwards, it gradually fell back to its "normal", faint level.

We obviously cannot know how much the comet brightened in relation with its discovery, but the evidence points to a large outburst, possibly rivaling the one in 2007. Comet Holmes also brightened suddenly by 3–4 magnitudes in January 2015, and such small outbursts might have happened during the time it was lost as well. Thus far, no specific physical mechanism has been proven to fully account for the behavior of comet Holmes.

Comet 29P/Schwassmann-Wachmann 1 has held a special place in the hearts of comet scientists ever since it was discovered in 1927. It has defied the simplest rules of cometary convenance in a spectacular way. Regarding its orbit with a perihelion distance of 5.7 AU and an aphelion distance of a little more than 6 AU, it was the only known Chiron-type comet long before Chiron was discovered. It is nowadays counted with the Centaurs, but before the Centaurs were discovered it was truly singular. Given the large perihelion distance, the activity of comet 29P cannot be driven by H_2O sublimation — contrary to the usual comets. To this comes its propensity for outbursts, which has been recognized since the beginning.

Some properties of the 29P nucleus have been more or less clarified. It is very large with a radius that can be estimated as $\sim$15–30 km. In the quiescent mode, a CO production rate from the nucleus of $\sim 10^{28}\,s^{-1}$ has been established by radio observations (see Sec. 2.6.6), and this is doubtless the driver of this activity. The grains that form the visible coma are likely icy, made to a large extent by H_2O. But the outbursts have not yet been monitored by mapping of the CO coma, and the information available is hence relatively sparse. It is certainly a question of a sudden, massive ejection of grains from the nucleus, but we do not know if this is due to a dramatic increase in the CO production rate. Led by earlier suggestions that crystallization of amorphous ice may explain comet outbursts, Froeschlé *et al.* (1983) found that the thermal conditions near the surface of comet 29P are particularly relevant for this explanation.

Some of the outbursts correspond to an increase of the brightness by a factor 10^4, while the majority are much smaller and occur several times per year. Thus, the outburst behavior of comet 29P is somewhat different from comet 17P, the events being on average smaller and occurring more frequently. However, we do not know if the mechanism is basically the same or not.

Comet 41P/Tuttle-Giacobini-Kresák was included in Table 4.1 like 17P/Holmes. However, in this case none of the three independent discoveries (Sec. 4.3.1) was due to an outburst. Instead, the comet is famous for only one extreme outburst — though a double one — during its 1973 apparition. In both cases the comet brightened by about 10 magnitudes (a factor $\sim 10^4$ in brightness) over several days. They occurred about 40 days apart, and in between the comet had dropped to its normal brightness. Another large but considerably smaller outburst happened during the 2001 apparition.

In a study of the 1973 outbursts, Kresák (1974) noted that the amount of radiation emitted by comet 41P during just a few days during an outburst was equivalent to that normally emitted during 80 revolutions, which means several centuries. He concluded that the extinction of comets by mass loss is largely caused by outburst events and that in between these, the evolution is slow.

That is certainly true for the outbursting comets, but the question remains which proportion of the comets at large belong to this category. The three comets just described may seem very few in comparison to the current sample of more than 300 Jupiter Family comets, but the vast majority of those have only been observed for a few apparitions, so some of them may well be of the outbursting kind, and we do not know how many these are. Concerning the relevant mechanism behind the outbursts, the issue is open. Supervolatiles like CO may play a role in causing the ejection of large quantities of grains, but this remains a speculation, and why it happens repeatedly in just a minority of the comets remains unexplained.

4.5. Aging of Comets

We have seen a few examples of observed comets experiencing phenomena of importance for their long term evolution, and some theoretical concepts dealing with this evolution have been introduced. Let us now face the question: How do comets evolve in terms of activity and observability in a statistical sense? This means to avoid studying the time dependence of individual comets, since this is known to be sensitive to the changes in observational techniques, so that comets can appear to be fading rapidly without evolving at all. We will instead take advantage of the fact that the observed comet population at large reflects a mixture of individuals in different evolutionary stages and may therefore record some features of the evolution.

4.5.1. *Long-period comets: The fading problem*

One of the first issues that Jan Oort noted, when he put forward his theory of the cometary cloud, was that over time, the returning comets would become so numerous that the relative height of the spike formed by the new comets would be much less than we observe. Of course, in his first investigation some comets were found to be returning, but there should have been more. One can use two related ideas to tackle this problem. One is that new comets like any comet fade so much that the returning comets tend to be missed, and the other is that the new comets may be special because they come directly from a storage in the Oort Cloud that may have lasted for billions of years. If so, they might be anomalously bright.

Regarding the second idea, one thing that makes the Oort Cloud environment special is the bombardment of comet nuclei by Galactic cosmic rays. This would form a refractory crust, which seems to go against the wanted effect of increased brightness. But if such a crust is combined with the presence of primordial, amorphous ice at a very shallow depth, one may speculate that a crystallization burst provides the necessary gas pressure to break the crust, eject the pieces and embed the nucleus in a thick shroud of dust, which provides the anomalous brightness. Another possibility that has attracted attention is that the Oort Cloud comet nuclei may accrete a surface frosting of supervolatiles, which sublimates fiercely on the inward orbital branch. Observational data suggest that dynamically new comets do behave in a special way when approaching the Sun by being particularly bright at large distances.

However, the fading that is needed to explain the orbital energy distribution of long-period comets does not only involve the first passage. Many papers have dealt with the problems of characterizing the gradual losses that seem to affect the long-period comets during their energy diffusion and of finding the true physical explanation. It is natural to think of surface differentiation as a contributing factor. The outgassing at each consecutive perihelion passage would leave a refractory residue at the surface (dust mantling as discussed in Sec. 4.2), which provokes an increasing amount of quenching of the gas production and growing inability to release dust grains.

In addition, comets may erode away by splits and surface fragmentation, as argued above. There is a very important distinction between the two mentioned processes although both of them would act to remove comets from the observable population. In the quenching scenario the comet enters into dormancy or, possibly, definitive extinction, leaving behind an object that may be observed as a long-period asteroid. This does not happen in the splitting scenario, where essentially nothing is left behind. In the models proposed by Paul Weissman, splitting plays a major role. Supporting evidence was found in a paper by Levison *et al.* (2002), where the abundance of extinct nuclei in the quenching scenario was confronted with the scarcity of observed long-period asteroids. Today, there are 29 long-period asteroids with perihelion distances $q < 5$ AU. All of these have periods less than 60 000 years. Since only three of these were known in 2002, a new look at the aging problem of long-period comets seems warranted.

Another legitimate question deals with the individuality of comets as far as aging is concerned. If two comets with equal-sized nuclei were injected from the Oort Cloud and experienced identical orbital evolutions, would they also age in the same way? Most scientists would likely answer no, because there are other parameters that may vary between different comets and would have an influence on the evolution. Some models include such variability and suggest that the Halley Type comets, which are statistically the oldest in terms of energy diffusion, tend to sample a kind of nucleus that is specially resistant against fading. This might be sufficient to explain their existence, even though the Halley Types could also have experienced particularly strong capture events that make them unrepresentative by being younger than they appear.

Comparison of the Halley Type comets and the Damocloid asteroids reveals an interesting complementarity. If we define both groups by orbital periods between 20 and 200 years, perihelion distances less than 5 AU and jovian Tisserand parameters less than 2, there are currently 67 Halley Type comets and 59 Damocloids. Among the Halley Types, there used to be a conspicious dearth of objects with perihelion distances $q > 1.5$ AU, but for the many recent discoveries this trend has disappeared. However, the Damocloids almost always have $q > 1.5$ AU. This indicates that both groups are captured Oort Cloud comets, and the perihelion distance influences the physical evolution, so that quenching is common for $q > 1.5$ AU and continued activity is the rule for $q < 1.5$ AU.

Finally, let us note that the simple models of aging that have so far been used for long-period comets imply a monotonous fading. This is self-evident in the splitting scenario[7] but less so when we deal with quenching. Rejuvenation from dormancy is a case in point as demonstrated by individual comets, and we shall soon see statistical evidence for this as well.

4.5.2. *Jupiter Family comets: The role of encounters*

There are hundreds of Jupiter Family comets and the discovery years range from the late 1700s to the present. Each new discovery in principle raises the question, why the comet was not discovered before. We have come across the cases of comets D/1770 L1 (Lexell) and 2P/Encke, which were among the first to be discovered and where

[7]The split may induce a brightness burst by exposing fresh, icy surfaces, but in the long run, the comet gets smaller.

the answers are different. Lexell's comet had just been captured by Jupiter into an orbit with much shorter perihelion distance, while Encke's comet had a stable orbit and was possibly missed by not being spectacular enough due to its closeness to the Sun at perihelion.

In our days, most of the new discoveries refer to comets with large perihelion distances, which simply could not have been discovered much earlier because of their faintness. With better telescopes, we are exploring new ground. Only few comets are discovered with small perihelion distances, and these tend to be recently captured, somewhat like Lexell's comet. This is the main criterion to consider the list of Jupiter Family comets with $q < 1.5$ AU as more or less complete.

Going through the pre-discovery histories of Jupiter Family comets, one is struck by the frequent occurrence of large decreases of the perihelion distance just a few revolutions before the comets were discovered. However, to some extent this may be because of the mentioned effect. If the relevant part of the Jupiter Family is already complete, each new discovery must be related to such a recent capture. Nonetheless, it is worth investigating, if these captures have any further consequence that facilitates discovery of the comets.

For this purpose it is best to consider new discoveries in a part of the Jupiter Family, where the coverage is far from complete. This means to look for comets that were discovered long ago with large perihelion distances. In Fig. 4.7 we see a plot of $\Delta q = q - q_{\rm disc}$ versus revolution number, where $q_{\rm disc} > 1.5$ AU is the perihelion distance at the discovery apparition, which has been given the number 0. The discoveries took place before 1950 in all cases. For the post-discovery apparitions we see a random behavior of q including both increases and decreases, but the pre-discovery apparitions are clearly different. More than one revolution before discovery, a significant fraction of the comets had $\Delta q > 1$ AU, and there is nothing similar for negative values of Δq.

How shall this asymmetry be interpreted? The most straightforward explanation is that recently captured comets are bestowed with an inherent property that makes them easy to discover. In turn, there is only one such property to think of, namely, an increased activity that makes the comets bright and spectacular. Let us refer

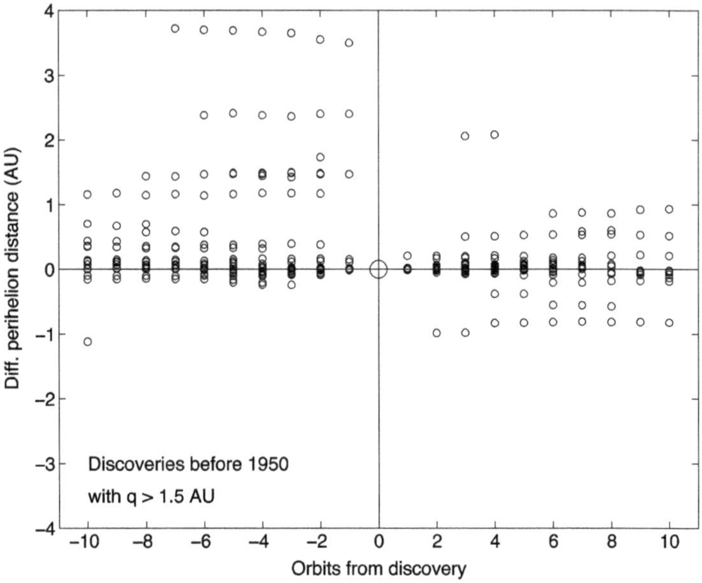

Fig. 4.7. Difference between current perihelion distance and the perihelion distance at discovery at the nine apparitions just before and after the discovery apparition. All these Jupiter Family comets were discovered before 1950 with perihelion distances $q_{\rm disc} > 1.5\,{\rm AU}$. Credit: H. Rickman, "Cometary Dynamics," in Lecture Notes in Physics, Vol. 790 (2010), pp. 341–399, © Springer-Verlag Berlin Heidelberg 2010. With permission of Springer Nature.

back to comet 65P/Gunn, which is seen in the right panel of Fig. 2.4 (Sec. 2.3.3). This very active comet was in fact captured into $q = 2.5\,{\rm AU}$ from $q_{\rm prev} = 3.3\,{\rm AU}$ at an encounter with Jupiter in 1965 shortly before its discovery (too recent for this comet to be included in Fig. 4.7). Licandro *et al.* (2000) took note of this fact and found a remarkable coincidence between those of their observed comets that showed distant activity, like comet Gunn, and those that had undergone recent encounters with subsequent decrease of the perihelion distance.

The presence of even more comets in Fig. 4.7 that were discovered without any recent capture does not contradict the above suggestion. When new comets are discovered in a large unexplored population, the main criterion is brightness, and some comets lived up to this criterion for other reasons — e.g., large nuclei or outbursts. Of course, the tendency for the captures to have occurred shortly before discovery implies that the increased brightness subsides relatively fast. There is another investigation that came to a similar result from a different aspect.

In Sec. 2.5.3, when dealing with the use of nongravitational effects for comet mass determination, we came across Eq. (2.14), which together with Eq. (2.15) expresses the observable perihelion delay ΔP in terms of the gas production curve $Q(t)$, the mass of the nucleus and other parameters including orbital elements. In papers by Festou *et al.* (1990) and Rickman *et al.* (1991), it was shown that a modified expression for the perihelion delay,

$$\Delta P'' \propto \Delta P\, a^{-5/2}\, Z(q)^{-1}, \tag{4.3}$$

could be derived, which is relatively independent of the orbit and thus isolates the physical trigger of the effect. Here $Z(r)$ is a dimensionless function approximating the sublimation flux of H_2O from an icy surface, and q is the perihelion distance. The triggering effect could thus be formulated as

$$\Delta P'' \propto \frac{f}{\rho R}(E + \sin\eta_e T), \tag{4.4}$$

where f and R represent the activity (see Sec. 2.4) and mean radius of the nucleus, respectively, E and T are parameters describing the shape of the gas production curve, and η_e is the lag angle between the effective jet force vector and the antisolar direction as projected onto the orbital plane.

Rickman *et al.* (1991) presented a statistical investigation of $\Delta P''$ as found from a large number of orbital linkages of short-period comets and associated E values, focusing on the variations with perihelion distance and the "age" of the linkage, i.e., the number of orbital revolutions spent before the center of the linkage since the last major decrease of the perihelion distance. The results are best illustrated by Fig. 4.8, which shows the absolute value of $\Delta P''$ versus the perihelion distance at the center of the orbital linkage for four subsamples. These differ according to the size of the preceding decrease of the perihelion distance.

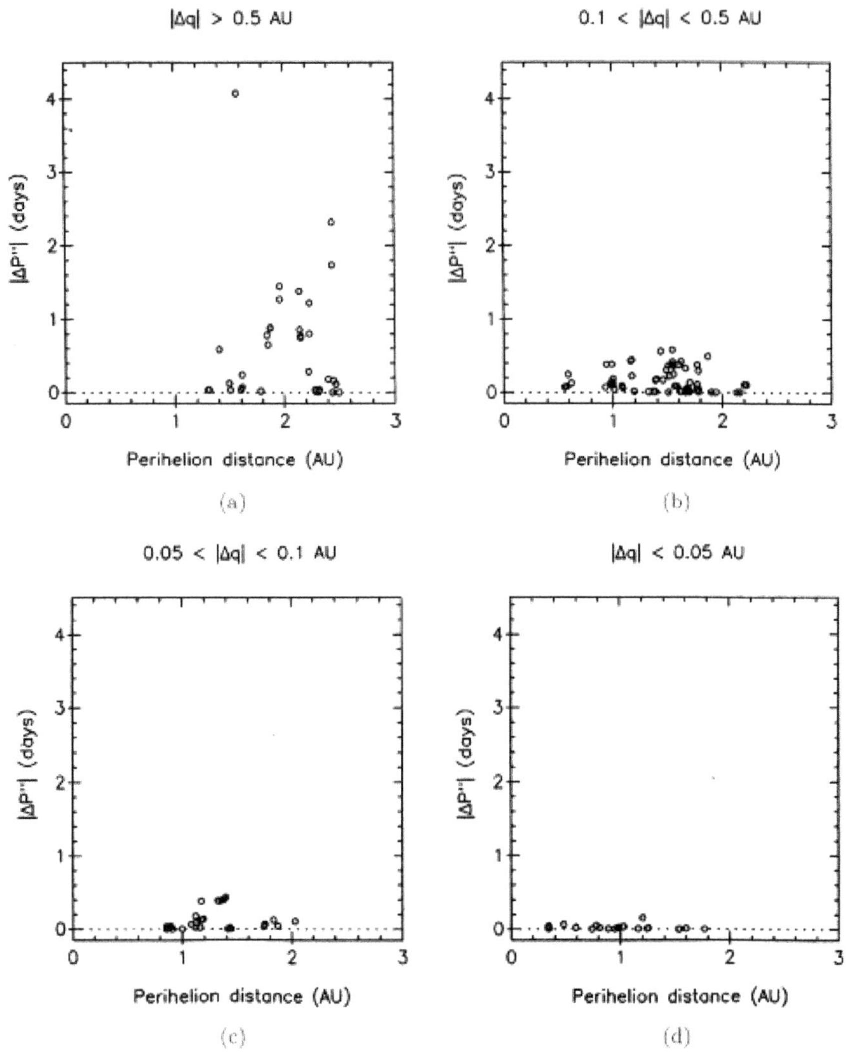

Fig. 4.8. The absolute value of a standardized nongravitational effect plotted versus the perihelion distance for four groups of linkages of short-period comets. These differ in the size of a preceding, sudden decrease of the perihelion distance caused by a close encounter with Jupiter. The time interval in question extends up to 25 orbital revolutions. From Rickman *et al.* (1991). © AAS.

Efficient triggers of the nongravitational effect include fresh nuclei (large f), small nuclei (small R), and large perihelion asymmetries of the gas production curves (large $|E|$). The only one of these, which can reasonably be expected to correlate with the size of the preceding orbital perturbation is f, and thus Rickman *et al.* concluded (in the widely-held picture of those days) that the larger decreases of perihelion distance cause larger mantle blow-offs, leading to fresher nuclei. In today's emerging picture as well, it is necessary to think in terms of a partial disappearance of the dust that covers the uppermost icy layer and quenches the outflow of gas.

According to Rickman *et al.* an age effect is also implied. This is because the comets that did not have any of the largest perturbations recently enough would likely have had larger perturbations earlier. But they do not seem to remember the freshness they once had, so there is an aging effect that operates on a time scale of $\sim$10 orbital revolutions.

We are thus faced with quite a rapid aging in spite of the fact that studies of light curves and absolute magnitudes of periodic comets do not corroborate this conclusion. Probably, given the above-described difficulties of characterizing secular brightness changes, the tools provided by such studies are too blunt compared to the reliable measures of nongravitational effects. When Rickman *et al.* performed their investigation, the sample of reliable f values based on direct measurements of gas production rates was by far too small, so a new look at the problem using the currently available f values may be an interesting pursuit.

A picture that arises naturally from the mentioned studies is that the physical evolution of comets is conditioned by the dynamical evolution via the activity level. This may typically be low but increases after a decrease of the perihelion distance and then falls back to a low value again upon a temporary stabilization of the perihelion distance. One may thus speculate that high activity levels only occur before the comet has reached its overall minimum of perihelion distance, so that after this time comets tend to remain at low activity levels or even dormant.

Chapter 5

Capture of Comets

In Sec. 3.6 we briefly sketched how the observed comets are linked to their distant source regions, namely, the Oort Cloud and the Scattered Disk. To be able to trace the comets back to their origin, it is of vital importance to understand some characteristics of these sources — the number of objects, their size distribution, and the distribution of orbital parameters.

For the Oort Cloud, there are no direct observations available to constrain these characteristics. We have to work our way from the only messengers we have — the new comets. The method is dynamical modeling of the cloud and the transfer mechanism. The quality of the results will depend on how well we understand the dynamics, and how well we can characterize the new comets. For the Scattered Disk, the situation is more complex. On the one hand, we do have some observational coverage of this population, which is of course extremely important. On the other hand, working our way from the observed messengers — the Jupiter Family comets — is hampered by specific problems that need to be resolved. In this chapter, we shall describe the current state of knowledge concerning all these issues with some historical background for better understanding.

5.1. The Oort Cloud

About a century ago, essential work was done, which laid the foundations for the concept of the Oort Cloud. Elis Strömgren and his colleagues at Copenhagen Observatory and Gaston Fayet at Paris Observatory made numerical studies of the motions of known long-period comets, integrating the orbits backwards in time to find the trajectories along which these comets had approached the planetary system. This study was of obvious interest in order to settle the issue, whether comets came from interstellar space or truly belonged to the solar system. The results confirmed the latter to be the case.

However, these calculations were extremely demanding and time consuming with the primitive computing aids of those days, and only a handful of comets could be accurately studied. A few decades passed, before the data came to use in a work that would overthrow the old concepts and set a new paradigm for the origin of comets. This happened at Leiden Observatory in Holland. One of the first studies of energy diffusion for comets (see Sec. 3.5) was made as a Doctoral work by Adrianus Van Woerkom (1948) with the result that a population with any initial energy distribution would evolve into a flat distribution extending to the parabolic limit, beyond which they escape into interstellar space. His supervisor, the celebrated Galactic astronomer Jan Hendrik Oort, made use of this conclusion in a study of the observed distribution of original orbital energies of comets as known from the works of Strömgren and Fayet, and he came to a remarkable result (Oort 1950).

Oort had a sample of only 19 original orbits with reliable values for the inverse, barycentric semi-major axis ($1/a_o$), which is proportional to the gravitational binding energy of the comet to the solar system. He noticed a conspicuous pile-up in a narrow energy interval next to zero, i.e., the parabolic limit. Ten of the comets had $1/a_o < 5 \times 10^{-5}\,\mathrm{AU}^{-1}$, which means semi-major axes larger than 20 000 AU. Four others were in the following bin with a_o from 10 000 to 20 000 AU, and the rest were scattered over a much wider range of energies with $a_o < 10\,000$ AU. This energy distribution is not flat at all, like it would be in a steady state, and Oort concluded that there must be a source of comets with $a_o > 10\,000$ AU that provides for the spike in the distribution. He identified this with a distant comet reservoir that was not observed due to the large perihelion distances and the fact that the comets would spend most of the time

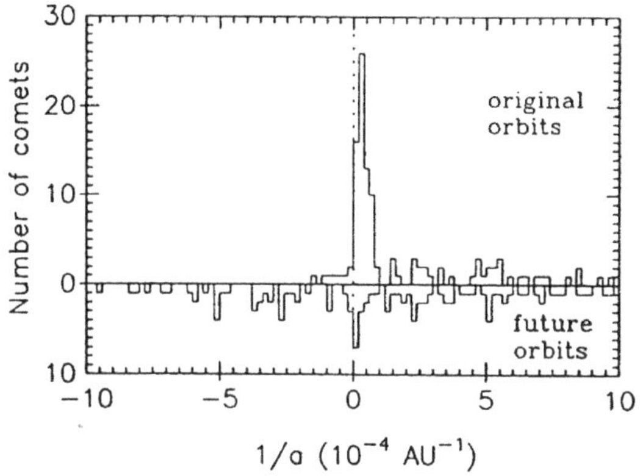

Fig. 5.1. Double histogram showing the distributions of inverse, original and future semi-major axes for a sample of long-period comets with high-quality orbits, available in the early 1990s. Reproduced from Rickman (2014) with permission. © The Meteoritical Society, 2013.

extremely far away. This reservoir has since been nick-named *the Oort Cloud.*

Nowadays many more comets are available for the kind of analysis that Oort performed. In Fig. 5.1 we see a double rendering of the energy distribution based on the sample available in the early 1990s. On the top there is a histogram of the original inverse semi-major axes, and the inverted histogram below refers to the future values for the same comets, reached after leaving the planetary system. The spike seen for the original orbits is essentially the same as that Oort had noted. However, for the future orbits this spike is almost entirely wiped out by the orbital perturbations suffered in connection with the passages through the planetary system. The scatter in $1/a$ is in agreement with the typical perturbations caused by Jupiter on random comets with small perihelion distances (see Sec. 3.5).

One consequence that is clearly demonstrated is that about half the long-period comets are ejected from the solar system by planetary perturbations. These will never return but will join a population of interstellar comets, of which nothing has yet been seen but which may be very numerous, in case many solar-type stars eject comets like our solar system does. The comets that are not ejected will come back to the subsequent perihelion passage some time in the future — the sooner, the further to the right they are in the diagram. They will then exhibit a distribution of original inverse semi-major axes similar to the one seen for the current comets that do not belong to the Oort spike. These are the returning comets, while those in the spike are called the *new comets*, i.e., the newcomers from the Oort Cloud.

Given that the orbits of the observed new comets are so unstable against ejection from the solar system, one may ask how large the perihelion distance has to be in order to avoid this instability. Relevant information is provided in Fig. 5.2. This shows the standard deviation of the perturbation distribution for inverse semi-major axis, considering random long-period comets that do not enter within the

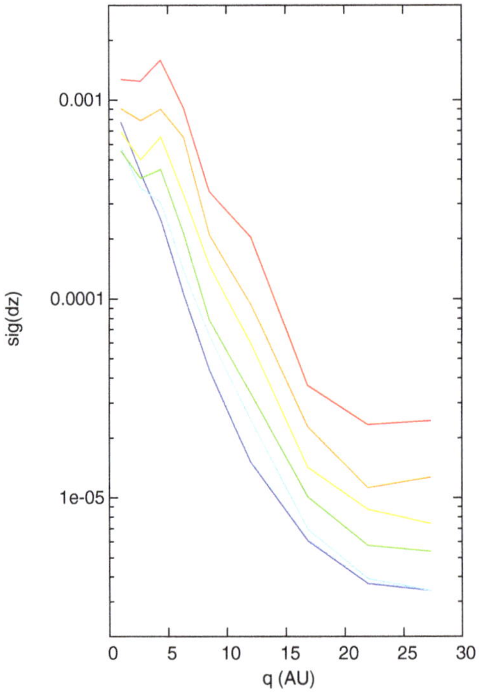

Fig. 5.2. Standard deviation of orbital energy perturbations as a function of perihelion distance and inclination for randomly oriented orbits with a_o = 20 000 AU. The colors indicate six inclination ranges of equal width (30°). All four giant planets are treated in the approximation of circular and co-planar orbits with $i = 0$. Reprinted from Fouchard, M. *et al.*, *Icarus* **222**, 20–31 (2013), with permission from Elsevier.

Hill sphere of any planet.[1] The six curves apply to six inclination ranges, all of the same width, and the red curve is for the lowest values ($0 < i < 30°$) while the dark blue curve is for the other extreme ($150° < i < 180°$). In the units employed (AU^{-1}) the width of the range of new comets measures 0.0001, and we see that this is exceeded by the typical perturbations for perihelion distances less than 5–15 AU, depending on inclination.

Thus, if q_c denotes the minimum perihelion distance, for which the typical perturbations do not remove a comet from the new comet range, we can expect the Oort Cloud to suffer a certain depletion for $q < q_c$ within the energy range of new comets. This depletion concerns the angular momenta (L) according to

$$L = r\,V_t = \sqrt{GM_\odot q(1+e)} \approx \sqrt{2GM_\odot q}, \tag{5.1}$$

where r is heliocentric distance and V_t is the transverse velocity component. As seen from a vantage point far away in the Oort Cloud (e.g., $r = 10\,000$ AU), the depletion of orbits with $L < L(q_c)$ can hence be pictured as a scarcity of comets with very small V_t. Using cylindrical coordinates in velocity space, this means a *loss cylinder* surrounding the radial velocity axis — a concept introduced by Oort (1950). In modern literature, the term *loss cone* is mostly used for the same concept.

Clearly, the observed new comets with $q \lesssim 5$ AU are situated in the depth of the loss cylinder. The easiest way for them to arrive there is by a large enough decrease of q during the orbit preceding the observed perihelion passage, so that the previous perihelion was outside the loss cylinder. The range of $q \sim$ 5–15 AU is troubled by the risk of ejection due to Jupiter or Saturn, and this is hence often called the *Jupiter–Saturn barrier*. From this perspective, new comets have been thought to be the lucky ones that jumped across the Jupiter–Saturn barrier in one single orbit. However, as we shall see in Sec. 5.2.2, this picture is oversimplified.

[1]This way one avoids the statistical instability of the standard deviation in the presence of the extremely rare but dominating close encounters.

It is remarkably easy to make a crude but still reasonable estimate of the number of comets in the Oort Cloud. Restricting ourselves to comets of sizes similar to those of the observed new comets, we can estimate that about one new comet passes perihelion every year within an interval of 1 AU in perihelion distance. If we take 25 000 AU as an estimate of their semi-major axis, the corresponding orbital period is four million years. Hence, assuming a steady state, there are four million comets in total occupying the orbits in question. Now comes the tricky and somewhat uncertain part. The observed new comets are deep within the loss cylinder, so their flux should be depleted compared to the majority of Oort Cloud comets with much larger perihelion distances. Taking this depletion to amount to a factor 2, and using a distribution of perihelion distances within the cloud that corresponds to a thermalized phase space distribution,[2] we can estimate the total number of comets in the Oort Cloud as

$$N_{\text{tot}} = 4 \cdot 10^6 \times 25\,000 = 1 \cdot 10^{11}. \tag{5.2}$$

This agrees with the estimate that Oort made in his seminal paper. Of course, it is possible to derive the total population of the Oort Cloud by a strict analytic formula that exhibits the role of each assumption made. One way to formulate this is

$$R_p(q_o) = N_{\text{tot}} \times \int_{a_{\text{min}}}^{a_{\text{max}}} \varphi(a) \cdot f_{lc}(a) \cdot \frac{2q_o}{a} \cdot a^{-3/2}\, da, \tag{5.3}$$

where $R_p(q_o)$ is the number of perihelion passages of new comets per year with $q < q_o$, $\varphi(a)$ is the relative frequency function of a in the Oort Cloud, whose integral from a_{min} to a_{max} is unity, and $f_{lc}(a)$ is the depletion factor of the Oort Cloud for $q < q_o$ (real number relative to that of a thermalized distribution).

However, while it is easy to write an equation, it may be much more difficult to put it into practice, and this holds for the above example. Some critical items are as follows. When deriving $R_p(q_o)$,

[2]Without the depletion, the number of observable comets in orbits with $q = a(1 - e)$ per AU in q would be 8×10^6. The thermalized frequency function of e is proportional to e, so the integral from 0 to 1 is 1/2.

one has to use a well-defined criterion on the physical properties of the included comets. Making mistakes about this may lead to serious errors. Second, the frequency function of semi-major axis is *a priori* unknown, and one can either make guesses or simulate the origin and evolution of the cloud using uncertain assumptions. Thirdly, the determination of the depletion factor as a function of semi-major axis is of critical importance and can be achieved with a good enough dynamical model for the cloud, but this is far from being a trivial task.

Moreover, one has to remember that N_{tot} is not the total number of comets in the entire Oort Cloud but only in the external part of the cloud, from which the new comets arrive. This is often called the *classical Oort Cloud*, but the classical Oort Cloud may just be a halo surrounding a more important *inner core* with $a < a_{\text{min}}$, which is not revealed by the Oort spike, since the depletion factor is close to zero. Hence, the conclusions to be drawn about the entire Oort Cloud will always be biased, until we know not only how the cloud works today, but also how it may have formed and evolved over time since the solar system was young. This will be a major issue to discuss in the following chapters.

5.2. Injection of New Comets

Facing the problem of how to explain the injection of comets into orbits with small enough perihelion distances to enable discovery (so-called observable orbits), Oort (1950) was able to show that the encounters of the solar system with Galactic field stars was a viable mechanism. This would remain for more than thirty years as the standard explanation of comet injection.

In Secs. 3.3.1 and 3.4, respectively, we got somewhat familiar with the workings of, first, the Galactic tide, and second, the stellar perturbations. Both can in fact do the job of injecting new comets from the Oort Cloud, and this became clear during the 1980s. The first paper to describe the action of the Galactic tide in changing the angular momenta of Oort Cloud comets was written by Byl (1983). The theory of Heisler and Tremaine (1986) was introduced in Sec. 3.3.1, and we saw that comets may perform oscillations of angular momentum (and perihelion distance) of very large amplitude. Moreover, during the course of such an oscillation, the largest change of perihelion distance between two consecutive perihelia occurs, when the Galactic latitude of perihelion obeys $\beta_G = \pm 45°$.

Therefore, transferring comets deep into the loss cylinder in one orbital revolution occurs most easily, when the comets have these preferred values of β_G, in case the Galactic disk tide is the dominant mechanism. By contrast, if stellar perturbations would dominate, β_G would not matter at all. Armand Delsemme (1987) showed that, indeed, there is a clear signature of β_G among the new comets in agreement with the prediction of the tidal theory. In Fig. 5.3 we see a more recent rendering of the β_G distribution for new comets, and the case is obviously quite clear.

This is not the only demonstration of the important role that the Galactic disk tide plays for comet injection. It has also been shown by analytic and numerical comparisons that this is the most important player in the game. Consequently, there came a time when

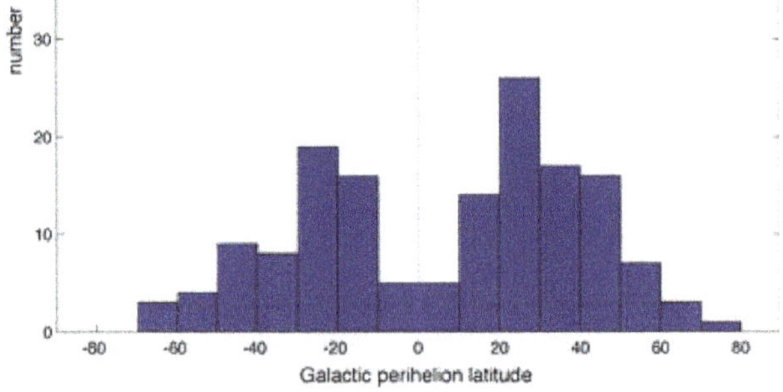

Fig. 5.3. Distribution of Galactic latitudes of perihelion for new comets, defined by elliptic original orbits with $a_o > 10\,000$ AU. The sample includes high quality orbits of comets observed until 2011. Reproduced from Rickman (2014) with permission. © The Meteoritical Society, 2013.

less and less attention was paid to the stellar encounters, and perhaps this sometimes went too far. But the role of the stars was to be investigated in detail about twenty years after the predominance of the tide was realized. Somewhat surprisingly, it was found that the role of stars is fundamental, even though the actual injections are mainly controlled by the tide.

5.2.1. *The role of stellar encounters*

The provision of new simulation tools for both stellar encounters and Galactic tides made new developments possible. Rickman *et al.* (2008) presented simulations of Oort Cloud dynamics covering the lifetime of the solar system, including both the mentioned effects at high accuracy. A salient feature was that not only the full model was treated but also unrealistic models including only the stars or only the tides. The comparison of these three models, as shown in Fig. 5.4, was revealing.

In each case, the cloud was started from a thermalized state. The black histogram shows how the injection rate drops significantly over about one billion years, when only the tides are included. This is because the disk tide is essentially integrable, and thus the injectable comets occupy an isolated phase space domain that gets depleted, as more and more comets enter into the loss cylinder and get ejected from the solar system. After 4 Gyr the injection rate is a bleak shadow of what it was in the beginning. Hence, the Galactic tide on its own cannot maintain a significant flux of new comets over the age of the solar system.

The gray histogram shows, on the average, a much more gentle decrease, which fits well with the decrease of the total number of comets in the cloud. However, there are many spikes that appear quite limited in time. These are so-called *comet showers*, which will be discussed in Sec. 5.3. Comparing with the white histogram, representing the full model, the stars dominate the injection rate during the showers but are less important otherwise. The main lesson to take home is that the full model yields a flux, which near the present time (more than 4 Gyr from the start) by far exceeds the sum of the separate tidal and stellar fluxes. Hence, the present flux

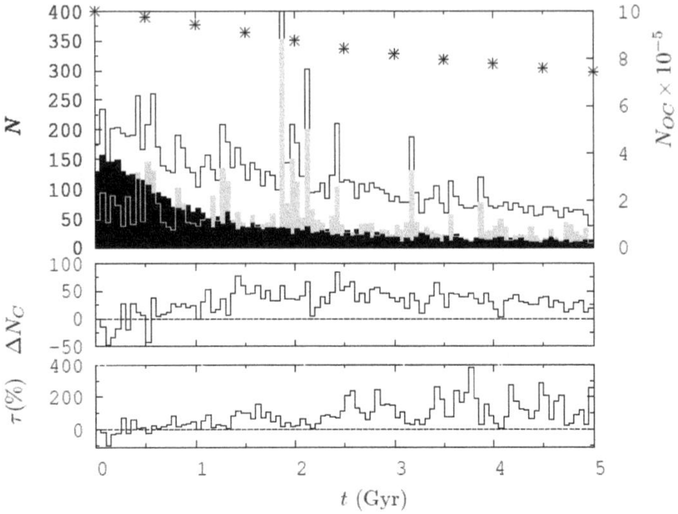

Fig. 5.4. Histograms of comet injection rate versus time from Oort Cloud simulations covering 5 Gyr (top panel). The white histogram is for the full model, while the black one is for the tides-only model and the gray one is for the stars-only model. The asterisks represent the full population of the Oort Cloud in the full model. The lower panels show the excess injection rate of the full model compared to the sum of the two restricted models. Reproduced from Rickman *et al.* (2008) with permission.

is controlled by tides (since the stellar flux is so much smaller) but at the same time due to a long-term stellar-tidal synergy that defeats the emptying of the phase space domain of tidally injectable comets. So, in a way, comet injections can be likened to the scoring of goals in international soccer: the tides are like the top attackers (currently Messi, Ronaldo, etc.) who score lots of goals, while few goals would actually be scored if it were not for the back-up of the entire team (the stars).

As indicated above, this long-term synergy has to do with replenishing the set of injectable comets, which occupy a special domain of phase space. Marc Fouchard and his colleagues have described this domain as the "tidally active zone" or the *TAZ* (Fouchard *et al.* 2011a). In a thermalized cloud, like the initial conditions for the simulations of Fig. 5.4, the TAZ contains a certain fraction of all the comets. If only the Galactic tide is taken into account, except for very large values of the semi-major axis, this perturbation is integrable and the orbital elements change periodically (see Sec. 3.3.1). They hence do not affect the status of a comet, i.e., if it belongs to the TAZ or not. However, the planets can remove comets by ejecting them or capturing them into orbits that are no longer perturbed by the tide to an appreciable extent. This is how the TAZ is emptied, and in the long run the only injectable comets are those of very large semi-major axis, so that the tide has lost its integrability and is able to feed new comets into the TAZ.

If stellar perturbations are also included, the situation changes radically. Normally, it does not take an extremely large stellar perturbation to shift many comets in various parts of the cloud into the TAZ from positions outside it. The TAZ can hence be replenished at the semi-major axes, where the tide is most efficient. This keeps the injection rate at a higher level, and the TAZ stays filled to an appreciable degree for the age of the solar system. It is of interest to know, generically, that "stars" are responsible for the TAZ filling, but also which kinds of stars are most efficient. Figure 5.5 shows the

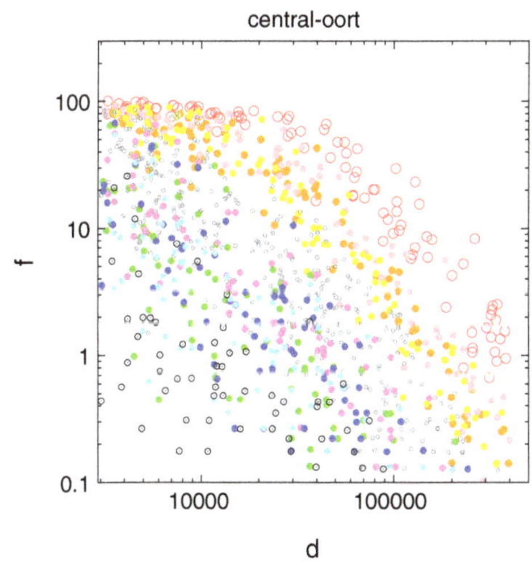

Fig. 5.5. TAZ filling efficiency as function of stellar encounter distance for extensive samples of stars, divided into 13 categories that are plotted with different colors/symbols. See the main text for more information. Reprinted from Fouchard, M. *et al.*, *Icarus* **214**, 334–347 (2011), with permission from Elsevier.

results of simulations that were targeted to answer this question. The experiment was described in detail by Fouchard *et al.* (2011a).

The plotted quantity f is the number of comets that are sent into the TAZ by each encounter relative to the population of the TAZ if it were full, i.e., in a thermalized cloud. Only semi-major axes between 20 000 and 50 000 AU are considered. The 13 stellar categories considered span from red dwarfs at the low mass end to OB stars at the high mass end. Low-mass stars tend to have higher encounter velocities than high-mass stars. The diagram exhibits a clear trend of TAZ filling efficiency, where the lowest stellar masses occur to the lower left and the highest masses to the upper right. This shows that massive stars are able to induce substantial TAZ filling (i.e., $\gtrsim 10\%$) even for distant encounters, while this is barely possible at all — and then only for extremely close encounters — when low-mass stars are considered.

The discussion of stellar encounters in Sec. 3.4 is of relevance, although the present topic is about the TAZ affinity instead of the effect on angular momentum or perihelion distance. The size of the heliocentric impulse imparted by the encounter was seen to be proportional to the ratio of stellar mass to stellar velocity, and this ratio varies considerably between the different stellar categories. An efficient filling of the TAZ does not primarily follow from a star passing very close to many single comets, thus giving them strong impulses, but rather from a star that can act on a large fraction of the Oort Cloud comets at the same time. The latter is the case for a massive star passing relatively far away, since the tidal effect will be quite strong on lots of comets in the densely populated, inner parts of the cloud. It is also the case for a star of any mass that passes very close to the Sun, because the impulse imparted to the Sun acts indirectly on all the comets in the cloud.

As a consequence, since low-mass stars are much more common than the high-mass ones, the average TAZ filling over billions of years is about equally caused by both kinds of stars. However, the TAZ filling due to high-mass stars is more transient. It happens rarely, but the TAZ may be filled to a high degree. Thus, the injection rate of new comets also increases and stays enhanced for a long time — until the TAZ gets depleted again. We will return to this issue in Sec. 5.3.1.

Now, let us stop for a while and look at the preceding discussion from another aspect. The stellar perturbations introduce a stochastic element into the Oort Cloud dynamics, thereby causing a reshuffling of orbits in phase space. This in turn can be viewed as a *thermalization*. In the papers so far described, the Oort Cloud was arbitrarily initialized in a thermalized state, which means that the TAZ was full. Thus, the role of the stars was to counteract the TAZ depletion induced by the tide and the planets, thereby keeping the injection rate of new comets at a higher level in the long term. However, the assumption of an initially thermalized cloud has no physical basis. In reality, the role of the stars may rather be the opposite. Their thermalizing influence may in fact counteract an initial overpopulation of the TAZ, thereby lowering the new comet injection rate.

5.2.2. *The role of planetary perturbations*

There is obviously a close relationship between the degree of TAZ filling and the rate of injection of new comets — alternatively, the amount of loss cylinder depletion. The more the TAZ is filled, the larger the rate of injection, and the less the loss cylinder will be depleted. However, this relationship is not one-to-one. There are several reasons. One is that stellar perturbations can inject comets on their own, including comets outside the TAZ. They can also add their own decrease of the perihelion distance to that of the tide, so that comets can be injected even though neither the tide nor the stars would have accomplished this on their own. The latter is a kind of real-time synergy, which is not very strong but cannot be neglected.

Another complication arises from the recent clarification of the role of planetary perturbations in the comet injection process. The Jupiter–Saturn barrier was often treated as totally opaque in the sense that no Oort Cloud comets would spend more than one perihelion passage with $q < 15\,\text{AU}$, since these would not stand a chance to avoid removal by planetary perturbations. In the quoted works by Rickman *et al.* (2008) and Fouchard *et al.* (2011a), this

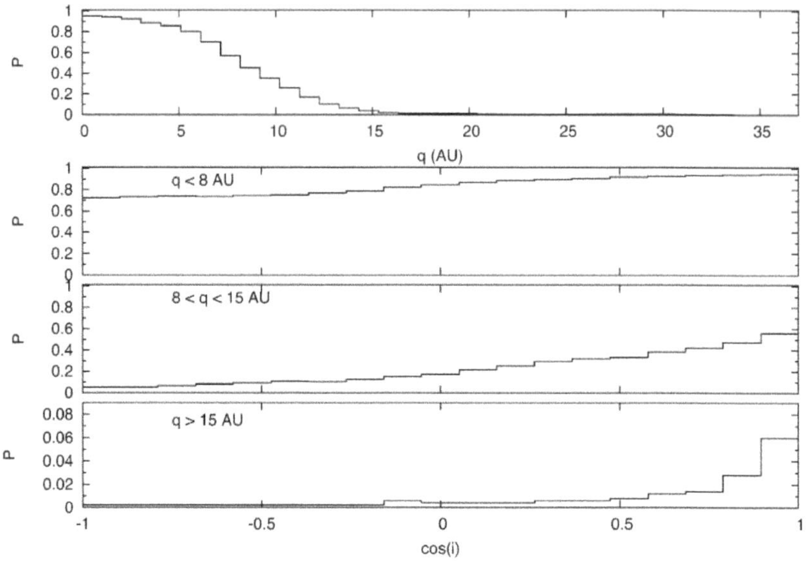

Fig. 5.6. Opacity parameter P (probability for a random comet with a_o = 20 000 AU to be kicked out of the energy range of new comets (a_o > 10 000 AU) by planetary perturbations. The upper panel shows P as a function of perihelion distance q, averaged over all inclinations i. The three lower panels show P as a function of $\cos i$ for three ranges of q. From Rickman (2014) with permission. © The Meteoritical Society, 2013.

assumption was made to simplify the calculations. However, it was also generally realized that this was an extreme point of view. Numerical experiments to reveal the details of the behavior were performed by Fouchard *et al.* (2013) and are reported in Fig. 5.6.

For $q < 5$ AU, which holds for the majority of observed comets, the assumption of complete opacity is seen to be rather good, but when q increases toward 15 AU, it becomes more and more erroneous for all inclinations. This partial transparency to planetary perturbations has deep consequences. Some of the new comets still evolve from $q > 15$ AU into $q < 5$ AU in one orbit, so they effectively jump across the Jupiter–Saturn barrier, when they become observable. However, others will make a stop-over inside the barrier including one or more perihelion passages, thus performing two or more smaller steps in q as a route into observability. The former comets may be called *jumpers* and the latter *creepers*. This terminology was introduced by Fouchard *et al.* (2014a) and also includes another aspect.

A fundamental discovery was made by Nathan Kaib and Thomas Quinn (2009). They treated comet injection from the Oort Cloud by a dynamical model that in general neglected stellar encounters but included Galactic tides and planetary perturbations. The Jupiter–Saturn barrier was modeled realistically and was thus partially transparent. As a result, creepers were common, and the creeping process was found to involve an important transfer of comets between different ranges of semi-major axis. The mechanism causing this is the energy diffusion induced by the planetary perturbations (see Sec. 3.5). Figure 5.7 illustrates how this works by one of their sample comets.

This comet initially belongs to the inner core of the Oort Cloud with $a \approx 6\,000$ AU. Its perihelion reaches Neptune's orbit after a slow decrease due to the Galactic tide. This decrease continues and is supplemented by a slow energy diffusion, until the perihelion

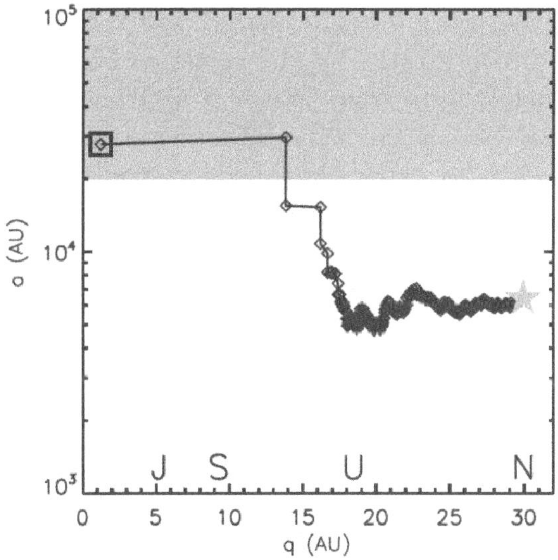

Fig. 5.7. Simulated evolution of a comet in the (q, a) parametric plane. The beginning and end are marked by a star and a square, respectively. The classical Oort Cloud is marked by the shaded area. From Kaib, N. A. and Quinn, T., *Science* **325**, 1234–1236 (2009). Reprinted with permission from AAAS.

penetrates inside the orbit of Uranus. While still outside the Jupiter–Saturn barrier, the comet now sees its semi-major axis increased during a sequence of orbits though still by a random walk process. A decisive kick by Uranus finally places the comet in an orbit whose semi-major axis is significantly larger than 10 000 AU, and the Galactic tide now moves the comet into the Jupiter–Saturn barrier with $q \approx 13$ AU. At the first perihelion in this orbit, the comet receives another decisive energy kick, after which it has $a \simeq 30\,000$ AU, and the Galactic tide brings it into an observable orbit with $q \simeq 1$ AU.

The main conclusion is that, if the Oort Cloud has an inner core with semi-major axes less than 10 000 AU, housing the majority of the comets, then these comets have an efficient way into observable orbits by means of the Kaib–Quinn mechanism. From their model, Kaib and Quinn (2009) estimated that about half the new comets would originate in the inner core. However, it is important to remember that there is no water-tight proof of the existence of the inner core. In particular, the new comets do not carry any observable marks or "flags" indicating their ultimate origin. Moreover, there is no way that a backward integration of the comet in Fig. 5.7 would reveal where it ultimately came from, in case it were one of the new comets observed today. Even if the stars did not exist, so that the Galactic tide would suffice to extrapolate back to the previous perihelion passage with $q \approx 13$ AU, it is way beyond present capability to identify the energy kick that actually occurred due to the planets at that time and thus to trace the comet to the inner core. The presence of additional perturbations, even though small, that are due to unknown stars and thus cannot be accounted for, makes the endeavor entirely hopeless.

It is fair to say that few if any comet scientists today doubt the existence of the inner core, but this is based on indirect evidence. The Oort Cloud itself has not been seen and remains a theoretical construct, but its existence cannot be put in serious doubt. The rest follows from our understanding of how such a reservoir should have formed and evolved, as will be seen in Sec. 6.3. Thus, under the assumption that the Oort Cloud is indeed equipped with an inner core, as mentioned, the contribution this offers to the new comet flux was estimated by Kaib and Quinn. It was also investigated using a model including stellar perturbations in the above-mentioned work of Fouchard *et al.* (2014a), where the additional terms *KQ jumpers* and *KQ creepers* were used for comets that experienced an increase of orbital energy $(-1/a)$ larger than $10^{-5}\,\mathrm{AU}^{-1}$ during the last unobserved perihelion passage — in recognition of the Kaib–Quinn mechanism.

Figure 5.8 from Fouchard *et al.* (2014a) shows a histogram of the energy distribution for the original orbits of new comets (first perihelion with $q < 5\,\mathrm{AU}$), where the contributions of the four classes are identified by different colors. The conclusion by Kaib and

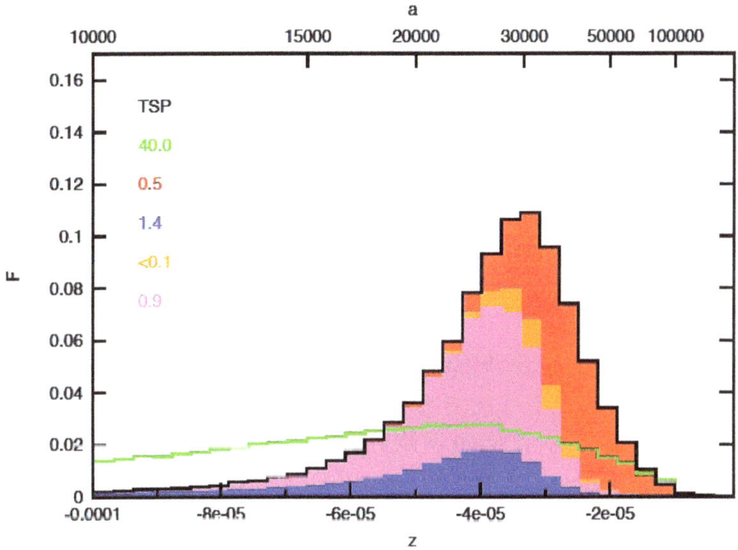

Fig. 5.8. Simulation results for the shape of the Oort spike, based on a model of Oort Cloud evolution from the beginning until the present time. Jumpers are marked in red, creepers in blue, KQ jumpers in orange and KQ creepers in magenta. The green histogram shows the distribution of initial orbital energy $(z = 1/a)$ at the start of the simulation. The plotted quantity F is normalized so that, when it is summed over all energy bins, the result is unity. The total contributions from all orbits with $a < 10\,000\,\mathrm{AU}$ are given by colored numbers on the left side. Reprinted from Fouchard, M. *et al.*, *Icarus* **231**, 110–121 (2014), with permission from Elsevier.

Quinn (2009) that about half the new comets come from the inner core via their mechanism is broadly verified, since 40% of the new comets had initial semi-major axes less than 10 000 AU. In addition, it is seen that the larger semi-major axes are preferred by the jumpers, while the smaller ones are dominated by the creepers, as is natural to expect from the steep increase with a of the average perihelion distance perturbation by the Galactic tide or stellar encounters (see Secs. 3.3.1 and 3.4). The fall-off of the flux with increasing a around and beyond 40 000 AU is due to overshoot, since the expected $|\Delta q|$ is so large that it is unlikely for the comet to hit the narrow range of 0–5 AU at the exact time of the perihelion passage.

Returning finally to the issue of TAZ filling and loss cylinder depletion, we have now seen that the comets passing perihelion with $q < 5$ AU and $1/a_o < 1 \times 10^{-4}$ AU^{-1} in fact sample a wider energy range of the Oort Cloud, if we focus on the orbits they had, when they first entered into the planetary system. Equation (5.3) in Sec. 5.1 is hence inappropriate to derive the total number of Oort Cloud comets, since it assumes that a is constant during the injection of new comets. In fact, armed with a simulation of the entire cloud accounting for its complete dynamics, the total population should rather be derived directly from the simulation by fitting the rate of injections to the observed flux of new comets.

5.2.3. *The observed Oort spike*

The Oort spike seen in Fig. 5.8 is based on dynamical theory and thus needs to be confronted with observations. However, this is easier said than done. As was seen above, establishing the presence of a spike at original, inverse semi-major axes $1/a_o < 1 \times 10^{-4}$ AU^{-1} is possible even with a small number of well-determined orbits and using purely gravitational dynamics. But now the issue is to find the detailed shape of the spike, and this is a much more challenging task.

In fact, the rendering of the Oort spike in Fig. 5.1 shows a few comets whose original orbits are apparently hyperbolic. In the current paradigm of comets as solar system members, this does not make sense. All the orbits in this diagram are purely gravitational, although they were mostly determined after the advent of the standard model for the nongravitational acceleration (Sec. 2.5.3). Suspicion has been cast on the presence of unmodeled, nongravitational effects as the reason for the strange results, but it was not possible to prove that this must be the case. Nongravitational effects are relatively easy to determine by linking several apparitions of a periodic comet — especially concerning the perihelion delay — but when we deal with new comets, there is only one apparition to work with, and mostly, only part of this is well covered by astrometric observations.

Nevertheless, without a determination of the actual nongravitational effects on new comets there is no hope of establishing the real shape of the Oort spike. Figure 5.9 gives a background to the following arguments. The comet orbit is nearly parabolic. The largest component of the jet force is the radial one outward from the Sun. On the pre-perihelion orbital branch the comet is decelerated, and after perihelion it is accelerated. This will slightly counteract the normal, Keplerian trend for the speed of the comet to increase with decreasing heliocentric distance. In a purely gravitational solution, the way to emulate this is to increase the orbital eccentricity, i.e.,

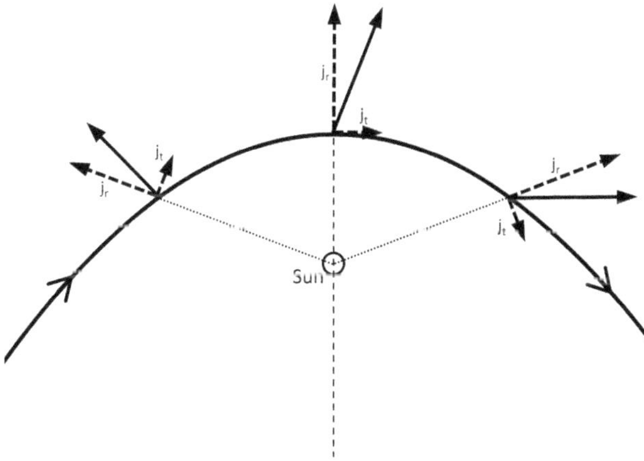

Fig. 5.9. Schematic representation of the orbit of a new comet as a parabola with the sense of motion indicated by arrows. The two in-plane components (radial and transverse) of the nongravitational jet force **j** are indicated at three orbital positions. Courtesy T. Wiśniowski.

to decrease the value of $1/a$ without changing the well-determined perihelion distance.

Hence, if a new comet is indeed affected by a significant nongravitational force, a gravitational solution will tend to bias its original orbit, making it look more hyperbolic than it really is. This may well be the reason behind the apparently hyperbolic orbits in Fig. 5.1. However, it is essential to establish the presence of this nongravitational force by the orbital solution and to estimate it correctly, because an overestimate of the force would imply an overestimate of $1/a_o$. It is thus fair to say that the realism and quality of the observed Oort spike hinges on very good astrometry and very careful determination of the nongravitational effects.

The latter has been provided during the last decade by Małgosia Królikowska. She was thus able to show quantitatively that the above suggestion as to the reason for apparently hyperbolic orbits is right (Królikowska 2006). Together with Piotr Dybczyński, she also investigated the shape of the Oort spike by orbit determinations including possible nongravitational forces for Oort spike comets with $q > 3$ AU (Dybczyński and Królikowska 2011). This study also involved backward integrations of the original orbits until the previous perihelion passage using huge swarms of comet clones. The large, current perihelion distances were chosen to minimize the nongravitational effects, but still, nearly 25% of the comets turned out to exhibit such effects, which were included into the determination of the original orbits.

Considering the absence of significant nongravitational effects for the majority of comets and the smallness of these effects for the rest, the resulting observational Oort spike seems to be the best one currently available. It is shown in Fig. 5.10 (upper panel). There is a rather good agreement between this spike and the one based on dynamical modeling in Fig. 5.10, which lends some confidence to the model. The general agreement also extends to the separation of the new comets into jumpers and creepers. In the observed spike, comets with previous perihelion distances $q_{\rm prev} < 15$ AU dominate for $1/a_o > 0.00004$ AU^{-1} ($a_o < 25\,000$ AU), and in the modeled spike this transition occurs at $1/a_o \approx 0.000033$ AU^{-1} ($a_o = 30\,000$ AU).

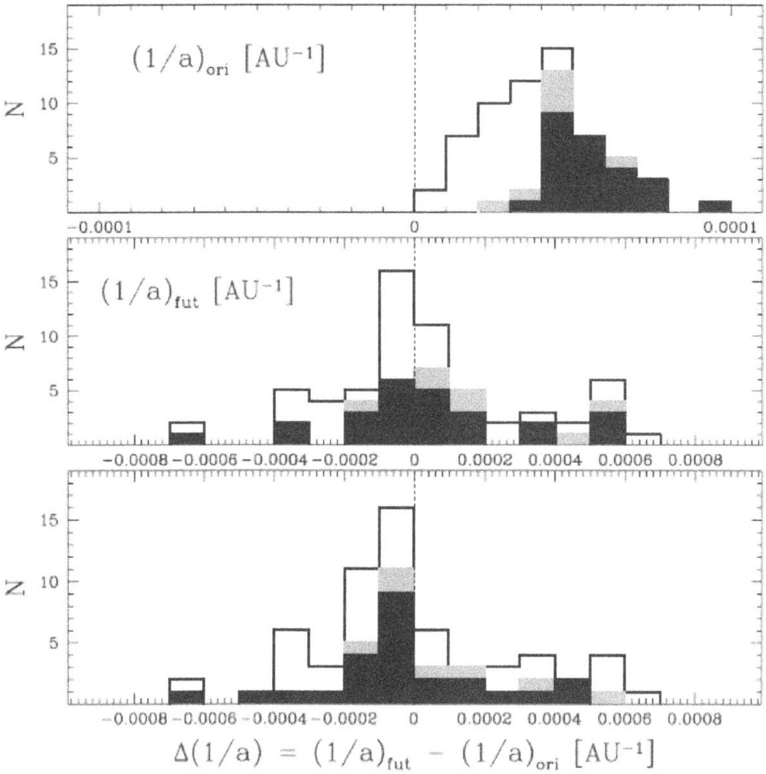

Fig. 5.10. Histograms of original and future $1/a$ values of Oort spike comets with $q > 3$ AU (upper panels), and of their differences (lower panel). Comets with preceding perihelion distance $q_{\mathrm{prev}} < 10$ AU are marked in black, those with $q_{\mathrm{prev}} > 15$ AU in white, and the intermediate ones in gray. Reproduced from Dybczyński and Królikowska (2011) by permission of Oxford University Press on behalf of the Royal Astronomical Society.

According to the model, the observed creepers are mostly of the KQ type, which means that they have diffused in energy from orbits originating in the inner core (semi-major axis less than 10 000 AU).

5.3. Comet Showers

From Fig. 3.8 we can see that, during intervals of 50 million years or more, stellar encounters at less than 10 000 AU from the Sun are expected. Which consequences would these have for the injection of new comets from the Oort Cloud? If we ignore the existence of an inner core and only consider the classical Oort Cloud with semi-major axes in excess of 20 000 AU, there would not be any remarkable consequences. All the stars can do, whatever the approach distance, is to thermalize the orbit distribution of Oort Cloud comets — when they reshuffle these orbits, they cannot aim at the loss cylinder rather than any other domain. For the classical Oort Cloud, the loss cylinder is filled to a significant degree under normal circumstances, and no star will fill it to much more than 100%. Thus, there would be a transient peak in the injection rate, but this peak would not be dramatic.

The situation is quite different, if the Oort Cloud energy distribution continues to semi-major axes of just a few thousand AU with a positive slope, so that the inner core dominates by numbers. In this case, more than 90% of all the comets in the cloud will have $a \lesssim 10\,000$ AU, and neither the Galactic tide nor the usual passing stars will be able to inject a significant number of comets into observable orbits. The loss cylinder is practically empty, until a very close stellar encounter occurs and it gets filled to a high degree. Thus, a very high or even dramatic peak in the injection rate occurs, and the inner solar system gets flooded with new Oort Cloud comets. This phenomenon, which was theoretically predicted by Jack Hills (1981), is usually called a *comet shower*.

Long-term Oort Cloud simulations including an inner core and stellar encounters never fail to exhibit comet showers, and an example was seen in Fig. 5.4. The high peaks featured by the stellar and combined stellar-tidal models are comet showers. Yet, the height of the peaks is strongly suppressed, since the bin width used for the histogram is 50 Myr, while the actual shower lasts for only about 1 Myr.[3] However, this only refers to the arrival rate of new comets, while in reality there is also an aftermath consisting of long-period and Halley Type comets. This continues for a few million years (Fernández and Ip 1987) and determines the duration of the interval

[3]This is easily realized, since almost all the shower comets come from the inner core, from which the free-fall time to the Sun is less than 10^6 years.

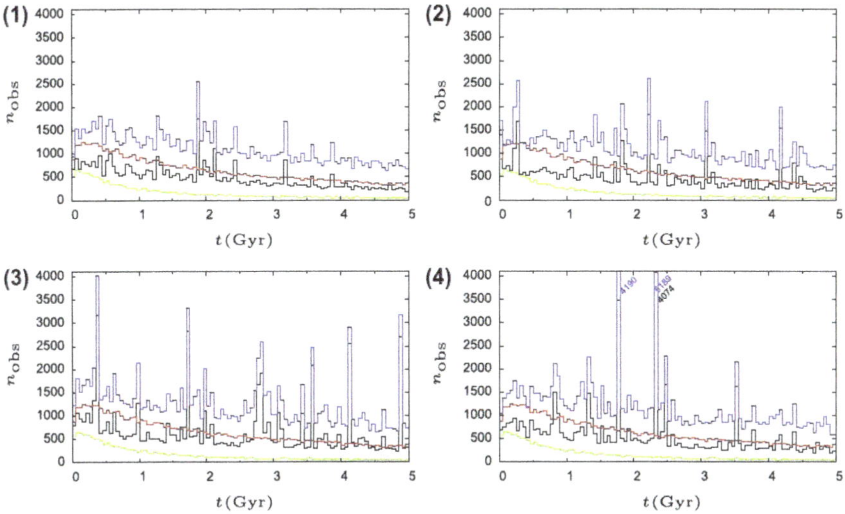

Fig. 5.11. Number of injected comets from a simulated Oort Cloud with four million initial members during consecutive intervals of 50 Myr. In each panel, green denotes a model with only Galactic tides, red denotes a model with tides and planetary perturbations, black denotes a model with tides and stellar encounters, and blue denotes a full model including all effects. Each of the four panels uses a separate sequence of random stellar encounters. Reprinted from Fouchard, M. *et al.*, *Icarus* **231**, 110–121 (2014), with permission from Elsevier.

of increased rate of perihelion passages of comets of Oort Cloud descent. A more recent example of simulated Oort Cloud injection histories is seen in Fig. 5.11. Each panel features four histograms, again with a bin width of 50 Myr. The red and green ones are smooth, since stars were not simulated, while the blue and black ones show results including stars, which means that comet showers are present. The blue histogram is the only realistic one, since both tides, stars and planets were included.

The comet showers are superimposed on a background, which is very similar in all panels. However, the shower peaks reflect the random selections of the strongest stellar encounters, and they obviously differ from each other. Actually, we see an example of picking at random a large number of events from a distribution with a very wide, low-probability tail, so that this tail is typically sampled just a few times. Even in its bulk features, the outcome is then purely a matter of chance. We simply cannot say, how strong the strongest comet showers of solar system history have been except in terms of statistical expectancy. As will be seen in Sec. 6.3.2, this also affects our knowledge of the long-term evolution of the Oort Cloud.

5.3.1. *An ongoing shower?*

As far as we can tell, comet showers must be part of the history of injections from the Oort Cloud. An unavoidable question is then: Are we currently experiencing one of those showers? Even though the answer seems quite clear (*No, we are not!*), the arguments behind this and the relevant discussion are worthy of attention.

Let us first estimate the peak injection rate during a shower. Only crude numbers will be derived, but they probably suffice. Let us use the results of Fouchard *et al.* (2011a). For a thermalized Oort Cloud with a flat energy distribution, containing one million comets, the authors found that the number of injected comets under the assumption of single stellar encounters as the only perturbing mechanism (N_*) could be fitted by power-law expressions in terms of stellar mass (M_*), stellar velocity (V_*), and approach distance to the Sun ($d_\odot$) as follows. For low-mass stars, the proper regression is

$$N_* = \left(\frac{M_*}{V_*\sqrt{d_\odot}}\right)^{1.82} \times 16.23, \tag{5.4}$$

while for high-mass stars the relevant formula is

$$N_* = \left(\frac{M_*}{V_* d_\odot^2}\right)^{0.89} \times 12.83. \tag{5.5}$$

The units are solar masses for M_*, 40 km/s for V_*, and 20 000 AU for $d_\odot$. It was found that Eq. (5.4) is to be used for $M_*/V_* < 0.60077$ and Eq. (5.5) otherwise.

Now, consider a stellar encounter yielding $N_* = 500$. From the estimate of the Oort Cloud population in Eq. (5.2), the real number of injected comets would be 5×10^7. If we take the duration of the shower to be one million years, we get an average injection rate of 50 new comets per year with $q < 5$ AU. This is roughly ten times the observed rate. If we would insist that the current Oort spike is due to a comet shower, the cloud population must be much less than in Eq. (5.2) to beat down the quiescent comet flux due to the Galactic tide. Ten times less would be relevant, and then the shower flux would also be in agreement with the observations. However, the considered kind of stellar encounter is extremely rare.

Fouchard *et al.* (2017) constructed ten random sequences of passing stars during an interval of approximately 4 Gyr and used the above equations to compute N_* for each individual encounter. They found on average about one encounter per sequence with $N_* > 500$. The random probability for the solar system to be currently experiencing a 1 Myr event that happens less than once per Gyr is less than 0.001, and thus we can safely discard the hypothesis of an ongoing shower on statistical grounds.

In other words, the new comets cannot be predominantly shower comets on the basis of simple statistics of stellar encounters. It is not so easy to prove the opposite, i.e., that the new comets are predominantly of the quiescent type. However, there are good indications, as we shall see.

We first make a new statistical estimate. Let us prescribe that 90% of the new comets are quiescent and 10% are shower comets. In this case, the shower-producing encounter should have $N_* \approx 5$. Such encounters were analyzed by Fouchard *et al.* (2011b). In their simulations, they found an average rate of appearance amounting to about one encounter per 12 Myr. Figure 5.12 shows approach data for stellar encounters used in Oort Cloud simulations, effectively covering 9 Gyr, highlighting the encounters in question by special symbols. To estimate the statistical chance to be inside such a gentle shower (Fouchard *et al.* preferred the name "enhancement" in this case), we should allow for the fact that the encounters are on average more distant (see Fig. 5.12), which may increase the duration relative to the strong showers from the inner core. If we take the duration to be 3 Myr, the chance of an ongoing enhancement is about 25%.

It thus seems statistically likely for the current new comets to be very little affected by injections due to any single star. Can we sharpen this statement by additional, observational evidence?

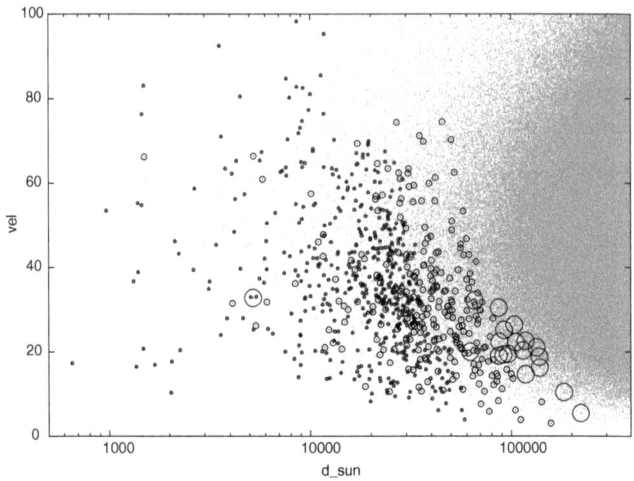

Fig. 5.12. Velocity at infinity versus solar impact parameter for stars used in Oort Cloud simulations. Tiny, gray dots denote stars with $N_* < 5$. The other symbols show stars with $N_* > 5$, separated into three different mass ranges. Credit: M. Fouchard *et al.*, *A&A* **535**, A86 (2011), reproduced with permission © ESO.

The good agreement between the theoretical and observational Oort spikes that we have seen followed from a theory devised to ignore all enhancement-making stars, but it does not seem likely that it would be ruined, if the theory were modified to include an ongoing enhancement. The same can be said about the distribution of Galactic latitudes of perihelia for new comets in Fig. 5.3 — while this shows that the injections are predominantly of tidal origin, a 10% contribution by one single star cannot be ruled out. A related argument has been discussed in previous literature and deals with the expectation that the perihelion directions of comets injected by close encounters with a passing star would be concentrated along a great circle on the sky (defined by the plane containing the Sun and the trajectory of the star). It was shown by Dybczyński (2002) that such a concentration should exist. However, to establish this feature in the observed sample, if only $\sim 10\%$ of the comets are involved and the encounters are not very close, is rather unlikely to succeed.

In summary, a current comet shower strong enough to dominate over, or even compete with, the flux of quiescent comets is excluded by the rarity of such events. In addition, it would likely have observational imprints that are not seen. However, a current enhancement of the injection rate at the 10% level or less cannot be excluded — neither from the statistical nor from the observational point of view.

As mentioned above, there is a phenomenon that bears resemblance to comet showers but is different in nature. This is the relatively efficient TAZ filling caused by moderately close encounters with massive stars, which allows for an increased flux of new comets for an extended period of time, until the depletion of the TAZ causes a return to normal conditions. This was mentioned by Fouchard *et al.* (2011a), who called it *comet drizzles* in analogy with the showers. In this case, the increase of the new comet flux is not at all dramatic. The degree of TAZ filling may increase by a factor 2–3 in exceptional cases but is mostly lower.

The question if we are currently experiencing a major comet drizzle is unfortunately impossible to answer. By definition, there is no signature in the statistics of comet orbits. Furthermore, the responsible star — even though of high luminosity — will likely have moved to such a large distance since the encounter took place, that it stands a very high chance to avoid recognition, even using the upcoming Gaia catalogue.[4] This problem is compounded by the fact that the star moves almost radially away for a considerable time and thus has a very small proper motion. As a result, it is very difficult to determine just how close the encounter was, and thus how important the drizzle would be, even if there is a good determination of the star's radial velocity.

5.4. The Trans-neptunian Objects

The discovery of lots of trans-neptunian objects is certainly one of the greatest advances in solar system science in recent times. Even though these are not counted with the comets, we need to pay

[4]This means a vast catalogue of parallaxes and proper motions for about one billion Galactic stars as measured with the aid of the *Gaia* astrometric mission, performed by ESA.

attention to them because of the close links that obviously exist. In particular, many of the well known, observed comets may have been trans-neptunian objects for billions of years in the past, and some may become trans-neptunian objects in the future as well.

In contrast to the Oort Cloud, the trans-neptunian populations have thus been subject to observations, so their characteristics need not be entirely inferred from models of their transfer into the comet populations. We have first hand information about their numbers and orbit distributions, and even physical properties of many members. However, one needs to keep in mind that the observations are hampered by the faintness of these distant objects. Most of the data concern objects with diameters ~ 100 km or more, while the observed comets are so few that, practically, this size range is not sampled at all. Moreover, what we observe on the surfaces of giant trans-neptunian objects is of limited use for comparison with the typical, km-sized comet nuclei, whose surfaces have been cooked in solar heat repeatedly at past perihelion passages.

Therefore, the following description of trans-neptunian objects (TNOs) will be focused on those aspects that undoubtedly offer insights about the origin and evolution of comets. This bias may be somewhat subjective but is intentional.

5.4.1. *The Edgeworth–Kuiper Belt and resonant objects*

Pluto, whose minor planet serial number (134340) will never be remembered except by the few specialists and afficionados, was the first to be discovered in this group and is also the largest with a radius of 1187 km. In 1930, by the time Pluto was discovered, there were only vague ideas about what the solar system might host beyond Neptune's orbit. It was of course natural to speculate about Pluto's nature — was it unique, as its recognition as the ninth planet would imply, or could there be many more of its kind? It would not take long, before new astrophysical concepts started to be applied to the problem of formation of the planetary system, and with this came ideas and hypotheses about the possible inhabitants of trans-neptunian space. In particular, small body populations were imagined, for instance, by Kenneth Edgeworth (1949) and Gerard Kuiper (1951).

Much later, as we shall discuss in Sec. 5.5.2, indirect evidence started to assemble concerning a flattened, belt-like structure hosting icy objects similar to comet nuclei beyond the orbit of Neptune. This was named the Kuiper Belt even before its first member was discovered, and in recognition of Edgeworth's contributions it has later often been named the *Edgeworth–Kuiper Belt* (EKB). Strictly speaking, Pluto is not a true member of this population because its eccentric orbit brings it to heliocentric distances less than that of Neptune, when it is near its perihelion. It is saved from collisions or close encounters with Neptune by a combination of a 2:3 mean motion resonance and a stable libration of the argument of perihelion around 90° (Milani *et al.* 1989). The true members, on the other hand, have orbits way beyond Neptune's orbit that are stable on very long time scales without any need for resonant protection.

In Fig. 5.13 we show orbital elements of TNOs with $a < 50\,\mathrm{AU}$ as provided by the IAU Minor Planet Center. This includes the EKB and resonant objects together with many others. A distinction has been made between objects with perihelion distances larger or smaller than 40 AU. The former can be assumed to be protected from Neptune as stated above, while the latter are likely to approach Neptune's orbit sooner or later, except if they are protected by a resonance. Among these protected objects, those in the 2:3 resonance are particularly abundant. Since this group includes Pluto, the rest of its members are often called *Plutinos*. The EKB objects show a striking dearth of semi-major axes beyond those associated with the 1:2 resonance with Neptune at $a = 47.8\,\mathrm{AU}$.

The plot of inclinations versus semi-major axes immediately catches attention, because there is a bimodal distribution of EKB objects with one group forming a very flat structure and the other being more excited, appearing like a halo with significant thickness. This means that the EKB is formed by two populations, usually called the cold and hot groups, respectively. Interestingly, the members of the two groups have different properties also in other regards. The cold objects exhibit reddish colors and are often binaries

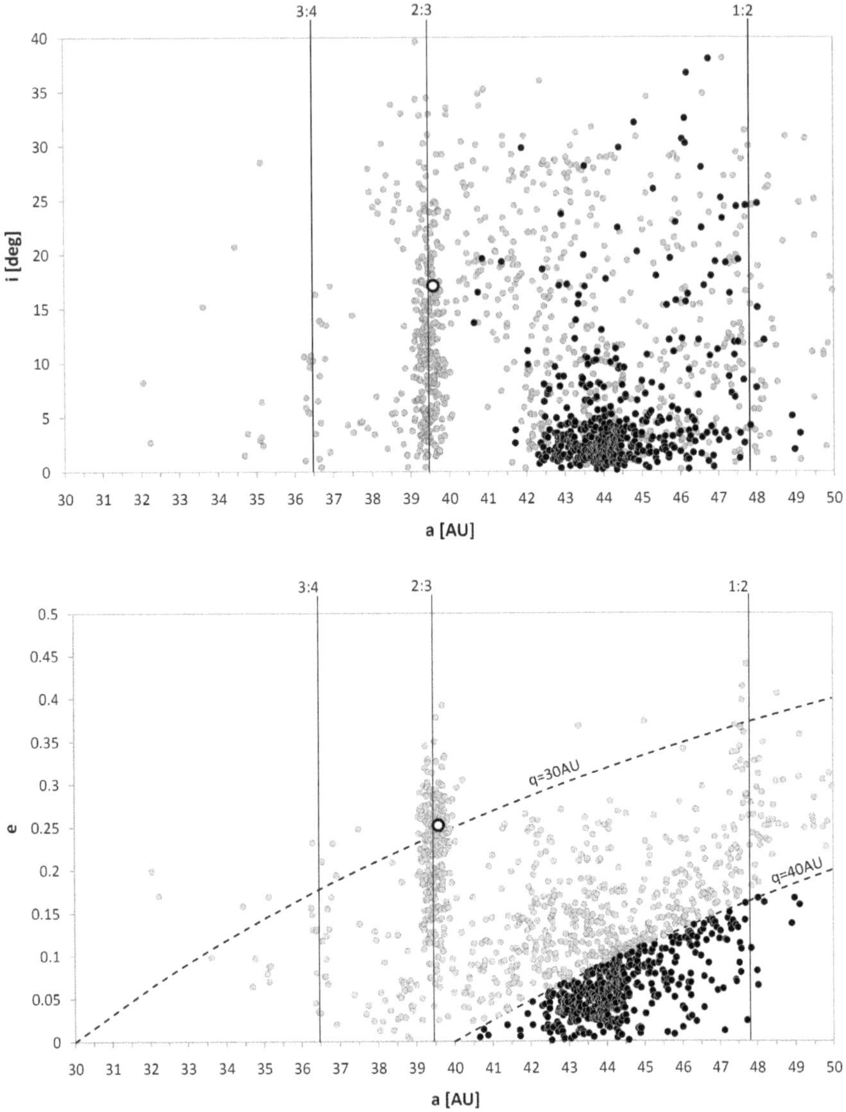

Fig. 5.13. Inclinations and eccentricities plotted versus semi-major axes for TNOs with $a < 50\,\mathrm{AU}$, listed in the IAU/MPC data base of August 2016. Only objects observed at more than one opposition or otherwise for more than one month have been included. Black dots denote EKB objects with perihelion distances $q > 40\,\mathrm{AU}$, and gray dots are used for $q < 40\,\mathrm{AU}$, whether the orbits are resonant or not. Pluto is identified by a white, open circle. Courtesy T. Wiśniowski.

with components of similar brightness, while the hot objects are spectrally diverse and binarity is less common. The resonant objects, in particular the Plutinos, seem related to the hot group rather than the cold one. In terms of total mass, recent estimates (Fraser *et al.* 2014) are that the hot group dominates over the cold group by a factor ~ 30. This seems to conflict with Fig. 5.13, but the apparent predominance of the cold group is due to the fact that searches for TNOs have been strongly concentrated to the ecliptic.

The existence of two distinct types of EKB objects has implications for the history of the TNO population. In particular, the cold EKB is thought to be a remnant of a primordial planetesimal disk, representing its most distant domains. Such planetesimals would be formed in nearly circular orbits very close to the ecliptic plane. The hot group, on the other hand, seems to have been emplaced by means of some dynamical process that occurred later on and involved scattering into more eccentric and inclined orbits. This process will be discussed in Sec. 6.3.1.

The absolute magnitudes of the objects on record are generally bright enough to tentatively assign diameters larger than 100 km, based on relatively low albedos. Concerning individual objects, the first one to be discovered (in 1992 by Dave Jewitt and Jane Luu) is (15760) 1992 QB_1, still unnamed. Its diameter can be estimated as somewhat larger than 100 km though with a large uncertainty. Among the largest, Pluto is outstanding with a radius of 1187 km, followed by (136472) Makemake with a radius of about 730 km and (136108) Haumea, which appears to be slightly smaller. The team of Caltech astronomer Mike Brown revealed the existence of both these objects. Both of them belong to the dynamically hot group.

Obviously, the populations in question are not limited to the big objects typically on record. However, reaching the smaller ones is extremely difficult due to their faintness. Hence, establishing a comprehensive size distribution is a formidable task. It is clear that no single power law can be used over the whole size range. On the other hand, a broken power law seems to be a workable model. Recall the discussion in Sec. 2.1.3, where the cumulative distribution of cometary nuclear radii was described by Eq. (2.4). Such a function is used also for EKB objects but with different values of the slope index (α) for the large and small objects. The transition is taken to be located at a diameter less than or about 100 km.

The index of the large objects has been found by analyzing the magnitude distributions of observational surveys. According to Gladman *et al.* (2001), the best fit is for $\alpha = 3.4$. A new determination by Fraser *et al.* (2014) gave $\alpha = 4.35$ for the hot group and even larger for the cold group. Thus, currently it seems preferable to use $\alpha \simeq 4$ for dynamically hot EKB objects. This is a very steep distribution, and the total mass would diverge toward infinity, if it were used down to a diameter equal to zero. However, it has been argued that collisional evolution in the EKB will reshape the size distribution of smaller objects, leading to a smaller α (Davis and Farinella 1997).

If the small objects are precursors of Jupiter Family comets (see Sec. 5.5.2), one may expect the α values of these groups to be similar. In Sec. 2.1.3 it was tentatively concluded that the Jupiter Family has α in the range from 1.5 to 2.5. For the small EKB objects, some direct observational indications have been available. In particular, Bernstein *et al.* (2004) used the Hubble Space Telescope for deep imaging to find EKB objects with radii ~ 10 km and found essentially none, while a steep size distribution would have predicted many. Some uncertainty over the exact slope index remains from such observations, but the close-up imaging of Pluto and Charon by the NASA *New Horizons* mission has yielded new insight. The size distribution of impact craters on Charon in the relevant range has a slope of $\alpha \simeq 2.3$ (Jutzi *et al.* 2017),[5] and this is a good indication that $\alpha \simeq 2.3$ for the small EKB objects too. In addition, it is certainly consistent with the observational estimates for the Jupiter Family.

5.4.2. *The Scattered Disk*

The largest known Scattered Disk object is (136199) Eris, which was discovered by Mike Brown and his team in 2005. Eris is just a little

[5]The authors refer to data by Singer *et al.* (2015).

smaller than Pluto with a radius of 1160 km, but due to a higher density, its mass surpasses that of Pluto by nearly 30%. However, Eris was not the first discovery of this population. This was instead (15874) 1996 TL_{66}, discovered in 1996 by Dave Jewitt and his team[6] but still unnamed. Contrary to the EKB, and in spite of the work by Fernández and Ip to be described in Sec. 5.5.2, the Scattered Disk had not become rooted in people's minds, so the discovery was unexpected.

In Fig. 5.14 we show orbital elements of Scattered Disk objects and Centaurs as provided by the IAU Minor Planet Center. The perihelion distance is plotted against the log of the semi-major axis. It is hard to draw an undisputable dividing line between the two populations and they are believed to be closely related, so both are

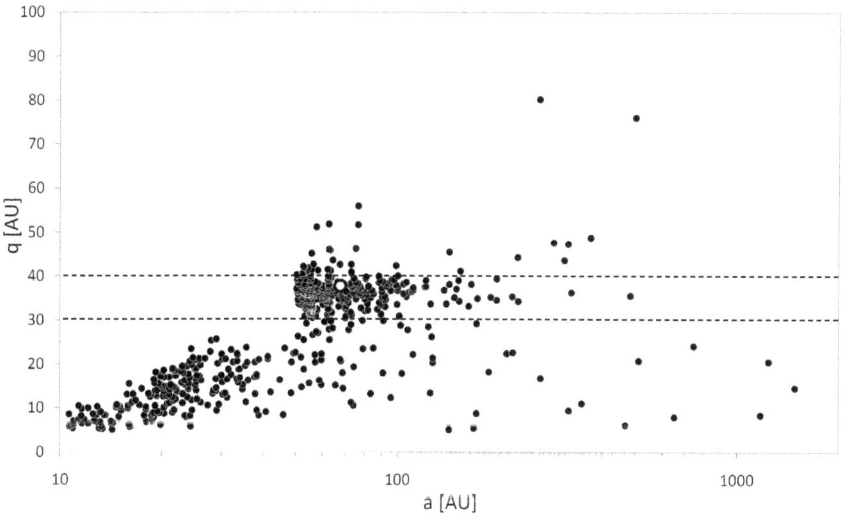

Fig. 5.14. Perihelion distances plotted versus semi-major axes for Centaurs and Scattered Disk objects with $a > 50$ AU, listed in the IAU/MPC data base of August 2016. Only objects observed at more than one opposition or otherwise for more than one month have been included. Eris is identified by a white, open circle. Courtesy T. Wiśniowski.

[6]Another object, (48639) 1995 TL_8, was actually discovered a year earlier but was not recognized as a Scattered Disk object until later.

included in the diagram. Even so, objects with $q < 25$ AU are clearly Centaurs, while all those with $q > 30$ AU belong to the Scattered Disk. The semi-major axes in Fig. 5.14 are restricted to $a > 50$ AU, but this limit is arbitrary and by comparison with Fig. 5.13 we see that the Scattered Disk actually includes many objects with $a < 50$ AU as well.

On the other hand, a few Scattered Disk objects have perihelia far beyond the planetary orbits, and these are usually called *detached*. They are of great interest and will be discussed below, but they conflict with the common property of Scattered Disk Objects (SDOs), namely, to be able to have close encounters with Neptune sooner or later without the intervention of external perturbing agents.

One difference between the EKB objects and SDOs is obvious from the large semi-major axes that some of the latter exhibit. While the former can in principle be observed all around their orbits, this is not the case for all the latter. Some can, if they are large enough, but the aphelia may be located several hundred AU away, and then the objects need to be close to their perihelia to be discovered. Since only a minor part of the time is thus relevant for discovery, the discovered SDOs may clearly represent only a small fraction of the whole population — apart from the fact that many that are currently discoverable are still awaiting discovery, like for the EKB objects.

As to their inclinations, the SDOs share more or less the same distribution as the hot group within the EKB. On the other hand, Centaurs sometimes have much higher inclinations (even retrograde orbits), and this may suggest an Oort Cloud origin in those cases. There are too few observations of SDOs, and discovery biases are too important, for a reliable determination of their size distribution. However, the similarity of inclinations between SDOs and hot EKB objects is thought to reveal a genetic relationship, in which case the α values of the latter can be extrapolated to the former. This is important when discussing the relationship to the Jupiter Family, because the Scattered Disk is believed to be the prime source region for the Jupiter Family with the EKB — and possibly the Oort Cloud too — of secondary importance (see Sec. 5.6).

Let us now turn to the detached objects. Among TNOs with $a > 50\,\mathrm{AU}$, in addition to SDOs, there are objects with $q > 40\,\mathrm{AU}$. Most of these have perihelia not far beyond the 40 AU limit and cannot be clearly separated from the SDOs, so they may be counted as a special kind of those. In addition, there is a group with $60 < a < 80\,\mathrm{AU}$ that stands out by more distant perihelia. These can be explained by the Kozai mechanism in association with mean motion resonances with Neptune (Gallardo 2006).

There is another group with $a \sim 300\,\mathrm{AU}$, which likely needs a different explanation, and with similar semi-major axes there are also a few that immediately stand out as peculiar. In particular, we refer to *(90377) Sedna* with $a = 506\,\mathrm{AU}$ and $q = 76\,\mathrm{AU}$, and *2012* VP_{113} with $a = 260\,\mathrm{AU}$ and $q = 80\,\mathrm{AU}$. Considering their orbits and the implied, severe discovery bias, these two objects should be the tip of an iceberg. This means that they signal the presence of a large population with similar orbits, whose perihelia are too far from the planetary system to be counted with the SDOs. These are detached, and an obvious question is how this came to be.

In fact, the property of detachment is shared with the Oort Cloud, and from this perspective the objects in question can be viewed as an extreme kind of inner Oort Cloud comets. Like the largest SDOs and EKB objects, they are of course much larger than the comets we are familiar with. Sedna's radius is estimated to be about 500 km, and its recently discovered colleague is about half as big. However, as seen in Sec. 5.1, the Oort Cloud is extremely poorly sampled, and it likely contains many members whose size is similar to the largest TNOs.

In connection with the report on discovery of 2012 VP_{113}, Trujillo and Sheppard (2014) noted that the orbits of SDOs with perihelia well beyond the orbit of Neptune and semi-major axes greater than 150 AU, including the two outliers, show a clustering of their arguments of perihelia close to the value zero. This means that these tend to pass their perihelia close to the ecliptic plane in northbound motion. While discovery bias could easily account for a coincidence of the perihelion direction with the ecliptic plane, there is no way to explain in such a way, why all the objects are northbound.[7] Batygin and Brown (2016) made the more far-reaching discovery that the considered group of distant, long-period TNOs share a preferred range of not only arguments of perihelia but also longitudes of perihelia, meaning that there is an alignment of the orbits in three-dimensional space.

These authors investigated possible reasons for such an alignment and found the likely reason to be a massive planet perturbing the TNO orbits in question. In a follow-up paper by Brown and Batygin (2016), the ranges of mass and orbital properties of this planet were narrowed down, and the conclusion was that the mass of the planet should be 5–10 Earth masses, its perihelion distance should be 150–350 AU, its semi-major axis should be 380–980 AU, and its inclination should be about 30°. This proposed planet has become known as *Planet Nine*. It has not yet been detected, but searches are under way.

We shall discuss the formation of the Oort Cloud in Sec. 6.3.1, and the possibly related origin of the detached population containing Sedna and 2012 VP_{113} will be one of the topics.

5.5. Capture of the Jupiter Family

The first Jupiter Family comets (JFCs) were discovered in the later part of the 18th century. That was a time, when comets had started to be actively searched for using telescopes and the newly discovered comets were mostly rather faint, so that there was no wonder they had not been discovered before. However, one comet caught attention. This was comet D/1770 L1 (Lexell), whose perihelion distance was only 0.67 AU and whose period was only about six years. It was bright enough to raise the question, why it had not been seen at earlier apparitions. The answer was provided by Anders Johan Lexell, who came from present-day Finland but was active in St Petersburg. He showed that the comet had encountered Jupiter in 1767 and,

[7]If this were not the case, there should also be a clustering of arguments of perihelia near 180°.

as a result, had been captured from a quite different orbit with $q \simeq 2.9$ AU. Given how rapidly comets brighten up when approaching the Sun, it would then have been too faint to be discovered, until its perihelion was moved much closer to the Sun.

Incidentally, Lexell also found another interesting result. The comet was in 2:1 resonance with Jupiter and was thus bound to encounter Jupiter again after two orbital revolutions. This would be like a mirror image of the first encounter, and the comet would be ejected outward again. It has been found that the actual outcome in 1779 was quite dramatic. Comet Lexell was perturbed into an orbit with $q = 5.2$ AU and $Q \simeq 80$ AU, where it is still performing its first revolution.[8]

This historic example is quite important, because it opened up the notion that comet orbits are very unstable and that short-period comets may have originated from long-period comets. This idea was also put forward in Pierre Simon de Laplace's *Exposition du système du monde*, following Lexell's results. As a consequence, all comets are related and could have a common origin — in principle, an attractive idea. It would, however, take a long time before this capture process was fully understood and a final verdict could be passed on its realism.

5.5.1. *Everhart's capture region*

As time went on, the number of Jupiter Family comets grew, and it also became clear that their lifetimes were limited. Comet 3D/Biela was seen to split and disappear, and several others were missed at their predicted returns — sometimes (as in the case of 5D/Brorsen in 1890) in spite of reliable ephemerides and a bright expected magnitude. This gave momentum to the hypothesis of *comet capture*. The short-period comets, in particular the Jupiter Family, would be a spin-off of the seemingly abundant flux of long-period comets via a dynamical mechanism relying on Jupiter's gravity, whereby the orbits were transformed and the periods shortened.

[8]This is an example of the fact that the letter 'D' in a comet designation does not imply the physical disappearance of the comet.

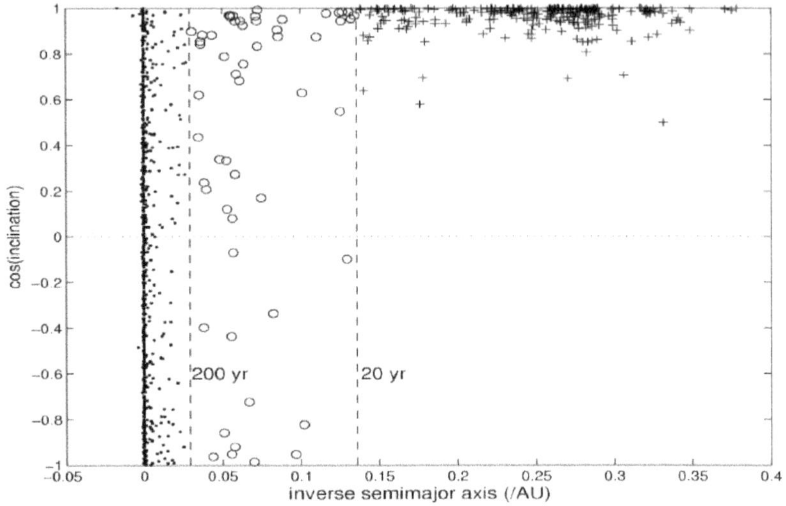

Fig. 5.15. The cosines of inclination versus the inverse semi-major axis for comets discovered until the end of 2011. Different symbols are used in different ranges of orbital period, divided by the vertical dashed lines.

In Fig. 5.15 we can see a problem with this idea. Using a modern sample of comets with good observational records, this is far more informative than the statistics available, say, one century ago, but the picture remains unchanged. The long-period comets are seen to the left (including the new comets at $1/a \approx 0$) and they exhibit a flat distribution of $\cos i$, which indicates isotropically distributed orbital planes. On their right side there are the Halley Type comets and these too sample the whole range of $\cos i$ though with a certain preference for prograde orbits. However, coming to the Jupiter Family, the situation changes completely. It is *a priori* remarkable, if not perplexing, that the inclination distribution should become totally different when passing the $P = 20\,\mathrm{y}$ limit. Obviously, this problem has to be solved.

The first mechanism to be investigated was single-encounter capture. This means that one single close encounter would transform a long-period orbit into a Jupiter Family orbit. It could be demonstrated that such an event is possible, but it was also realized more than a century ago that the probability is far too small and that the results are not in agreement with the observations (Newton 1893). We can see this by returning to Eq. (3.2) in Sec. 3.1, from which we see that the quasi-conserved quantity T can be written as $T = 2L_z - E$ in terms of the orbital energy E and the perpendicular component of the angular momentum L_z. If we start from a long-period orbit with $E \simeq 0$ and the encounter leads to an orbit with a semi-major axis smaller than that of Jupiter, it means that the comet reaches $E < -1$ in the units used. Thus, L_z has to decrease by a large amount due to the same encounter. Depending on the encounter geometry, this can be achieved by a large decrease of L or a large decrease of $\cos i$, or a combination of the two. In practice, this leads to many captured comets with high inclinations — even in retrograde orbits.

A question thus arose, if multiple-encounter capture like the energy diffusion of long-period comets, ending up in Halley Type orbits, could also explain the Jupiter Family and its characteristics. Is there a way from the Oort Cloud into the Jupiter Family by such a process? There was no easy answer by means of theory or studies of individual comets, and the problem thus had to be studied by numerical simulations. In the 1960s, Edgar Everhart started to implement such a method by investigating the probability distributions of orbital perturbations at close encounters with the giant planets for random encounter geometries. In 1972, he published a paper that appeared to provide the solution.

Everhart (1972) took a large sample of initial, parabolic comet orbits characterized by their inclinations and perihelion distances and, otherwise, with random orientations and integrated these — typically with only Jupiter as perturbing planet — until they were either ejected from the solar system or captured into short-period orbits. The number of orbits spanned by the integrations was anything from just one to thousands. He found that, even though all inclinations were considered along with perihelion distances up to 6.2 AU, practically all the captures occurred for $i_0 < 9°$ and $4 < q_0 < 6$ AU. He called this the *capture region*.

The inclinations remained low during the capture process, and the perihelion distances decreased with the orbital periods. When the perihelia had been moved into $q < 2.5\,\mathrm{AU}$ — corresponding to the orbits of nearly all the Jupiter Family comets that were then on record — the periods had dropped into a good agreement with the range then observed for the Jupiter Family. It had thus been shown that multiple-encounter capture from the near-parabolic flux offers a possible route into the Jupiter Family. In essence, this route is like an extension of the evolutionary curves in Fig. 3.1 (Sec. 3.1) to the limit of infinite aphelion distance. Everhart remarked that his captured comets spanned a range of 2.44–3.04 in T, thus indicating the range of the capture region.

However, Everhart did not say that there was an agreement between the necessary capture rate to maintain the Jupiter Family in a steady state and the flux of long-period comets; nor did he maintain that his capture mechanism was necessarily the one actually in operation. These issues would be the subject of later works.

5.5.2. *The need for an additional source*

Joss (1973) published the first paper to indicate the quantitative need for an additional source of Jupiter Family comets. He derived the steady state number of Jupiter Family comets by using estimates for the arrival rate of new comets from the Oort Cloud, the observable lifetimes of comets, and the capture probability as found from Everhart's paper, involving the measure of the capture region in terms of inclination and the conditional capture probability within this region. As a result, he found that the real Jupiter Family is much richer in comets than Everhart's capture mechanism could explain.

A similar argument was put forward by Julio Fernández and Wing Ip ten years later. They estimated that no more than ~1% of the Jupiter Family comets could be supplied from the new comet flux. But Fernández and Ip (1983a) also proposed an explicit solution to the problem. They had studied the scattering of planetesimals during the accretional growth of Uranus and Neptune and found that this would leave behind two distant reservoirs of such planetesimals, which would be potential source regions of the observed comets. One is the Oort Cloud and the other is a trans-neptunian "cometary belt", which in hindsight can be seen as a prediction of the Scattered Disk.

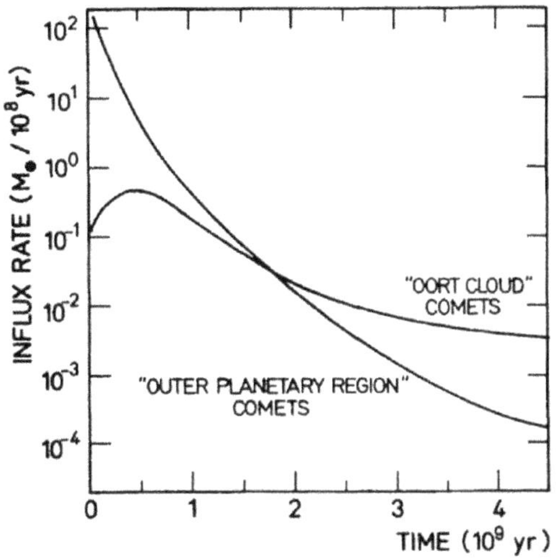

Fig. 5.16. Mass influx rates of comets into the terrestrial planet region as functions of time from the Oort Cloud and the Scattered Disk ("Outer planetary region") according to numerical simulations. Reproduced from Fernández and Ip (1983b) with permission.

The idea of such a comet belt beyond Neptune was not new — it had, for instance, been suggested by Whipple (1972a). But this was the first time that it was backed by a quantitative, dynamical explanation of its origin. Figure 5.16 illustrates the results concerning the influx rate of comets into the terrestrial planet region as a function of time according to numerical calculations (Fernández and Ip 1983b).

The Scattered Disk contribution was clearly dominant in the beginning but should then have dropped by several orders of magnitude following the dissipation of the disk. By now, the new comet flux from the Oort Cloud is expected to be ten times larger, but only 10^{-3} of this would contribute to the Jupiter Family in contrast to the entirety of the disk flux. The above-mentioned $\sim 1\%$ contribution by the Oort Cloud could thus be understood.

In the meantime there had been a somewhat different approach to the problem, which was initiated by orbit integrations performed by Elena Ivanovna Kazimirchak-Polonskaya (1972). Her work had shown that Jupiter Family comets have a dynamical link to the Centaur region and that the giant planets can change their orbits so that

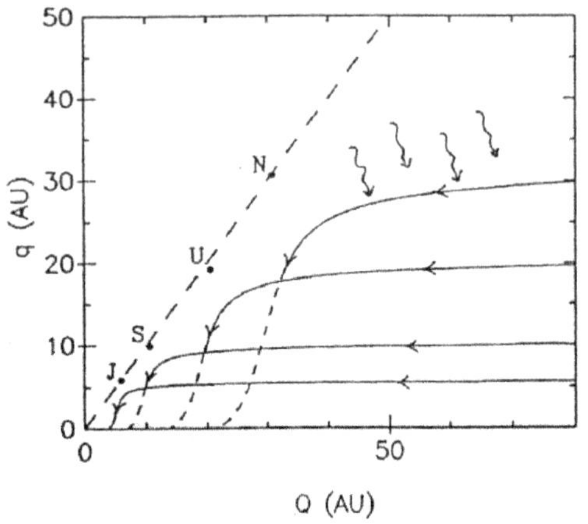

Fig. 5.17. Schematic picture of dynamical transfer routes for comets within and inside the giant planet region. Evolutionary curves for Tisserand parameters typical of Jupiter Family comets and low inclinations are shown separately for the respective planets. Aphelion and perihelion distances are plotted on the axes. The arrows on the curves indicate the direction of capture toward the Jupiter Family. The curly arrows indicate the influx from a trans-neptunian reservoir.

dynamical control is passed between them. This indicates a transfer mechanism between the Jupiter Family and trans-neptunian space, which opens up new possibilities of capture. Such transfer routes are indicated schematically in Fig. 5.17.

The lowest curve drawn in the figure shows how the Jupiter Family would be connected to Everhart's capture region, since the curve would continue in an essentially horizontal direction to infinite aphelion distance. The next curve counting upward belongs to Saturn, indicating that Saturn will have a capture region too. However, the captures do not form a Saturn Family in this case, since Jupiter takes control as soon as the perihelion moves into the vicinity of its orbit. Similarly, Uranus and Neptune have their capture regions, from which they can capture comets with initially larger perihelion distances. In these cases too, the evolution proceeds only until the perihelion reaches the inner neighboring planet. Such a multi-planet capture from the Oort Cloud was investigated and found to work by Everhart (1977).

However, the efficiency of this capture may still be too low to account for more than a minority of the Jupiter Family, and hence the figure also shows an expected, additional capture flux from a trans-neptunian comet reservoir. This may indeed be related to the belt that Fernández and Ip proposed.

In principle, it is possible that a multi-planet capture starting via Neptune or Uranus could work quantitatively, if the Oort Cloud has a very massive inner core from which many comets are still evolving into perihelia near the orbits of these planets by the slow Galactic tide. This scenario was proposed by Mark Bailey (1986), but it now seems that the real inner core does not possess the necessary population. One more argument against capture from the Oort Cloud was made by Duncan, Quinn and Tremaine (1988). These authors simulated captures from an isotropic Oort Cloud flux with perihelia in the 4–6 AU range into visible comet orbits with $q < 2.5$ AU to compare with Everhart (1972), and their results were different. While Everhart had found comets of the high-inclination type not to become visible at all, such comets were now in majority among the visible comets. Most such comets were long-period, most of the rest were Halley Types, and only a few were typical Jupiter Family comets that may have started with low inclinations.

In another experiment, Duncan *et al.* simulated multi-planet capture from the Oort Cloud and found that captures into orbits with $P < 200$ y and $q < 1.5$ AU, on the one hand, occurred only from initially prograde orbits, but on the other hand produced final orbits with a very broad inclination range. There is a controversial issue about the latter results, because to speed up the integrations, the planetary masses were increased by a large factor. It has been argued that this could lead to artifacts, for instance by introducing too many single-encounter captures, which are known to lead to many retrograde, captured comets (see above). However, the papers by Duncan *et al.* (1988) and Quinn *et al.* (1990) found wide support in the research community and were often seen as eye-openers, proving that the Jupiter Family has to have its own source region in a trans-neptunian belt, which the authors called the *Kuiper Belt.* This naming was in remembrance of a paper by Kuiper (1951), in which small bodies beyond Neptune were suggested to be left-overs from a trans-neptunian extension of the original solar nebula (Sec. 5.4.1). This is different from the disk that Fernández and Ip had imagined.

5.6. Steady-State Conditions and Lifetimes

We have seen that comets do not live forever. Let us take Jupiter Family comets as an example. The first thing that may happen to a Jupiter Family comet is that its activity ends or subsides, so that the comet becomes dormant. We know that the Jupiter Family consists of both active and dormant comets and that the active to dormant transition can be reversed (Sec. 4.3.1). Therefore, these two kinds should be treated together. However, active comets lose mass and may thus shrink below a limiting size, where they effectively cease to exist as observable objects, and from this there is no return.

There is hence an ongoing loss of comets from the Jupiter Family. The longer a comet stays in the Jupiter Family, the more likely it is to meet the disintegration end state, although this also depends on how big the nucleus was initially and, perhaps, to which extent it avoids the kind of physical-dynamical evolution that stimulates the quenching of its activity and slow-down of its mass loss. However, for dynamical reasons, disregarding the weak nongravitational effects, all Jupiter Family comets are bound to face the end state of evolution into long-period orbits and eventually ejection from the solar system. This may be in the active or dormant state and the only escape is disintegration.

Thus, given the properties that Jupiter Family comets have, including their size distribution, their population is endowed with a certain distribution of lifetimes — meaning the time spans during which they exist as active or dormant Jupiter Family comets. Our conclusion may either be that some process acted in the recent past to set up a Jupiter Family population that is now in a state of decay, or that the process of initializing new Jupiter Family comets is acting continuously, thereby keeping the population in a steady state. In the absence of a natural candidate mechanism to form the Jupiter Family at once, the latter alternative is usually considered and capture from the source population is supposed to secure a long term steady state.

This means that the following equation has to be satisfied in order to guarantee the steady state:

$$N_{\rm JF}/\tau_{\rm JF} = N_{\rm sou} p_{\rm cap}, \tag{5.6}$$

where $N_{\rm JF}$ is the number of Jupiter Family comets (active and dormant together), $\tau_{\rm JF}$ is their average lifetime facing the risk of disintegration or ejection, $N_{\rm sou}$ is the number of comets in the source population, and $p_{\rm cap}$ is the probability per unit time for the average source comet to get captured into the Jupiter Family. It does not matter how long the capture takes, as long as the comets are not physically destroyed in the meantime.

Equation (5.6) is well suited to investigate the constraints on a trans-neptunian source population for the Jupiter Family, based on observational estimates of the Jupiter Family population size and lifetimes. However, the treatment of the left-hand member has usually been different from what we described. For instance, only active comets may be counted in $N_{\rm JF}$, and $\tau_{\rm JF}$ may be taken to be the active lifetime, i.e., the time spent by the average comet in the active state.

Hal Levison and Martin Duncan (1997) performed a comprehensive study of the capture of Jupiter Family comets from the Kuiper Belt by the multi-planet mechanism that we described above. The thrust of this paper was that a large number of test objects were integrated over a very extended time scale, so that relevant statistics was available to characterize the capture process and the captured comet population in some detail. In the picture they used, a Kuiper Belt object remains such until it has a close encounter with Neptune. It is then sent into something we may call an "adventure", which may temporarily lead into the Jupiter Family but never back into the Kuiper Belt. The integrations by Levison and Duncan were limited to the tracing of these adventures.

Orbits with $2 < T < 3$ and $q < 2.5\,$AU were taken to define the *visible Jupiter Family*. Comets that entered into this group were considered to stay active for a certain "fade time" t_f and then to

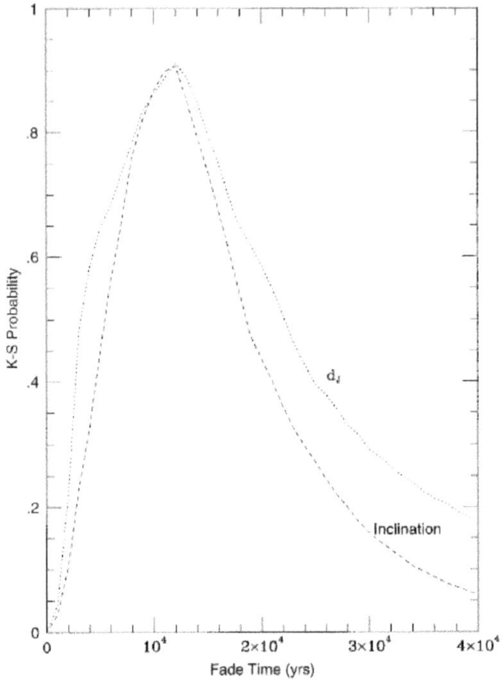

Fig. 5.18. Kolmogorov–Smirnov probability of agreement between the Levison–Duncan steady state distributions of orbital elements for visible Jupiter Family comets and the observed distributions versus the assumed fade time. The inclination is represented by the dashed curve and the least difference between the nodal distance of the comet and Jupiter's semi-major axis (d_J) by the dotted curve. Reprinted from Levison, H. F. and Duncan, M. J., *Icarus* **127**, 13–32 (1997), with permission from Elsevier.

become extinct, meaning an irrevertible dormancy. It was found that a good fit to the orbits of observed, active Jupiter Family comets was obtained by putting $t_f = 12\,000$ years (see Fig. 5.18). The reason for this result is that the close encounters with Jupiter that Jupiter Family comets experience cause a gradual increase of the inclinations and scatter of the Tisserand parameters ("dynamical heating"), which may eventually lead to disagreement with the observed orbits, if the comets are assumed to live too long. With the fade time preferred by Levison and Duncan, the ratio between the steady state numbers of extinct and active Jupiter Family comets was 3.5. Based on an estimate of the number of Jupiter Family comets by Fernández *et al.* (1992), Levison and Duncan thus came to the result that the visible Jupiter Family should contain about 500 comets — active and extinct together.

Since such comets turned out to be about 4.1×10^{-5} of all the "adventurous" comets at any time, the number of the latter should thus be 1.2×10^7. Their average dynamical lifetime was estimated as 4.5×10^7 years, so the necessary rate of renewal is 0.27 per year. The statistical chance for a Kuiper Belt object to encounter Neptune per year had been estimated as 4×10^{-11}, so the Kuiper Belt has to contain 6.7×10^9 members of the relevant sizes in order to provide the steady state source for the Jupiter Family.

The "relevant sizes" means nuclear radii larger than one kilometer. It is still very difficult to estimate how many such objects there are in the trans-neptunian populations. At the time of the work by Levison and Duncan in the mid-1990s, there was no way to tell how realistic the above estimate would be. It was rather a prediction to be verified or rejected by future observations. In this regard, the discovery of the scattered disk was important, because it was quickly realized that this reservoir might host as many comets as the Edgeworth–Kuiper belt. But the adventurous comets can be expected to behave in the same way independent of which source they come from, and scattered disk objects have a closer connection to Neptune and are more likely to encounter the planet. Hence, one may conclude that most Jupiter Family comets are captured from the scattered disk rather than the Edgeworth–Kuiper belt (Duncan and Levison 1997).

The necessary number of scattered disk objects is just one tenth of that found for the Kuiper Belt. Later works have tended to agree on this rough estimate. Among those, let us note the one by Di Sisto *et al.* (2009), who introduced a model for the physical destruction of Jupiter Family comets. The free parameters of this model basically concerned the frequency and magnitude of random splitting events, which act to erode the nuclei in addition to the ice sublimation. Di Sisto *et al.* simulated the orbital distribution of the steady state Jupiter Family and looked for the best fit to the observed distribution by tweaking these parameters. They found that comets have to split very frequently and substantially — more so than observations seem to suggest.

This way the lifetime of visible Jupiter Family comets was brought down to 3 600 years, and a good fit to the cometary orbits was obtained. This is shorter than the 12 000 years found by Levison and Duncan (1997), probably because the source population used by Di Sisto *et al.* was more excited — a natural consequence of the fact that more recent observational surveys have revealed more high-inclination members of the trans-neptunian populations. Of course, the higher the starting inclinations of the captured Jupiter Family comets, the shorter the lifetimes have to be to comply with the observational evidence and avoid an overheated steady state population.

In a recent investigation by Rickman *et al.* (2017a), this problem was pushed to the extreme. The ultimate source of the Jupiter Family was not the currently observed scattered disk but an early scattered disk consistent with the Nice Model (Sec. 6.1). While there is no inconsistency between this model and the current discovery statistics of SDOs, the implication is a source that is so puffed up that the comets have the inclination distribution of the observed Jupiter Family already the first time they reach orbits with perihelion distance $q < 2.5$ AU. This is illustrated in Fig. 5.19. As a consequence, even if the observed Jupiter Family comets were assumed to have ages less than one thousand years, they would already have acquired too large inclinations, and such a picture is clearly unacceptable. It would be impossible to explain both the perihelion distances and the inclinations of the Jupiter Family at the same time.

This is a serious conundrum, whose solution may be sought along two different lines. In one case, the true scattered disk is considerably flatter than the Nice Model prediction, but this may be at odds with observations. The other possibility is that the concept of fade time that Levison and Duncan introduced is oversimplified and needs revision. Some key issues were discussed in Chap. 4. Comets become dormant and more or less immune to physical evolution, and they may get rejuvenated, e.g., in connection with perturbations reducing

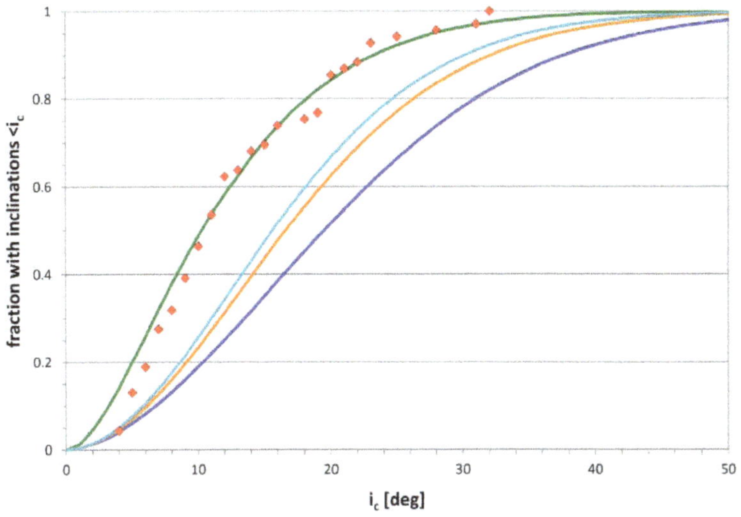

Fig. 5.19. Cumulative inclination distributions for Jupiter Family comets. The red symbols refer to observed members of the Jupiter Family with nuclear radii >0.8 km. The green curve refers to simulated comets at their first perihelion passage with $q < 2.5$ AU. The other curves show simulated steady state inclination distributions for $q < 2.5$ AU derived by assuming infinite lifetimes (blue) and lifetimes limited by ice sublimation (orange and cyan). Credit: H. Rickman *et al.*, *A&A* **598**, A110 (2017), reproduced with permission © ESO.

their perihelion distance. The observed Jupiter Family may therefore be a biased selection out of a much larger, true population. The bias consists of having recently experienced encounters with Jupiter that decreased their perihelion distances and caused reactivation. This will favor low-inclination orbits with a nodal distance close to 5.2 AU.

If this is the explanation of the observed orbital distribution, the aging of comets is slower than believed. The Jupiter Family is larger than usually thought, and the predominance of dormant or almost dormant objects is more pronounced. However, this does not necessarily affect the capture rate from the scattered disk, so the current estimate of the scattered disk population may not need any revision.

Chapter 6

Formation of Comet Reservoirs

We now come to how the comet reservoirs were formed. This discussion will concern comet nuclei that have been formed by mechanisms to be discussed in Chap. 7. The main topic of this chapter will be the dynamical processes, whereby these nuclei were brought from their place of origin into the more or less quiescent environments, where they were stored for most of the solar system history and from where they are currently arriving.

6.1. The Nice Model

This term stands for a set of rather different models, proposed by one team of scientists since 2005 with the fundamental papers dating from 2005–2011, and describing the long-term evolution of the planetary orbits in the Solar System. The most recent one is of course favored, but it is of great use to know how the ideas developed throughout the project, so we will examine several versions of the model here. The Nice Model deals primarily with planetary orbits, but there is hardly any category of solar system objects that is not concerned, and comets are no exception.

The name *Nice Model* has been used since the beginning and refers to the fact that the work behind the first version took place at the Nice Observatory in France. It involved the local astronomer Alessandro Morbidelli and the US visiting scientist Hal Levison plus the long-term visitors Kleomenis Tsiganis (Greece) and Rodney

Gomes (Brazil). Later versions and applications have involved other people too, for instance Ramon Brasser and David Nesvorný.

The common features of all versions are (1) long-term gravitational interaction between the giant planets and an outer, massive planetesimal disk; and (2) a resulting instability of the giant planet system involving mutual close encounters, whereby the ice giants (Uranus and Neptune) get ejected into the disk and migrate into their current orbits by scattering the planetesimals all around the Solar System. The short time scale of the evolution following the instability (in reality, $\sim$10 Myr) makes it appear almost like an explosion when illustrated by computer animations. The term *planetesimals* stands for the macroscopic building blocks, out of which the terrestrial planets and the cores of the giant planets were accreted. This will be further described in Chap. 7.

Comets play a central role, because the planetesimals of the outer disk are thought to be similar to — or even identical to — comet nuclei. Thus, the Nice Model offers one explanation for the formation of the comet reservoirs, but there are also alternative ideas that will be referred to in this chapter.

6.1.1. *The initial model*

This was published in three adjoining *Nature* papers (Tsiganis *et al.* 2005, Gomes *et al.* 2005, and Morbidelli *et al.* 2005). As a prerequisite to the actual model, the giant planets were assumed to have formed along with lots of stray planetesimals during the existence of the short-lived, primordial *solar nebula.*[1]

The four giant planets were assumed to move initially on quasi-circular and coplanar orbits at specific heliocentric distances (in a typical case, 5.45, 8.18, 11.5 and 14.2 AU for Jupiter, Saturn and the two ice giants, respectively) by the time the solar nebula had dissipated. All stray planetesimals in this quite narrow zone

[1]This term stands for the accretion disk, through which the gas and dust of the presolar cloud were assembled into the Sun, mostly during the earliest stage. Somewhat later, it also formed the environment for the growth of planetesimals and subsequent growth of the giant planets.

were shown to have very short dynamical lifetimes and thus were assumed to have already been ejected. However, about 1 AU beyond the outermost planet orbit and further out, all initial planetesimals would have survived. Thus, a disk with a total mass of about 30–50 Earth masses and extending out to 30–35 AU was considered and represented in numerical simulations by a few thousand roughly Moon-sized particles.

The planet orbits were not in mutual resonances. However, Saturn and Jupiter typically had an initial period ratio of about 1.84 — thus, relatively close to the 1:2 mean motion resonance. At present, these two planets are at 9.54 AU and 5.20 AU, so their period ratio is approximately 2.48. This means that, if the system has evolved smoothly from the assumed initial condition into the current situation, the two planets must have passed their 1:2 resonance, and the mutual perturbations then suffered constitute a great risk of instability for the whole planetary system.

The computer simulations showed a slow heating of the innermost part of the planetesimal disk (i.e., increase of orbital eccentricities), leading to a slow trickling of objects into close encounters with the outer ice giant. Actually, this would not necessarily be Neptune, because Uranus and Neptune may change places during the forthcoming instability. Figure 5.17 in Sec. 5.5.2 shows what may then happen by means of an illustration of the multi-planet capture of Jupiter Family comets from a trans-neptunian disk, because the dynamics is very similar in the two cases.

In the present case, the planetesimals meet the first ice giant near their perihelia. Hence they overtake the planet at close encounters, and their motion relative to the Sun is decelerated by the hyperbolic deflection, as explained in Sec. 3.2.1. The ensuing evolution leads toward the lower left in the (Q, q) diagram of Fig. 5.17, following curves of constant Tisserand parameter. When the perihelia reach the next ice giant, the same happens, and the planetesimals are handed over to Saturn, which in turn hands them over to Jupiter. The reason for the efficiency of this handing-down process is that none of these giant planets is likely to eject the objects from the Solar System. In principle, after the first decelerating encounter, this might happen as a result of the ensuing random walk along the evolutionary curve at repeated encounters, so the issue is whether the process is ended by ejection or handing down to the next planet, and the latter is much more likely than the former.

However, Jupiter is massive enough to easily eject the planetesimals into interstellar space and at the same time has no inner neighboring giant planet to hand them over to. Thus, eventually, nearly all of them get ejected by Jupiter. The above-described handing-down process of course has dramatic consequences for the planetesimals, but each planetesimal is so small that its reaction on the planets is completely negligible. However, when similar evolutions are experienced by billions of planetesimals, there is a significant net effect on the planetary orbits. In the present case, this is called *planetesimal driven migration* and was first studied by Fernández and Ip in the 1980s. The result is a slow expansion of the giant planet system, since Jupiter migrates inward and all the others migrate outward. After several hundred million years, Saturn and Jupiter may thus reach their mutual 1:2 mean motion resonance, and the system is doomed.

What actually happened in the simulations is that Jupiter and Saturn drifted across the resonance without being trapped in it. Orbital mechanics then implies a sudden increase of their eccentricities. Thus, for the first time, the two gas giants achieved eccentricities similar to their current ones. Their orbits started to precess due to their mutual gravitational perturbations, and secular perturbations (see Sec. 3.3) started to act between all the four giants. These excited the ice giants to even larger eccentricities, and at this point — due to the compactness of the system — the ice giants started to encounter both each other and the gas giants.

This implies a period of chaotic evolution, which lasts only for a million years or so. Both ice giants get ejected into the planetesimal disk and thereby scatter its objects in a chaotic fashion. Their orbits start out with large eccentricities (acquired from the preceding close planet–planet encounters), but due to the myriads of encounters with planetesimals, these eccentricities are damped (a mechanism known as *dynamical friction*), and then the orbits expand due to the

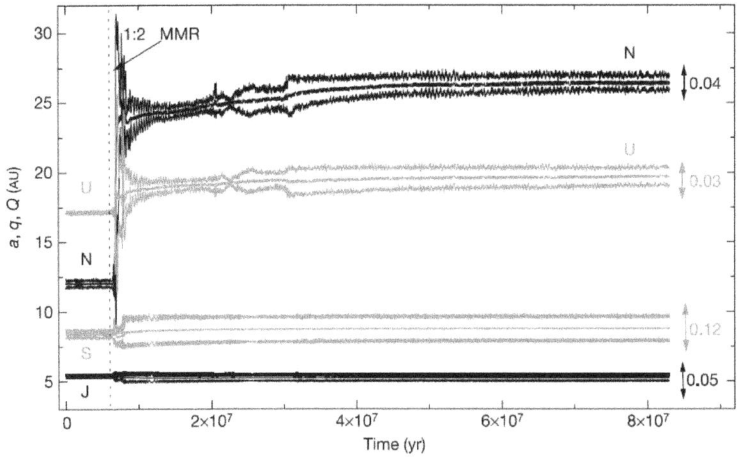

Fig. 6.1. Evolution of giant planet orbital elements during the instability period of the Nice Model. The three curves plotted for each planet show, from above to below, the aphelion distance, semi-major axis and perihelion distance. This integration starts with Saturn very close to the 2:1 resonance with Jupiter, and the crossing of the resonance — marked by a vertical dotted line — occurs after 6.6 Myr. Adapted by permission from MacMillan Publishers Ltd: Tsiganis, K. *et al.*, *Nature* **435**, 459–461 (2005), © 2005 Nature Publishing Group.

scattering of the objects that remain in the exterior parts of the disk. This outward migration may push some objects ahead by means of locking into the 1:2 or 2:3 resonances with Neptune, thus populating the resonant groups of TNOs (Sec. 5.4.1).

Figure 6.1 illustrates the time evolution of the semi-major axis and the perihelion and aphelion distances of the giant planets in a typical model run. Note that in this case Uranus started out as the most distant planet, but this is not a general trend in the simulations at large. Thus, the actual initial order of the ice giants cannot be predicted. These calculations were taylored to show in detail the consequences of the resonance crossing and reconfiguration of the giant planet system, but the question remains of how long time it would take from the dissipation of the solar nebula until these events would occur.

The answer clearly depends on the detailed choice of initial giant planet orbits (i.e., the values of the semi-major axes), the mass of the

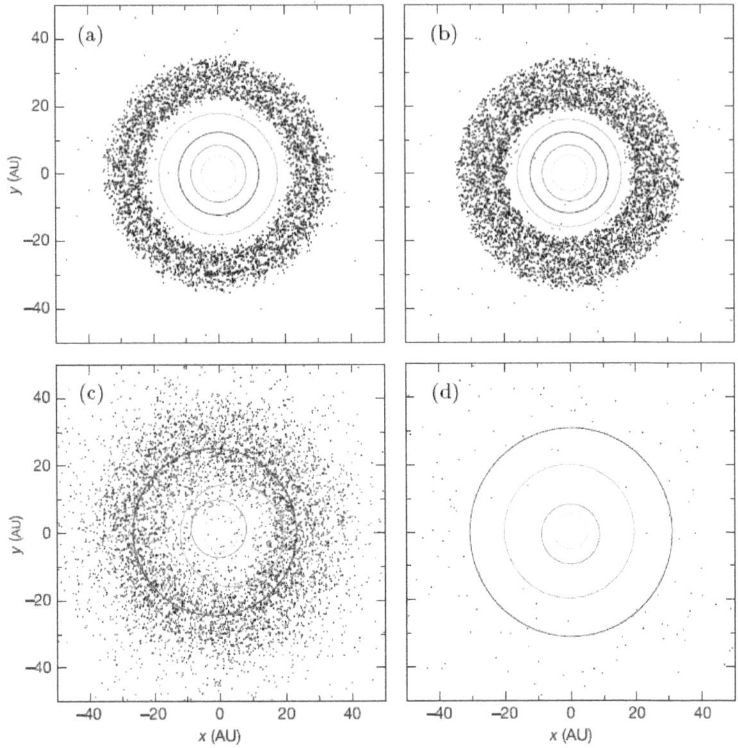

Fig. 6.2. Planetary orbits and positions of disk particles in a reference Nice Model simulation, where the mass of the initial planetesimal disk is 35 solar masses and the initial gap between planets and disk extends from 14.2 to 15.5 AU. The elapsed times from the start of the simulation are 100, 879, 882 and 1 082 Myr for panels (a), (b), (c) and (d), respectively. Adapted by permission from MacMillan Publishers Ltd: Gomes, R. *et al.*, *Nature* **435**, 466–469 (2005), © 2005 Nature Publishing Group.

planetesimal disk and the exact distance from the outermost planet orbit to the inner edge of the disk. In Fig. 6.2, four snapshots of the dynamical evolution are shown representing one such choice. These show the planet orbits and remaining planetesimals at (a) 100 Myr after the beginning; (b) a time just before the instability breaks out; (c) a time shortly afterwards; and (d) 200 Myr later, when everything has settled.

This simulation, like those similar to it, was favored for a particular reason. While the evolutions shown in Fig. 6.1 did not differ much between different choices of model parameters, the timing of the critical events could vary a lot and was strictly set by those choices. The reason why the timing in Fig. 6.2 was found attractive is that this allowed one to find an explanation for the so-called *Late Heavy Bombardment* (LHB). The latter is the popular name for an episodic, strong increase of the lunar cratering rate that appears to have taken place roughly 4 Gyr ago. In 2005, the most common estimate was that about 700 Myr passed from the birth of the solar system until the LHB. Since it was shown to be possible for the Nice Model instability waiting time to be equivalent to this delay, it could be surmised that the scattering of planetesimals from the primordial outer disk formed a convenient source of lunar impactors.

6.1.2. *The first improvements*

The initial version of the Nice Model was widely acknowledged by the international research community but was not found as satisfactory by its authors. An important reason was the ad-hoc choice of initial radii for the planetary orbits. This did not take into account modern theoretical developments related to how the emerging giant planets interact dynamically with the gas disk (solar nebula) in which they are immersed. Such theory is beyond the scope of this book, but for the understanding of the Nice Model we need to briefly introduce a few aspects.

The necessity to account for planet-disk interactions was made acute by the observations of exoplanets in the late 1990s. In particular, the *hot Jupiters* immediately raised a dilemma, because these were giant planets often exceeding the size of Jupiter, which orbited around their parent stars in the immediate vicinity of the stellar surfaces. This is so different from what we see in the solar system that explanations had to be sought. One avenue of research was the evolution of the planetary orbit due to the exchange of energy and angular momentum with the native gas disk.

This kind of evolution was known from a different context, namely, the above mentioned interactions with a planetesimal population and was thus termed *migration*. Two basic mechanisms have been identified. In *Type I migration* the mass of the planet is relatively low and it excites spiral density waves in the disk, which exert a torque on the planet thereby changing its orbital angular momentum. The resulting migration may be inward or outward depending on the circumstances. If the planet is massive enough, which is the case for Jupiter, its gravity clears a gap in the disk surrounding its orbit, where the gas density falls to zero. In this case there may be *Type II migration*, which means that both the inner and outer parts of the disk (with respect to the gap) undergo the inward motion associated with the accretion onto the star. Barring this phenomenon, the planet would find its equilibrium position within the gap, so that the torques from the inner and outer gas would cancel out. But the gas accretion makes the inward torque from the external gas predominate, and thus the planet is dragged toward the star.

Type II migration offers a natural way to explain the existence of hot Jupiters, but one nagging question is why this did not happen in the solar system. The answer to this question might also offer a good proposal for the initial spacing of our giant planets. A relevant scenario was introduced by Masset and Snellgrove (2001) and is illustrated by Fig. 6.3. It goes as follows in the case of the solar system. There is a pair of giant planets, where the inner one (Jupiter) is massive enough for Type II migration, while the outer one (Saturn) is left with Type I migration as the only way to move radially. But the latter type is faster, so Saturn may have started by approaching Jupiter from the outside. Eventually, it would be locked in a mean motion resonance with Jupiter, whereupon the two planets had to migrate together. At this time, the two gaps would overlap, but since Saturn's gap was not cleared, gas was able to stream past the planets into the Sun. In this situation, the inward motion might be reversed, so that the planet pair would start migrating outward.

Morbidelli and Crida (2007) found that, under some conditions regarding the viscosity and scale height of the disk, the inward drift of the Jupiter-Saturn pair would be halted but not reversed, so that a stable 2:3 resonant configuration would be reached. The predominance of hot Jupiters among exoplanets and the lack of systems like ours would likely mean that our disk, unlike most others, fulfilled the criteria for stability. This is one indication that our solar

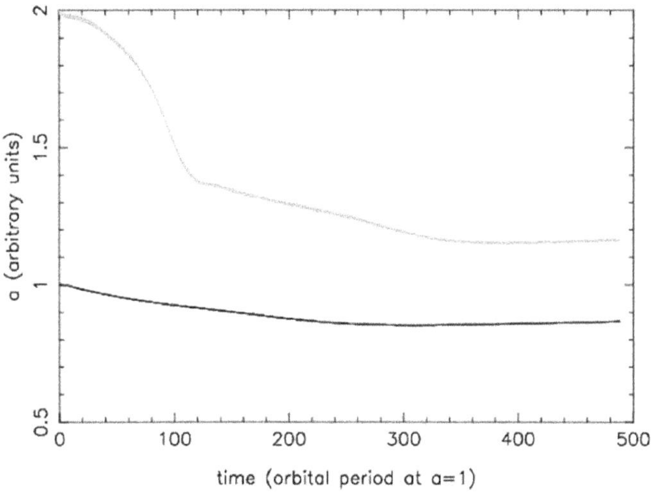

Fig. 6.3. Orbital radii of two migrating planets as functions of time, resulting from their interactions with a modeled gas disk. After the capture into mutual resonance, the two curves are parallel. Reprinted from Morbidelli, A. and Crida, A., *Icarus* **191**, 158–171 (2007), with permission from Elsevier.

system may be special because of a particular property that is shared by only a minority of the planetary systems.

A similar outcome would be likely for the ice giants. The first one to approach Saturn from the outside by its Type I migration would get stuck in the 2:3 or 3:4 resonance with Saturn, and the second one would again land in such a resonance with the first ice giant. Thus, the whole system of giant planets was imagined to initially arrange itself into a multi-resonant structure, where all the mean motions were locked into 2:3 or 3:4 commensurabilities.

Irrespective of any exterior planetesimal disk, such a system was found to be violently unstable on a short time scale as soon as the gas disk was gone — simply due to the mutual gravitational interactions. This is because the system is very compact, so the planets are in dangerous vicinity of each others' orbits. But a minority of constellations was found to be stable, and this must then have been the one that the Solar System got into. Planetary systems around other stars might often have landed in unstable configurations and got destroyed at an early stage. Most of these exoplanets would have been ejected and the survivors left on eccentric orbits, as is often seen for exoplanets.

In the stable case of the solar system, Saturn and Jupiter would have started in the 2:3 resonance, implying a period ratio of 1.50. If this ratio would increase due to the interactions with the planetesimal disk, the first major resonance to be encountered would be 3:5, for which the period ratio is 1.67. According to the new simulations that were performed, the scattering of planetesimals making their way into the region of planetary orbits occurred in the same manner as before. However, the planetary orbits were locked in resonances and thus could not directly start to migrate — instead, the eccentricities increased. This, in turn, led to the break-up of the resonance locks, and the system was finally free to drift toward new resonances. Once again, the waiting time for this to occur was typically several hundred Myr.

The ensuing passage of the 3:5 resonance by Saturn and Jupiter was found to excite their eccentricites similar to what was found in the initial model, and once again, the ice giants were perturbed onto eccentric, crossing orbits leading to close encounters. The resulting, chaotic scattering pf planetesimals was similar to that of the initial model. But a problem remained, since Jupiter and Saturn continued to drift rapidly toward the 1:2 resonance, and this crossing would be dangerous.

It had already been found in the initial model that there were inconveniences connected with the terrestrial planets and asteroid main belt, if the Nice Model instability took place so late that it could explain the LHB. A gentle drift of the two gas giants across the strong 1:2 resonance would stir up the asteroids and excite the terrestrial planet orbits too much. There was even a threat of the Earth being expelled from the solar system. In the improved model, the instability of the giant planets had already been provoked by the 3:5 resonance crossing, so in a sense there was no longer any useful role to be played by the 1:2 resonance.

The preferred scenario was found as one, where one of the ice giants first gets "captured" by Saturn into an orbit that reaches into the realm of Jupiter, whereby Saturn is kicked outward. Then another

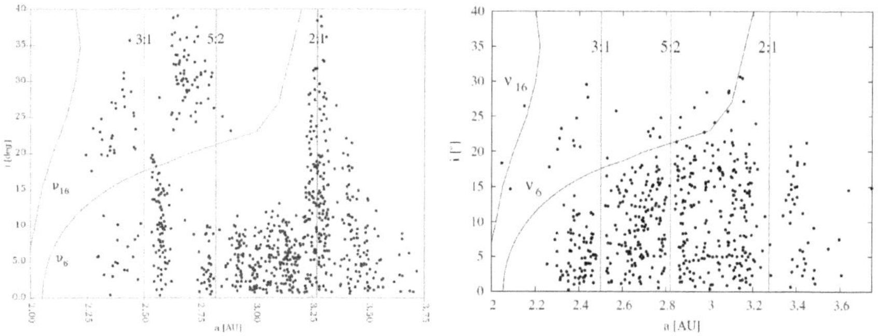

Fig. 6.4. Simulated, final orbital distributions in the asteroid main belt as found by integrating fictitious objects with perturbations by the migrating giant planets. Left panel: smooth migration of Jupiter and Saturn across their 1:2 resonance. Right panel: jumping Jupiter scenario. From Morbidelli *et al.* (2010). © AAS. Reproduced with permission.

close encounter happens — this time with Jupiter, whereby the ice giant is ejected into the planetesimal disk and Jupiter gets an inward kick. This is known as the *jumping Jupiter scenario.* It is preferred, because the system of terrestrial planets and the asteroid belt avoid the large perturbations that a slow drift of the two gas giants across the 1:2 resonance would imply.

Figure 6.4 shows results by Morbidelli *et al.* (2010) concerning the simulated distribution of final orbits in the asteroid belt by plotting the proper inclinations[2] versus the semi-major axes — resonance gaps are clearly visible. In the case of a slow drift across the 1:2 Saturn–Jupiter resonance (left panel), a sweeping of the belt by secular resonances would excite far too many asteroids into high-inclination orbits, while in the jumping Jupiter case (right panel) a very good agreement with the observed orbit distribution is obtained.

An interesting consequence of this scenario has to do with the explanation for the LHB. Of course, the scattered planetesimals from the trans-planetary disk represent a source of projectiles for lunar

[2]Usually, the inclinations and eccentricities of asteroid orbits undergo periodic oscillations, as the orbits precess due to the planetary perturbations. By filtering out those oscillations, one obtains stable elements called proper inclination and proper eccentricity.

cratering in both the initial Nice Model and the revised, jumping Jupiter model, since the disk gets dispersed in the same way. But in the initial model, the 1:2 resonance crossing by Jupiter and Saturn stirred up the asteroid belt so much that another very important source was provided by asteroids escaping from the belt. Thus, the revision of the model meant that the latter of the two sources more or less disappeared, so that the LHB might have been caused predominantly by the trans-planetary objects — in other words, by comets.

6.1.3. *The current model*

The revised version of the Nice Model was a great step forward, but some problems remained. In the initial model, where the planets were not in mutual resonance during the phase of planetesimal driven migration, the outer ice giant would indeed migrate outward, thus keeping pace with the erosion of the disk's inner edge, and the process was therefore self-sustained. However, in the improved model this could not occur, since the planet was locked into a resonance with its inner colleagues. Therefore, there seemed to be a risk that the process would stall, and the planets would be left forever in their resonant configuration.

Another worry is that the planetesimal disk must have some internal dynamics that is not considered in either of the two previous versions of the Nice Model. Actually, the disk cannot be made up of only very small bodies, but there must have been thousands of Pluto-mass objects and some probably larger still. This is because we know that there are a few Pluto-mass objects in the known trans-neptunian population (Pluto and Eris), and if the Nice Model is correct, the current population must be a bleak shadow of the very massive scattered disk that existed in the beginning (see Sec. 6.4.2).

One consequence is that encounters with the massive objects will stir up the eccentricities of the small ones, and values of about 0.1–0.2 may be expected. If one takes account of this so-called *viscous stirring*, there is a way to avoid the stalling of the planet orbit evolution, since new disk objects are fed into the realm of the planets by scattering off the massive members. But another consequence of the significant eccentricities, which had not been expected beforehand, was also discovered by the new simulations.

Levison *et al.* (2011) showed that there is a secular exchange of energy between the planetesimal disk and the planets, and this was found to be related to the orbital eccentricities. When the disk particles have clearly eccentric orbits due to viscous stirring, this exchange becomes noticeable. The drainage of energy from the planet orbits is not uniform but mostly involves the inner ice giant, whose orbit has the largest eccentricity among the planets. Therefore, this planet should migrate with respect to the others, but it cannot because of the locking into resonances. Hence its eccentricity is pumped up, and this leads to changes in the precessional frequencies of the planet orbits. As a consequence, there is a possibility of secular resonance in some pair of planets, which disrupts the resonance locks and induces chaos into the system. When this happens, the planets find themselves in the same situation as in the preceding model, and the outcome is bound to be the same.

This was found to happen in typical cases, and the time scale from the start of the simulations was found to be between 300 and 1000 Myr. Contrary to earlier Nice Model versions, there is no particular correlation between the length of this time and the width of the initial gap between the planets and the disk — the latter does not matter very much due to the viscous stirring of the disk.

In the mentioned version of the Nice Model, which may currently be called the standard version, the orbital evolution in the compact system of giant planets is due to a collective action of the disk rather than individual objects intruding into the planetary system, even though such intruders must still exist. The lesson to be learned is that the existence of a massive disk of objects with a wide size distribution outside a compact planetary system formed by gas driven migration into stable resonances is a formidable threat to the long-term stability of the planet orbits, and a break-down is to be expected in a time much shorter than the age of the solar system. The initial configuration of giant planets that Levison *et al.* (2011) explored has Jupiter's orbit centered at 5.4 AU, Saturn's at 7.3 AU, and the two ice giant orbits at 9.6 and 11.7 AU.

The significance of the Nice Model for the creation of the main comet reservoirs will be evident from the discussion in Sec. 6.3. Basically, the scattering of planetesimals by the outward migrating ice giants does not only send comets into the terrestrial planet region, where these may hit the Moon and planets, but it also forms an exterior, scattered disk that may be seen as the direct predecessor of both the current scattered disk and the Oort Cloud. In fact, if the Nice Model is true, the whole trans-neptunian part of the solar system was set up at the time of the instability according to a plan that was already inherent in the primordial structure of the solar system during the earliest accretion phase.

It should, however, be emphasized that there is significant freedom in the Nice Model concerning details. Different variants have been proposed within the standard version, where even the initial number of giant planets can be different from four (Nesvorný and Morbidelli 2012). Batygin and Brown (2016), when proposing the existence of Planet Nine, suggested as a likely explanation the ejection of a fifth giant planet from its formation zone, and the Nice Model would likely provide a means for this to happen. Moreover, the timing of the instability is to some extent free. It is not a model parameter but a result that can change depending on the choice of initial conditions.

A late instability occurring about 4 Gyr ago would offer the only natural explanation for the LHB and thus, if the LHB is taken as a fact, this is strongly preferred over an early instability. Another reason may be that the early delivery of water to the Earth works best in case the orbits of Jupiter and Saturn are nearly circular (O'Brien *et al.* 2006). However, the question whether or not there was a LHB has been under debate since a long time. This cannot be answered by lunar crater counts and dating of lunar surface strata — nor can the issue be settled by the Nice Model in itself. It is rather a question of unraveling the early bombardment history of the Earth and Moon based on geochemical data and the dynamics of the left-over embryos and planetesimals from the formation of the terrestrial planets.

Independent of which way this open issue is resolved, the Nice Model will survive because of its remarkable ability to account for many different observed properties of the solar system. Among these we find the orbital properties of all the planets, the irregular satellites of the giant planets, the Trojans associated with Jupiter and Neptune, and all the trans-neptunian small body populations. The latter include both the Scattered Disk and the Oort Cloud, as we shall see in Sec. 6.3.1.

6.2. The Solar Birth Cluster

Among the few things that we can actually claim to know about comet origin are that it was very long ago, when the Sun was young or even infant, and that it was relatively far from the Sun. In particular, all the current reservoirs of comets are distant, and the dynamics involved in forming them must be the dynamics of the outer solar system. Considering these two aspects, the issue of the environment in which the solar system was born is quite important, when discussing comet formation and mechanisms of storage into the reservoirs.

This issue deals in particular with the stellar environment. The flux of energetic radiation that may have eroded the outskirts of the solar nebula depends on the number of nearby, hot and massive stars. The amount of radioisotopes that was dumped into the material from which planetesimals formed depends on whether a supernova explosion occurred in the vicinity at the relevant time, for which the probability in turn depends on the number density and mass distribution of nearby stars. Moreover, if the Sun was born in a dense stellar environment, this must have affected the dynamics of objects that orbited at large distances (see Sec. 3.4).

At the same time, we have to realize that stellar clusters are common in the Galaxy and a stellar cluster is a very likely place for the Sun to have been born. Obviously, we have to investigate all the consequences this may have for the early history of comets.

6.2.1. *Stellar clusters and star forming regions*

The Sun is a typical Galactic disk star. As a result of long standing efforts to probe the contents of the Galactic disk and its subcomponents, we know that most disk stars are single, binaries or belong to multiple systems but are not members of clusters, which typically contain hundreds of stars or more. However, most disk stars have lived for much longer than the typical life span of most stellar clusters. They may hence be escapees from long since dissolved clusters, and there is no saying that stars in general are not formed in clusters. In fact, the observational evidence points in the opposite direction. Studies of star forming regions show that the nascent and emerging stars belong to aggregates that may be referred to as *birth clusters*.

In terms of dynamics, a stellar cluster is a gravitational many-body system. If the number of stars is large enough, a statistical approximation is relevant. This means that a star belonging to the cluster moves in a gravitational field with two components. One is the global cluster potential, which defines the orbit that the star will follow. Since the cluster is close to a steady state, this potential is time invariant and thus conservative. The other component is a fluctuating potential due to the moving nearby stars, which causes a permanent change of the orbital energy of the star as well as its angular momentum.

Thus, the orbits of the cluster stars are in permanent change due to the interactions at close encounters. As a consequence, from time to time it happens that a star is pushed to a velocity exceeding the velocity of escape and leaves the cluster. This means that the cluster is subject to a slow evaporation process, which limits its lifetime. There is also a mass segregation, since the lighter cluster stars are more prone to escape, and the more massive ones tend to collect near the cluster center.

Other dynamical phenomena within stellar clusters include the formation and dissolution of binaries, and very close, low-velocity encounters. The relative motions of stars within a cluster are much slower than those of stars in the solar environment in the Galaxy, and thus the mentioned phenomena are of importance inside clusters, while they are extremely unlikely outside clusters. If we take for granted that the Sun was born as a cluster member, it is thus natural to ask if it has always been single or spent part of its youth as a binary component. If the latter happened, it might have had ramifications for solar system dynamics.

The higher the mass of a stellar cluster, the less important is the fluctuating component of its gravitational potential. Consequently, in relative terms the evaporation is slower and the cluster lives longer. The Galactic disk hosts some rich clusters, whose age may even exceed that of the solar system. We then observe the remnants of much more massive, initial clusters, even though these remnants still stand out by their number of stars and total mass. But most clusters that are apparent for visual observers of the Galaxy are much more modest. Their typical ages are less than 100 Myr, and thus they contain some main sequence stars of early spectral types and high luminosities. But their observed memberships are limited to a few hundred stars.

Of course, the youngest clusters are found in the star forming regions. These regions are mostly connected with giant molecular clouds, which were mentioned in Sec. 2.6.4 as exceedingly dark environments and the homes of protostellar objects. This darkness means that no visual light emitted by the protostars can emerge from the clouds, but the infrared radiation can be studied to some extent. There is a good observational material of this kind, allowing one to map out the clustering tendency for stars that have yet to break through their cocoons of dust. These clusters are referred to as *embedded clusters* and are of course the youngest of all. To find an unbiased size distribution characterizing stellar clusters at their birth, the embedded clusters form an invaluable data source.

In a review of the topic of embedded clusters, Lada and Lada (2003) concluded that the initial mass function of such clusters has a slope such that the decrease of the number of clusters formed per unit time with the number of members is just about compensated by the increasing number of involved stars per cluster. Using this result with some extrapolation, one may conclude that Galactic disk stars have about equal chances to be born in embedded clusters with masses of 10^2, 10^3 and 10^4 solar masses. For each cluster of the last kind, 100 clusters of the first kind are born and even more of the even smaller kind that do not survive for long after the natal molecular cloud has been dissolved (this means very short lifetimes). Thus, almost all embedded clusters are very small and short-lived, but the few large and long-lived ones contain an important fraction of all stars.

6.2.2. *The cluster size*

The suspicion that the Sun could have been born in a cluster of any size does not tell us much about the history of the solar system. The situation would be different, if we had evidence from the solar system itself, pointing to a certain size of the Sun's birth cluster. There are indeed some relevant criteria, discussed in a review by Fred Adams (2010), which we will now present.

One important reason to believe in a big birth cluster is that we need to explain the presence of large quantities of short-lived radioisotopes in materials contained in planetesimals. The evidence comes from chondritic meteorites — a sort of stony meteorites, whose elemental composition is very similar to that of the Sun. These are hence thought to represent the original solid material of the solar nebula. They must come from parent bodies that have preserved this material. The chondrites are not homogeneous as to structure or chemistry, but they contain a few typical components that are present in more or less all of them.

A component of particular interest hosts the so-called *calcium-aluminum-rich inclusions* or CAIs (Sec. 2.6.2). These have a common age (4 567 Myr) that exceeds all other ages of solar system materials, so their formation is taken as the marker of the origin of the solar system. The whole-rock ages of the chondrites containing the CAIs are typically a few Myr younger. Concerning the radioisotopes, there are several of them, but the most famous is aluminum-26 (^{26}Al) with a half-life of only 0.72 Myr. This isotope decays into ^{26}Mg, which is a stable isotope. In the *Allende meteorite*, which fell in Mexico in 1969, an excess abundance of ^{26}Mg was found in good correlation with the local aluminum abundance — in particular, in the CAIs (Lee *et al.* 1976).

From the amount of this correlation one can deduce the time elapsed from the creation of the radionuclides until the formation of the CAIs. This turns out to be very short, so that the two events were practically simultaneous. The origin of the solar system can be identified with the collapse of a dense core inside the presolar cloud (see Sec. 2.6.4), but what about the origin of the unstable isotopes? The favored mechanism is the explosion of a supernova, which not only produces those isotopes in abundance but also sends them out into space as part of a dense, quickly propagating gas shell. Hence the idea that a nearby supernova went off at a time just before the solar system started to emerge from its collapsing cloud. It does not take much imagination to see a possible logical connection in that the collapse could have been triggered by the shock front signaling the supernova ejecta.

However, this scenario raises an urgent question. The presolar cloud would not have lived very long until the solar system had formed, and supernova explosions do not occur very frequently in the vicinity of a random place in the Galaxy, so how can we explain the near simultaneity of the two events? This seems like a strange coincidence. The problem is most easily solved by assuming that the birth of the Sun occurred in a rich stellar environment, where many stars of different masses were born during a short time. In the present case, according to the theory of supernovae, a progenitor star of 25 solar masses or more is necessary to produce the radioisotopes that were present in the infant solar system. Such massive stars are so rare that, statistically, this is a likely outcome only if the solar birth cluster contained at least several hundred stars.

The other reason has to do with the detached TNOs (Sec. 5.4.2) and, in particular, Sedna. This concerns the topic of Sec. 6.3.1, but briefly, the issue is how to detach the perihelion of a scattered disk object with a semi-major axis around 500 AU into a distance of more than 70 AU. A diffusive process involving many small perturbations is not an option, because it would too easily be ruined by a return of the perihelion to the planetary region. One large impulse conveyed by a close stellar encounter is the favored explanation.

The star in question must have passed within about 1 000 AU of the Sun, preferably with a low velocity (Rickman *et al.* 2004). As seen from Fig. 3.8 in Sec. 3.4, considering the usual Galactic field stars, only one such encounter can reasonably be expected during the age of the solar system. The problem is that it could have happened at any time, and the scattered disk was rich enough to provide a chance for the extraction of Sedna-sized objects only for a relatively short interval. Therefore, the favored environment is the solar birth cluster (Morbidelli and Levison 2004), and the timing must be early enough for this cluster to survive.

Clearly, the size of the birth cluster is important in this regard. The larger it is, the longer it will survive and the more frequent the close stellar encounters will be. According to Adams (2010), the likelihood to satisfy the Sedna constraint is significant only for clusters with about 1 000 members or more.

However, there is also a down side of close stellar encounters. These have the potential to upset the planetary orbits to an extent that is incompatible with the current structure of the solar system. Looking at the encounters that occurred within the solar birth cluster, we likely have to deal with the multi-resonant orbits of the giant planets before the instability of the Nice Model. These orbits were small enough to be relatively insensitive to the effects of passing stars, but on the other hand, even slight effects might have destroyed the fine details of the pre-instability clockwork of this model.

Such issues have not yet been studied in detail, but simulations of the effects of stellar encounters on the planetary orbits have been made (e.g., Malmberg *et al.* 2007), which led Adams (2010) to conclude that the closest encounter must have had a miss distance of more than 225 AU. Carrying out the analysis like in the case above, one may then deduce an upper limit to the number of stars in the birth cluster. This limit was found to be about 5 000.

Finally, there is a constraint coming from the preservation of the solar nebula. When immersed into the far ultraviolet radiation field of nearby stars, this gas disk is eroded by hydrogen ionization. The resulting heating of the gas will lead to unbinding from the solar gravity beyond a certain distance, which depends on the uv flux. A question remains as to the seriousness of this criterion within

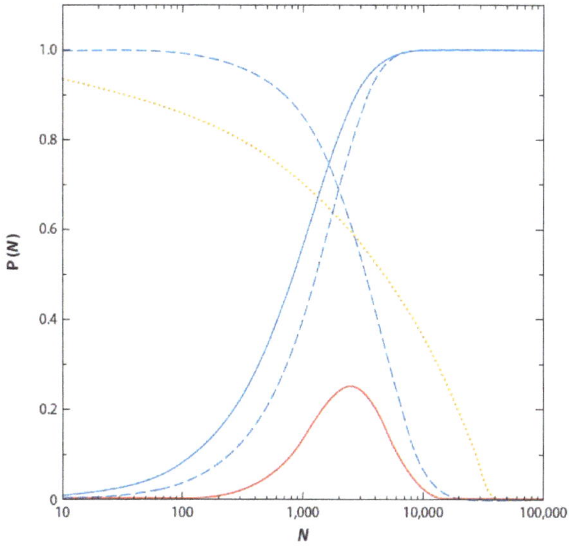

Fig. 6.5. Probability curves for the population size of the solar birth cluster. The solid blue curve shows the probability to have a supernova progenitor of relevant mass, and the rising dashed curve shows the probability to obtain a stellar encounter close enough to detach the orbit of Sedna from the planetary system. The waning dashed curve shows the probability not to have a stellar encounter that destroys the regularity of the planetary system, and the dotted curve shows the probability not to have the solar nebula too heavily eroded by far uv photons. Reproduced from Adams (2010) with permission of *Annual Review of Astronomy and Astrophysics*, Vol. 48. © by Annual Reviews.

the Nice Model, but Adams (2010) concluded that birth clusters with more than 10 000 stars would likely be too hostile to the solar nebula.

Figure 6.5 summarizes the above constraints with the overall result that the most likely range of the population size for the solar birth cluster can be expressed as $4\,300 \pm 2\,800$ (Adams 2010). However, Adams also pointed out that the applied constraints are necessary but not sufficient. For instance, the existence of a supernova of sufficient mass in the cluster does not mean that the explosion had to occur at the right distance from the Sun at exactly the right time. In this particular case, it might help to have more than one relevant supernova, which would raise the estimate for the population range.

6.2.3. *Explicit proposals*

Most of the short-lived radioisotopes that existed in the infant solar system can be explained either as products of stellar nucleosynthesis in the Galactic disk at relatively large distances, which is continuously ongoing, or as results of the corpuscular irradiation of the solar nebula by the solar wind, which was very strong when the Sun was formed. The above-described scenario of a nearby supernova is of relevance mainly for ^{26}Al and ^{60}Fe. Referring to just one star as the origin of the whole amount of these isotopes has been standard, but actually, there are serious problems.

Gounelle and Meynet (2012) pointed out that the delivery of the estimated amount of ^{26}Al requires a distance of only about 0.5 pc for the supernova explosion and that this leads to a dilemma, because the supernova progenitor would have ionized and heated the surrounding hydrogen gas (thus creating a so called HII region) out to a larger distance. Such a region is not a suitable place for the survival and collapse of the presolar cloud. Furthermore, only part of the ^{26}Al but all of the ^{60}Fe produced inside the progenitor would be released during the explosion, making it difficult to understand how the initial solar system ^{26}Al/^{60}Fe abundance ratio could be as high as the estimated 8.2.

Gounelle and Meynet therefore proposed a multi-generation scenario for the origin of the two radionuclides and the formation of the Sun. Their scenario is based on the popular picture of sequential star formation and considers three stellar generations. The first generation produces supernovae that enrich the surrounding gas in ^{60}Fe and ^{26}Al, but the former isotope wins out as time proceeds due to its slower decay. After about 10 Myr, the second generation of stars is formed in the vicinity under the influence of the expanding supernova ejecta shock fronts. The massive stars of this generation develop HII regions that are surrounded by expanding shells of compressed gas, and these are the sites of formation of the Sun and other third generation stars. The shells are enriched in ^{26}Al, which is produced inside the massive second generation stars.

The proposed stellar environment is quite massive — Gounelle and Meynet estimate about 5 000 first generation and more than 1 000 second generation stars, while the shell where the Sun was formed contained only a few hundred third generation stars. However, the dynamics of this large stellar population was not discussed and it is therefore difficult to conclude on the existence of a solar birth cluster or its possible population size.

In fact, there is one observed and well-known stellar cluster that has been considered to be a possible remnant of the solar birth cluster, namely, *Messier 67* (M67). This idea was based on a remarkable similarity of physical and chemical properties between the Sun and one of the M67 members (Önehag *et al.* 2011). The large distance of M67 from the Galactic plane causes a dynamical problem facing the suggestion of a genetic relationship with the Sun, and while the processes causing the secular heating of the Galactic disk do not rule out such a relationship (Gustafsson *et al.* 2016), the latter does not appear likely. On the other hand, the spectral similarity between the Sun and some M67 members may be taken to indicate that the Sun was indeed formed in a dense stellar environment (Önehag *et al.* 2014).

6.3. Oort Cloud Formation and Evolution

6.3.1. *Oort Cloud formation*

Basically, the principle of Oort Cloud formation is quite simple. Two steps are involved. In the first step, the giant planets scatter comets gravitationally into orbits with larger and larger energy, reaching further and further away. In the second step, when the orbits reach sufficiently far, an external perturbing agent adds angular momentum, thereby increasing the perihelion distances and decoupling the comets from the planets. This scheme holds for practically all mechanisms so far proposed. The variants concern which giants are mainly responsible for the scattering in step 1, and whether Galactic tides, field stars or cluster stars cause the decoupling in step 2.

Historically, the first ideas used the tacit assumption that the Sun was born alone, because the observations of star forming regions had not yet revealed the strong clustering tendency of new-born stars. Moreover, the influence of the Galactic tides had not yet been realized, so the decoupling was considered to result from encounters with field stars like the present ones. There is only a narrow range of orbital energy, where these stars are efficient in decoupling the comets — this range is the same as that of the Oort spike for new comets, which was discussed in Sec. 5.1.

This places a constraint on the planetary scattering. Saturn and, above all, Jupiter cause large perturbations of the orbital energy for typical cometary perihelion passages. A typical step in the random walk is thus larger than the width of the range of efficient decoupling and, hence, there is an overwhelming risk of overshoot, so that the comets are ejected from the solar system rather than being decoupled. Thus, for populating the Oort Cloud, Jupiter and Saturn are remarkably inefficient. Instead, Uranus and Neptune became known from the work of Viktor Safronov (1977) to be the main providers of comets during the formation of the Oort Cloud.

This scenario was explored in simulations by Fernández and Ip (1983b) — see Sec. 5.5.2. The Oort Cloud comets would be escaping planetesimals from the Uranus–Neptune region. After the role of Galactic tides was discovered a few years later, these found a natural place in the modeling of Oort Cloud formation. The new picture was first described by Duncan *et al.* (1987) in a highly influential paper. Since the Galactic disk tide is more efficient than the stars in decoupling the comets from the scattered disk, the critical energy range is broader and the scattering job of the planets is easier. The resulting picture of Oort Cloud formation and evolution is illustrated in Fig. 6.6. However, from the point of view of avoiding ejections from the solar system, the situation was not dramatically improved. Uranus and Neptune were still the main comet providers into the semi-major axis range $a > 3\,000\,\mathrm{AU}$, from which the cloud was populated. As the figure shows, the current Oort Cloud is predicted

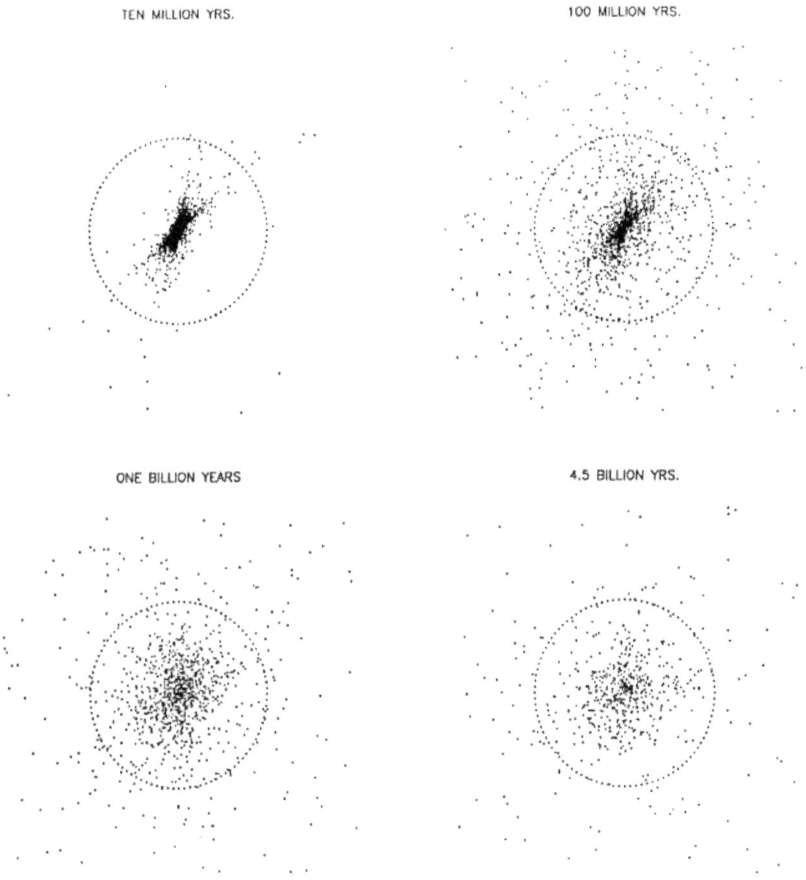

Fig. 6.6. Four snapshots of the comet distribution in the Oort Cloud resulting from numerical simulations. The cloud is projected onto a plane parallel to the Galactic midplane. The dotted circle indicates a radius of 20 000 AU. The flattening of the inner cloud (apparent in the upper panels) indicates the orientation of the ecliptic plane. From Duncan *et al.* (1987). © AAS. Reproduced with permission.

to be dominated by an inner core, where the orbits extend to less than 20 000 AU.

However, while there was no immediate reason to doubt this description, a dilemma was suspected well into the 1990s. This concerned the mass of the cloud. From the number of cloud comets large enough to produce the brightness of the new comets, as derived from injection models, and an assumed mass distribution among the cloud members, a rough estimate of the total mass could be made. While the large uncertainties did not allow full confidence, this mass was large enough to place a severe constraint on the efficiency of Oort Cloud formation. It thus appeared that the formation scenario under consideration was not efficient enough.

In the mid-1990s the time had come for the solar birth cluster to enter the picture. Julio Fernández (1997) pointed out that this might increase the efficiency of Oort Cloud formation significantly, because the increased flux of close and slow stellar encounters would allow decoupling to take place at smaller semi-major axes, and this would strongly increase the chances for Jupiter and Saturn to contribute to the delivery of comets. Given the large amount of planetesimals in their formation zone, the effect on the Oort Cloud population size could be large.

Models of Oort Cloud formation in an embedded cluster were produced ten years later after the discovery of Sedna. The semi-major axis of Sedna's orbit — about 500 AU — is too small to be compatible with decoupling except in a stellar cluster, and in Sec. 6.2.2 we saw that this even provides a constraint on the size of the cluster. This motivated Ramon Brasser, Martin Duncan and Hal Levison (2006) to study the conditions under which the solar birth cluster would allow the formation of an Oort Cloud that includes objects in Sedna-like orbits. They used a model of a short-lived embedded cluster with 200–400 stars and found that the constraint would be met for a high central density, as seen from Fig. 6.7.

We note a related and very important result, namely, that an Oort Cloud formed in a dense birth cluster is very small. This is because most of the comets in the scattered disk have orbits with small semi-major axes. When the cluster provides enough close stellar encounters to act efficiently on such comets, these tend to dominate the cloud by forming a very concentrated inner core. Of course, the cloud also contains comets in much larger orbits, but the latter are relatively rare.

The indicated range of central densities is not necessarily realistic for a small birth cluster, but this problem is minor compared to

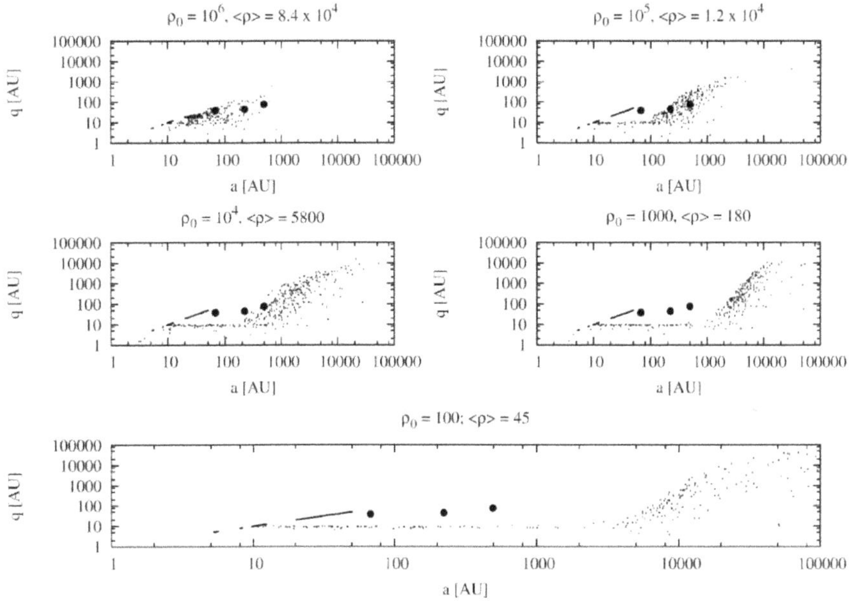

Fig. 6.7. Snapshots of the (a, q) distribution of comets at the end of dynamical simulations for planetesimals from the giant planet region. Each simulation uses a birth cluster with a given value of the central density (ρ_0), as indicated above the diagrams. The rightmost filled circle shows the orbit of Sedna. Reprinted from Brasser, R. *et al.*, *Icarus* **184**, 59–82 (2006), with permission from Elsevier.

what was found in a follow-up paper by Brasser *et al.* (2007). In this paper the dynamical model was made to include one more item — the drag force imposed on planetesimals by the gas of the solar nebula. The result was quite remarkable. The orbits of planetesimals with the size of typical comet nuclei were found to be subject to an efficient circularization that, effectively, prevented evolution into the scattered disk. On the other hand, much larger objects were so much less affected that these could easily be scattered like in the earlier case without the gas drag. The authors referred to this as *size sorting*.

This result is important, but the considered scenario with planets in stable orbits and a quiescent gas disk now seems obsolete, while it was relevant at that time. First, we have to refer back to Sec. 6.1 and our account for the evolution of the Nice Model. Consider the inward motion of the Jupiter–Saturn pair within a broad disk gap with a very low gas density. During this motion, planetesimals may have encountered these planets and been scattered outward with less intervention by the gas, and perhaps this was the case for small as well as large objects. However, the dynamics of the planetesimals during the gas-driven migration of the giants has not yet been addressed.

The question of gas-driven migration is even further complicated by the *Grand Tack* scenario (Walsh *et al.* 2011), which holds that Jupiter's inward migration was not stopped until the planet had reached a distance of about 1.5 AU from the Sun, at which point the Jupiter–Saturn pair turned outward. This scenario has its main virtue in providing a reasonable explanation for the relative masses of the terrestrial planets — in particular the fact that Mars has just about 1/10 the mass of the Earth. It involves scattering of planetesimals, but the further fate of these objects was not discussed. Again, the formation of a primordial scattered disk may be imagined.

Unfortunately, while we thus cannot rule out the existence of a primordial scattered disk, its formation efficiency and the dependence on the size of the planetesimals are essentially unknown. The same can be said about a primordial Oort Cloud — whether it exists depends on whether the primordial scattered disk existed, and its population size depends on the population size of the disk. However, it also depends on the decoupling efficiency, which in turn depends on the stellar environment.

Even in a dense birth cluster the stellar encounters do not allow a very high decoupling efficiency, but a way forward was found by Levison *et al.* (2010). This work considered a small birth cluster with only about 100 members. Primordial scattered disks were assumed to exist around all stars. The simulations showed a new phenomenon when carried on until the break-up of the cluster. Many comets that had escaped from their native stars were picked up by the same star or another star, as the velocities decreased during the cluster expansion. Hence, the resulting comet clouds were dominated by "stealth comets". If this scenario is applicable to the solar system, the efficiency of Oort Cloud formation would be increased by a large factor, and most of the Oort Cloud comets would have their origin in the protoplanetary disk of some other star in the Sun's birth cluster.

Thus far, we have discussed the formation of a primordial Oort Cloud due to planetesimal scattering by the forming giant planets. As suggested in Sec. 6.1.3, there is also a mechanism for delayed Oort Cloud formation within the Nice Model. If this model does describe reality, the delayed cloud is in fact a necessary consequence. The mechanism, as investigated by Brasser and Morbidelli (2013), means that the trans-planetary planetesimal disk was stirred up by Uranus and Neptune during the time of their outward migration. As a result, a large fraction of these planetesimals were scattered by Neptune into an exterior disk, much like the formation of the primordial scattered disk. Given the presence of external perturbers, some fraction of these scattered objects were decoupled and formed an Oort Cloud. Brasser and Morbidelli assumed that the solar birth cluster was long gone by the time of the LHB and only considered the Galactic tide and the field stars to play a role in this decoupling.

In Fig. 6.8 the simulation results by Brasser and Morbidelli are illustrated. The upper panel shows the mass fraction of the planetesimal disk that occupies the Oort Cloud as a function of time after the end of the planet migration. For typical disk masses considered within the Nice Model, the current Oort Cloud would measure about 1–2 Earth masses. In the lower panel we see the mass ratio between the Oort Cloud and the Scattered Disk, which converges around the value 12. We shall return to this in Sec. 6.4.1.

Interestingly enough, there is hence a possibility that the current Oort Cloud is a mixture of two parts, which were formed at different times by planetesimals from two different regions. The primordial part would ultimately originate from the planetesimals of the giant planet region, which according to the Nice Model extended from about 5 to about 15 AU from the Sun. The more recent LHB part would be made up of planetesimals from the trans-planetary disk, which started somewhat further away and continued out to about 30 AU. It is also possible that the formation of the primordial

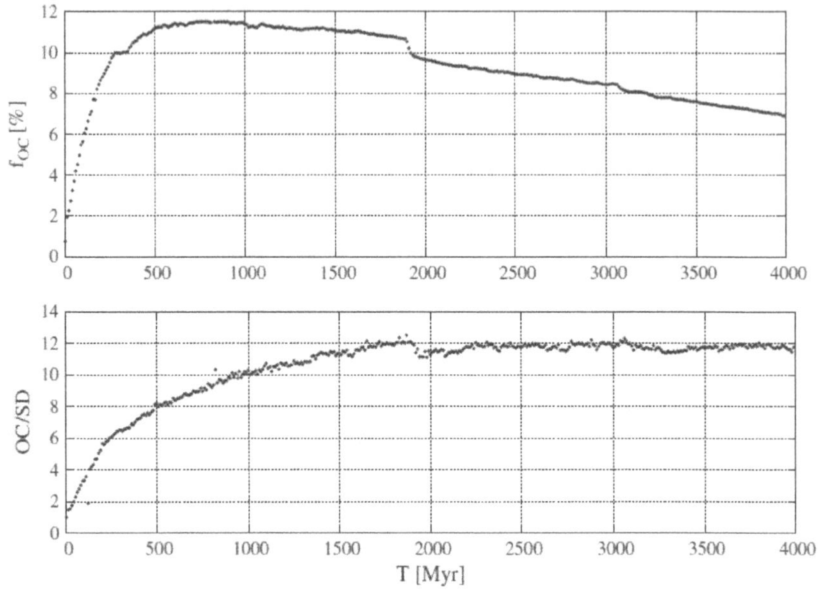

Fig. 6.8. Time evolution of the fraction of primordial disk members stored in the Oort Cloud (upper panel) and the ratio between the same fraction and the fraction populating the Scattered Disk (lower panel). The time is counted from the end of planet migration in the Nice Model. Reprinted from Brasser, R. and Morbidelli, A., *Icarus* **225**, 40–49 (2013), with permission from Elsevier.

scattered disk was so inefficient that the current Oort Cloud is completely dominated by its more recent part. Moreover, there is an issue concerning the survival of a primordial cloud in a long-lived birth cluster, as will be shown below.

Finally, as indicated in Sec. 6.1.3, it is possible that the planet instability of the Nice Model took place very early — more or less contemporary with the creation of a primordial Oort Cloud discussed above. In this case the Oort Cloud would be formed much earlier than in the case that Brasser and Morbidelli treated. However, Rome was not built in one day, and nor was the Oort Cloud. In the Nice Model scenario, as shown in Fig. 6.8, the formation took several hundred Myr. This would likely reduce the survival issue concerning the solar birth cluster but not remove it completely.

6.3.2. *Oort Cloud evolution*

If the solar birth cluster plays a role in forming the primordial Oort Cloud, it also poses a threat to the cloud, because if stellar encounters that are efficient in raising the perihelia of scattered disk objects occur frequently, there may also be encounters that kick comets away from the Sun or kick the Sun away from its comet cloud. In the case of a small and short-lived birth cluster, there is hardly any problem, but the situation is quite different if the birth cluster is rich and long-lived.

Let us consider a numerical example, using the classical impulse approximation as described in Sec. 3.4. From Eq. (3.23) we can estimate the impulse imparted to the Sun by a passing star. A typical encounter velocity in a birth cluster can be taken as 1 km/s, and the stellar mass is about the same as that of the Sun. It then turns out that an encounter distance of 400 AU yields an absolute value $I \approx 4.5$ km/s. This is about equal to the speed of escape from the Sun at a distance of 100 AU, showing that essentially the entire Oort Cloud would leave the solar system — or, more correctly, almost the whole Oort Cloud would be deserted by the Sun. The comets would thus become cluster vagabonds. Only comets in strongly bound orbits that obtain nearly the same vectorial impulse as the Sun would in this case follow the Sun and stay bound to it.

With a birth cluster of moderate size and lifetime, such close encounters are not very likely. Conversely, if the cluster contains several thousand stars (Sec. 6.2.2), there is a substantial chance of occurrence. Thus, a primordial Oort Cloud may not survive at all. Nordlander *et al.* (2017) have found by simulating Oort Cloud comets with semi-major axes from 5 000 to 20 000 AU inside a birth cluster of 2 000 initial stars or more that the chance of survival is practically zero, given the likely structure and lifetime of such a cluster.

Thus, if we take for granted that a primordial scattered disk could be formed, the size of the birth cluster has a strong influence on the population of the current Oort Cloud. If this cluster was very small, a substantial part of the primordial Oort Cloud comets may survive. But if it was of intermediate or large size, it is more likely that the primordial Oort Cloud comets survive only in the inner core, while most of the Oort Cloud may be of later origin in the case of a LHB-related Nice Model instability.

After the Sun left its birth cluster, whenever this occurred, there clearly had to be a loss of comets from the Oort Cloud for several reasons. One reason is the ongoing variations of angular momentum of the comet orbits caused by the Galactic tide and stellar encounters, which means that comets are sent into orbits with perihelia inside the planetary system. As we have seen, the planetary perturbations then ultimately eject the comets from the solar system. Another reason is the action of the closest stellar encounters experienced over time. The effects of such erosion are seen in the slow decline of the mass of the Oort Cloud after a maximum reached after 500–1000 Myr in Fig. 6.8.

The first effect is much more predictable than the second. The Galactic tide is essentially integrable, and the typical stellar encounters occur with a frequency that does not vary much as long as the solar orbit in the Galaxy does not exhibit large variations. However, the closeness and other circumstances of the few most efficient stellar encounters is largely a matter of chance. Just like in the case of the birth cluster, we can expect the loss of comets in this case to be due to the gravitational kick received by the Sun from a star passing at close range. Fouchard *et al.* (2014b) found that an initial Oort Cloud, when integrated in a full dynamical model for the age of the solar system, would face a cumulative loss of 24–41% of its members in four cases with different random sequences of passing stars. Only 2% were lost by tidal injection into the planetary system and the large variation of the other losses was interpreted to result from the random nature of the closest stellar encounters, showing that these are of great importance.

Of course, the effect of an extremely close stellar encounter cannot be limited to the ejection of Oort Cloud comets by the impulsive change of the heliocentric orbital energy. Many other comets will stay in the Oort Cloud though with strongly perturbed energies. Thus, they may migrate between the inner core and outer halo of the cloud. In fact, the mobility of Oort Cloud comets in terms of orbital energy found by Fouchard *et al.* (2014b) is a remarkable result, but in addition to the revolutionary changes induced by the kind of stellar encounter that is so rare that it may never have happened in reality, there is also the effect of planetary scattering, which is always bound to occur.

The initial conditions for a long-term Oort Cloud simulation are very important for the resulting dynamical evolution. In a recent study, Fouchard *et al.* (2017) discontinued the earlier use of a cloud that had already been made isotropic by stellar perturbations and considered instead a precursor scattered disk population. Thus, these simulations dealt with the entire history of the Oort Cloud from formation till the present time. The stellar influence was modeled using typical Galactic field stars, which makes the simulations particularly relevant for the LHB-related formation scenario. An important result is that the inner core of the Oort Cloud, containing most of its members, retains a memory of this anisotropic structure — in one sense, this is similar to the findings of Duncan *et al.* (1987) as illustrated in Fig. 6.6.

However, before the randomization of the orbits by stellar encounters has time to occur, the comets keep a memory of the fact that they started out with perihelia in the planetary region — the integrable Galactic tide cannot wipe out this memory. This means that, in phase space, the orbits remain close to the TAZ (see Sec. 5.2.1). They are hence vulnerable to planetary ejection, and hence the loss rate of such an Oort Cloud is higher than was found in the earlier models. The cumulative value is now $\simeq$70–80%. In addition, the new comets arriving today from the outer Oort Cloud were shown to sample the initial population rather uniformly in terms of semi-major axis. If the initial cloud was strongly concentrated, we mostly see comets that started out in the inner core.

The current Oort Cloud may thus only contain about 25% of its initial members, corresponding to the same fraction of the initial mass. In terms of numbers, Fouchard *et al.* (2017) carried out what was recommended in Sec. 5.2.2, namely, to derive the population size of the cloud from their simulations by using an estimate of the rate of arrival of new comets from observations — in this case, based on a paper by Francis (2005). The result is that the initial Oort Cloud should have contained roughly 1×10^{12} comets, and the current cloud thus counts $\sim(2\text{–}3)\times10^{11}$ comets. In terms of mass, it is unfortunately impossible to draw any precise conclusion, since the average mass of a new Oort Cloud comet is poorly known.

Finally, let us note that there is a threat to the long-term stability of the Oort Cloud, which was not included into any of the mentioned investigations. This comes from the rare passages of Giant Molecular Clouds in the vicinity of the solar system. As seen in Fig. 3.8 (Sec. 3.4), over several Gyr, one would statistically expect a closest encounter at a distance of $\sim$ 10 pc. For a very crude estimate, we may take the GMC to be a point mass with $M_* = 2\times10^5$ solar masses. With an encounter distance of 10 pc, we use the tidal approximation to the heliocentric impulse of a comet situated at $r_c = 20\,000$ AU as given by Eq. (3.25) in Sec. 3.4. The result is that such comets would receive impulses similar to the case of a one solar mass star passing at about 1 000 AU from the Sun with the same velocity.

This means quite a drastic effect, and the outer part of the Oort Cloud could be mostly stripped away — only to be replenished from the inner core. Clearly, GMC encounters should be included into a full model of the long-term evolution of the Oort Cloud. They should also be modeled realistically without the above point-mass approximation, but this poses problems that have so far prevented such encounters from being routinely included into Oort Cloud modeling. When the problem became realized, following a paper by Ludwig Biermann (1978), it was frequently noticed in the literature. Finally, it was partly alleviated, as Hut and Tremaine (1985) studied the problem in depth and concluded that the structure of GMCs was too poorly known to say for sure, if the problem was severe or not. Nowadays, much more is known about both structure and evolution of GMCs, and realistic simulations may be forthcoming.

6.4. The Number of Comets

6.4.1. *The current number*

The time has now come to summarize what we know about the number of comets in the solar system and how the estimates are made. In Chap. 5 we made a crude estimate of the number of comets in the Oort Cloud and sketched how one goes about making a better estimate based on dynamical modeling of the cloud. In Sec. 6.3.2, we have seen the result of such an estimate. One of the ingredients of this analysis is the arrival rate of new comets, which is derived from observations. This is in fact a difficult task, since severe biases are involved in comet discoveries, and the resulting rate has to refer to a certain limit in the size or brightness of the comets.

The mentioned estimate by Francis (2005) is in good agreement with other results and appears rather reliable because of the use of an observational material that is homogeneous and well characterized in terms of discovery bias. In this case, the rate is four new comets per year with perihelion distances $q < 5\,\text{AU}$ and total absolute magnitudes $H < 11$. The definition of H was given in Sec. 2.2.2. It approximates the heliocentric magnitude that the comet would have if placed at $r = 1\,\text{AU}$. According to Eq. (2.8), it also yields a rough estimate of the H_2O production rate of the comet at this standard heliocentric distance. As mentioned, according to Fouchard *et al.* (2017), the number of Oort Cloud comets with this outgassing capacity should currently be about $\sim (2\text{–}3) \times 10^{11}$.

The number of Scattered Disk comets is estimated differently. This estimate is based on a necessary capture rate into the Jupiter Family to keep this in steady state. We discussed the issue in Sec. 5.6 and referred to an estimate of about 6×10^8 SDOs by Duncan and Levison (1997). Rickman *et al.* (2017a) adjusted this number upward to about 1×10^9. It refers to objects that would develop into Jupiter Family comets with an outgassing capacity expressed by $H \lesssim 9$. From a statistical correlation between H and the absolute nuclear magnitude H_N for Jupiter Family comets, as found by Fernández *et al.* (1999), it appears that the relevant size limit is given by $H_N < 17$. Thus, the nuclear radius is at least about 1.3 km.

Returning to the estimate for Oort Cloud comets, we should find the value of the nuclear radius that corresponds to $H = 11$. This cannot be done using observed nuclear magnitudes, but Sosa and Fernández (2011) used nongravitational parameters (A_2) for a number of long-period comets to determine their masses. Assuming a density of $400\,\text{kg}\,\text{m}^{-3}$, they derived nuclear diameters and correlated these with the corresponding H values. Adopting their result, we find an average nuclear radius of 0.3 km for $H = 11$.

If we assume a common size distribution in the Scattered Disk and the Oort Cloud such that the cumulative number above radius R_N goes as R_N^{-2} (see Sec. 2.1.3), we find that the number of SDOs with $R_N > 0.3$ km is about 2×10^{10}. Hence, the population size of the Oort Cloud is about 10–15 times larger than that of the Scattered Disk. This is in perfect agreement with the simulation results for formation and evolution of the Oort Cloud and Scattered Disk by Brasser and Morbidelli (2013).

6.4.2. *The initial number*

Given the good fit between the observed OC/SD population ratio and the prediction of the Nice Model, we may also check if the absolute numbers are in agreement with the Nice Model scenario. As mentioned in Sec. 6.1, this model admits substantial freedom concerning several parameters, and one of these is the total mass of the primordial planetesimal disk. However, some observational constraints exist in this case too. Rickman *et al.* (2017b) recently simulated the LHB cratering of the Moon and terrestrial planets due to the cometary component that arose from the dispersal of the planetesimal disk. They considered two variants of the disk with total masses of 18 and 36 Earth masses, each coupled to an estimate of the fraction of the objects that would enter into the inner solar system.

A probability distribution was found for the size of the largest lunar impactor in each variant. Given that the probability to hit the Moon could be estimated as $\simeq 1/30$ of the probability of capture as a Jupiter Trojan, a comparison with the radius of the 30th largest Trojan could be made, and only the low mass disk model gave a good result. With the size distribution adopted by Rickman *et al.*, the number of disk objects with radii $R > 1$ km is about 2×10^{11} in this case. This is a little more than 100 times the current Scattered Disk population, which agrees very well with the expected 100-fold attenuation of the Scattered Disk from the time of its formation (see Fig. 6.8).

6.5. Collisional Evolution of Comets

Let us now discuss a threat to the physical existence of comet nuclei, which is independent of cometary activity and related phenomena. The issue is, if there are or have been regions of space, where the number density of comet nuclei is so high and the relative velocities so large that the nuclei suffer collisional destruction. Obviously, this cannot be the case for the Oort Cloud thanks to its enormous volume, which could host many more comets without any risk. The same can be said about the inner solar system, where Jupiter Family comets thrive along with Halley Types and occasional long-period comets. The number density of comets and meteoroids in this region is far too low for collisional destruction to be an issue.

6.5.1. *The current TNOs*

The trans-neptunian comet reservoirs, i.e., the Edgeworth–Kuiper Belt and the Scattered Disk, present a more interesting case. Judging from the observations, these two reservoirs have roughly the same population size. Since the SDOs occupy a far larger volume, their collisional frequency is far lower, compared with the EKB objects. Interestingly, while the SDOs offer a direct source for captures into the Jupiter Family, the delivery from the EKB requires the injection of objects into "escape routes", where resonances bring them into planet-crossing orbits. This injection is thought to occur by collisions in the EKB.

Before the existence of the Scattered Disk was realized, the EKB appeared as the necessary source for a large majority of the Jupiter Family comets (see Sec. 5.5). Alan Stern (1995) noted that a massive, initial EKB would be collisionally evolved. Don Davis and Paolo Farinella (1997) developed a model for the EKB, which had an initial mass of three Earth masses at the time of formation of the solar system. They computed the collision rate within this population and found that initial objects with diameters less than about 50 km would be heavily depleted at the current time due to collisional destruction. Currently, the comet-sized objects would be predominantly collisional fragments. The number of such objects was estimated to be about 5×10^9 and the rate of collisions was estimated to account for an injection rate into the escape routes large enough to explain the capture rate into the Jupiter Family.

However, the situation changed with the discovery of the Scattered Disk, as explained in Sec. 5.5. Captures from this source were shown by Duncan and Levison (1997) to be ten times more efficient in providing Jupiter Family comets. Hence, the number of comet-sized EKB objects should be less than 1/10 of the estimate by Davis and Farinella, and the previous collisional evolution of the EKB must be less severe. In fact, the recent estimate by Fraser *et al.* (2014) results in a total mass of only 0.01 Earth masses. Even comet-sized objects may then be partly primordial survivors.

For the current SDOs the collisional evolution is insignificant. However, the current population is but a bleak shadow of what it was, just after the disk was formed, as seen in Sec. 6.4.2. Moreover, before it was formed the risk of collisions must have been much larger still.

6.5.2. *The primordial disk*

The formation of the distant comet reservoirs involves planetesimal scattering according to all the current concepts. Obviously, vast numbers of km-sized planetesimals must have been involved. Before the scattering took place, these must have coexisted within a volume much smaller than they would later occupy. Thus, the issue of collisional evolution arises. If the scattering happened very early, for instance during the accretion of the giant planets, the amount of collisional evolution might be very small. However, the objects being scattered would collide with smaller debris to the point of being destroyed (Stern and Weissman 2001). Even the surviving nuclei would likely bear the scars of such collisional erosion.

Charnoz and Morbidelli (2007; cited by Morbidelli and Rickman 2015) modeled collisions within the population of objects being scattered into the Oort Cloud or Scattered Disk during the planetary formation epoch. They found that the Oort Cloud comets would be relatively little affected, while those staying around in the disk for longer times would be shattered to the extent that almost all the small nuclei would finally be collisional fragments. Thus, it seems that the comets that formed in the giant planet region were heavily processed, and chances are slight to find such comets in an undamaged state today.

In the Nice Model with a late instability, the situation is special because there would be a planetesimal disk with a total mass of more than 15 Earth masses, which existed for several hundred million years. This primordial disk is in fact an obvious candidate site for collisional evolution of the objects that later would become comet nuclei.

That collisional evolution may be expected is obvious already from the fact that this disk has a much smaller volume and larger mass than the EKB, for which Davis and Farinella (1997) found significant evolution to occur. Let us now present the same numerical estimate that was shown in a paper about comet 67P/Churyumov-Gerasimenko by Rickman *et al.* (2015). We consider a target planetesimal with radius R_c, orbiting within the primordial disk, and we estimate the number of disruptive collisions ($N_{\rm coll}$) that it will undergo during a time interval T. For this we use the so-called particle-in-a-box formula:

$$N_{\rm coll} = \frac{N_p u T A_p}{V}, \tag{6.1}$$

where N_p is the number of collision partners ("projectiles") in the disk of sufficient size to disrupt the target, V is the volume of the disk, u is the average relative velocity when two disk objects encounter each other, and $A_p = \pi(R_c + R_p)^2$ is the collisional cross section. Here, R_p is the radius of the projectile. Since we have a wide range of projectile sizes, we have to replace $N_p A_p$ in this formula by

$$N_p A_p = \int_{R_{\rm min}}^{R_{\rm max}} n(R_p) A_p(R_p)\, dR_p, \tag{6.2}$$

where $n(R_p)$ expresses the differential size distribution of projectiles and the integral is carried from the minimum radius of destructive projectiles (a function of R_c) to an arbitrary maximum, above which the number of collisions can be neglected.

For a rough estimate we take V to be the volume of a circular annulus with inner and outer radii of 15 and 30 AU and a constant thickness of 4 AU. In the standard version of the Nice Model (Sec. 6.1.3), the viscous stirring of the disk increases the orbital eccentricities and inclinations, implying an average encounter velocity of $\simeq 0.6\,\text{km/s}$ (Levison *et al.* 2011). For T we take the time from solar system formation until the LHB, which is about 400 Myr. We consider a target with $R_c = 2\,\text{km}$, not very different from the 67P nucleus. To break up such a target at a collision with the estimated velocity, the projectile should have a radius of at least $R_{\text{min}} = 0.5\,\text{km}$ as a conservative estimate.

Let us use a disk of relatively low mass (18 Earth masses) as modeled by Rickman *et al.* (2017b). In this case, the differential size distribution for small objects can be taken to be a power law with index -3 and the whole mass can be put into objects with $R > R_{\text{min}}$. We then get $N_p \simeq 8.3 \times 10^{11}$. Evaluating the above equations, we get $N_{\text{coll}} \simeq 6$. Clearly, a large majority of the targets with the chosen size would undergo at least one destructive collision, and the chance of survival is very small.

A detailed analysis of the collisional evolution in the primordial disk as well as during the dispersal stage, when the Scattered Disk was formed, was presented by Morbidelli and Rickman (2015). The method was based on the above-mentioned work of Charnoz and Morbidelli (2007). This means a Monte Carlo simulation, where the physical-dynamical evolution of a large number of test particles (representing comet nuclei) is traced step by step in time. Collisions are registered based on their probabilities, and the results are judged in terms of disruption or not using an assumed specific disruption energy coming from previous literature.

The result was that survival against collisional disruption is extremely unlikely for a target with $R_c = 2\,\text{km}$ (once more, roughly the size of the 67P nucleus). This is the case wherever in the disk the object is assumed to be placed, but the conditions are especially harsh in the inner parts. For the same assumed power law size distribution as above, essentially all the objects with $R < 50\,\text{km}$ would likely be disrupted. This concerns the assumed 400 Myr lifetime of the primordial disk. During the dispersal phase, this size distribution would imply an expectancy of more than one for the number of disruptive collisions facing a target of the same size, showing that the chances of survival are further decreased.

As a conclusion, if the 67P nucleus took part in a long-lasting primordial disk of the Nice Model scenario, it would hardly be possible to consider it as a collisional survivor, thus favoring an origin as a collisional fragment. The problem is then to understand, if this can be the case considering the measured physical and chemical properties of this nucleus. We shall return to this discussion in Sec. 7.4.4.

Chapter 7

Origin of Comet Nuclei

We have come to the ultimate question about comets: How were their nuclei formed? As we have seen in earlier chapters, all the current concepts about the reservoirs, from which we know that comets come, suppose that the nuclei existed very early in the history of the solar system. Hence, these must have been among the first objects to be formed, so the timing is clear: the period before any planets were formed or when the giants were very young. In any case, the solid material that assembled into comet nuclei was typical of the oldest solar system solids including some that have been preserved in meteorite parent bodies. We may thus be talking about solar nebula condensates or even presolar grains. In any case, the growth of comet nuclei started with the agglomeration of nebular grains, so let us consider how this growth may have proceeded.

7.1. Growth of Constituent Grains

7.1.1. *The solar nebula*

The solar nebula formed the environment, where grain growth took place. The concept of a solar nebula is old, dating back to the Kant–Laplace nebular theory of solar system origin in the 18th century. However, the version that is currently discussed has little to do with what scientists and philosophers of old times imagined. Above all, observations of galactic regions of ongoing star formation now play a central role, while previously the solar nebula was entirely a theoretical concept.

According to the current concept, the solar nebula started out as an *accretion disk*, through which material was channeled into the forming Sun. Its origin was the gravitational collapse of the presolar cloud. Due to the finite amount of angular momentum contained in this cloud, the collapse resulted in a rotating disk. Probably under the influence of a process known as *magneto-rotational instability*, torques arose in the disk transporting specific angular momentum outward and letting the deprived material spiral inward, finally to enter into the Sun.

Of course, the life span of the solar nebula should be comparable to the time it took to channel one solar mass into the center. For an order-of-magnitude estimate we may take the free fall time from the presolar cloud core that gave rise to the solar system. Typically, such cloud cores have radii $\sim$5 000 AU, which yields a free fall time of $\sim 10^5$ yr. If a major fraction of the Sun's mass accreted on that time scale, the rate would have exceeded 10^{-6} solar masses per year. Such high rates are not observed, as shown by Fig. 7.1, but protostars may not be observable at those very early stages. What we do see in this diagram is that stars continue to accrete material for millions of years but that the rates are very low after ten million years. The stars are called *T Tauri stars* and show clear evidence of surrounding disks at least when they are young, but for the older ones this evidence is gone, and we talk about "naked T Tauri stars".

Ten million years is likely an upper limit to the solar nebula life span, and several papers have argued for just a few million years. Let us now estimate the mass of the solar nebula. From Fig. 7.1 we can make an estimate. A disk that supports an accretion rate of 10^{-8} solar masses per year for an interval of $\sim$2 Myr must have a mass of at least $\sim$0.02 solar masses (Hartmann 2008). This value seems reasonable from the point of view of explaining the planetary masses of the solar system, considering that these were made of material from the solar nebula. Each planet has a bulk elemental composition that can be crudely estimated, and this represents a certain fraction of the solar nebula material from which it was built.

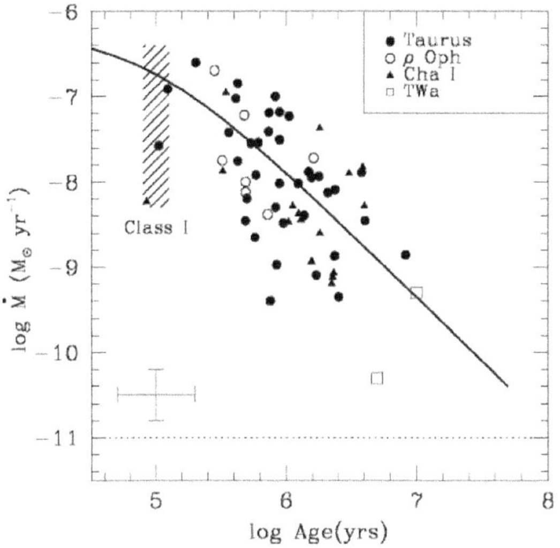

Fig. 7.1. Observationally determined mass accretion rates in T Tauri stars, plotted versus the estimated ages of the stars. Different symbols denote stars in different regions of stellar formation. The solid curve is a calculated model fit. Reproduced from Hartmann (2008) with permission from N. Calvet. © The Royal Swedish Academy of Sciences. Reproduced by permission of IOP Publishing. All rights reserved.

This fraction can be estimated by adding up the missing amounts of major, light elements that are needed to restore the solar elemental abundances. Thus, adding together the estimated nebular masses for all the planets, we get the mass of what is called the *minimum mass solar nebula* (MMSN). This mass turns out to be about 0.01 solar masses. However, it is only a minimum, since the planet-forming process certainly was not 100% efficient.

What was the radius of the disk? At first, one may be tempted to answer "at least, 30 AU", since this is the radius of Neptune's orbit. However, in Sec. 6.1 we saw that the ice giants, Uranus and Neptune, probably formed much closer to the Sun and may have migrated into their current orbits after a long time (the Nice Model). Within this framework, we have to ask instead for what reason Neptune's outward migration stopped at 30 AU. According to the Nice Model, this was because the planetesimal disk that made it migrate had an outer limit not much further than 30 AU from the Sun. Since there

Fig. 7.2. A region within M42 (the Orion nebula) imaged by the Hubble Space Telescope, showing several protoplanetary disks, shining by reflected starlight against a darker background. Credit: NASA/HST/WFPC 2/C.R. O'Dell (Rice University).

is no reason why planetesimals should not have formed throughout the solar nebula, the conclusion is that the solar nebula had a radius of about 30 AU.

Comparing with observations of current protoplanetary disks, such as the ones imaged in Fig. 7.2, one finds a reasonable agreement in many cases, while there are also many disks of much larger radii. The most popular explanation is that the Sun, like many observed young stars, was born in a harsh environment with a large flux of ultraviolet radiation coming from nearby, hot stars. This would cause ionization of the hydrogen gas, which pumps so much kinetic energy into the disk that its outer parts are lost into space. The process is called *photo-evaporation.* We note that, by invoking this to explain the relatively small size of the solar nebula, we also have an argument in favor of a rich birth cluster, which was discussed in Sec. 6.2.2.

Finally, we have to note that the gas of the solar nebula must have been turbulent, at least as long as the accretion went on. This involved a vigorous dissipation of the kinetic energy of bulk motion into heat, which caused a large temperature gradient between the mid-plane and the disk surface, from where the heat was radiated.

As a consequence, convection was set up. Turbulent eddies with a broad size spectrum arose. This highly dynamic state was bound to calm down with time, as the accretion rate decreased and the disk temperatures fell. But it was important for the growth of grains, as we shall see.

7.1.2. *Grain aggregate formation*

By the time the presolar cloud core collapsed and the solar nebula arose, the solid material existed predominantly in the form of submicron-sized grains with refractory cores and icy mantles (see Sec. 2.6.4). In the inner, hotter parts of the solar nebula the ices were lost by evaporation, while outside a certain distance from the center, called the *snow line* (see below), they might be retained.

The initial grains were part of the solar nebula and thus embedded in a gas that was not extremely rarefied. This means that the gas behaved like a continuous medium and moved according to the laws of hydrodynamics (or magneto-hydrodynamics) with a well-defined velocity field. However, grains much smaller than the mean free path of the gas molecules would interact with the individual impinging molecules rather than the continuous medium. With standard models of the solar nebula one can estimate the number density of hydrogen molecules at 5 AU from the center to be $n_g \sim 10^{14}\,\mathrm{cm}^{-3}$. Using a cross-section of $A_g \sim 10^{-16}\,\mathrm{cm}^2$ per molecule, we find the mean free path to be $\lambda \sim 1\,\mathrm{m}$. Clearly, grain aggregates up to pebble size meet the above condition. Their interactions with the gas are described within the *Epstein regime*.

In particular, we can derive an expression for the drag force on a grain moving through the gas at a speed V much lower than the sound velocity c_s. This force will be proportional to the gas density ρ_g, the specific momentum c_s transferred per collision, and the sweep up efficiency AV, where A is the cross-section of the grain. Hence, the rate of change of V is given by $-V/\tau_f$, where the *grain/gas coupling time scale* τ_f is

$$\tau_f = \frac{R\rho_{gr}}{c_s\rho_g}, \tag{7.1}$$

where R and ρ_{gr} are the radius and density of the grain, respectively. Another term often used for τ_f is *friction time*. Assuming this to be constant, V will decay exponentially with an e-folding time equal to τ_f.

For a crude estimate, let us use $\rho_{gr} \sim 10^3\,\mathrm{kg\,m^{-3}}$. For the above value of the gas number density at 5 AU from the center, the gas density is $\rho_g \sim 10^{-7}\,\mathrm{kg\,m^{-3}}$, and for hydrogen molecules at an estimated temperature of 100–200 K we get $c_s \sim 10^3\,\mathrm{m/s}$. Hence, letting R_{gr} vary from $1\,\mu\mathrm{m}$ to 1 m, we find that τ_f increases from about ten seconds to about one year. Grain/gas coupling occurs extremely fast for tiny particles and takes very little time even for meter-sized boulders.

Grain growth by accretion means that the grains need to come together and collide, so that they can stick. Starting from a monodisperse population of submicron grains, which are all perfectly coupled to the gas in which they reside, the only way for collisions to occur is via *thermal motions*. This means that these tiny grains act like giant gas particles sharing kinetic energy with the dominant hydrogen molecules. From the factor $\sim 10^{10}$ between the typical masses, we obtain a factor 10^{-5} in thermal velocity, which means that the grains have random motions at the level of $V_{gr} \sim 1\,\mathrm{cm/s}$.

If the collisional cross-section of such a grain is A_{gr} and the number density of grains is n_{gr}, the collision frequency ν_c is

$$\nu_c = n_{gr} A_{gr} V_{gr}. \tag{7.2}$$

We may estimate the cross-section to be $A_{gr} \sim 10^{-13}\,\mathrm{m^2}$. From the above gas density, using the fact that the mass fraction of solids in cosmic material of solar composition is about 1%, we estimate a mass density of $\rho_{solid} \sim 10^{-9}\,\mathrm{kg\,m^{-3}}$. If this is represented by grains of mass $m_{gr} \sim 10^{-17}\,\mathrm{kg}$, the number density is $n_{gr} \sim 10^8\,\mathrm{m^{-3}}$. These estimates yield $\nu_c \sim 10^{-7}\,\mathrm{s^{-1}}$, i.e., on the average a few collisions per year for a given, individual grain.

Thus, in a few years, the initial submicron grains will have collided several times at very low speeds. Since these collisions normally imply sticking, initial aggregates up to several microns in size can be expected to form. However, the number density of these aggregates is much lower than that of the initial grains and their motions are slower, so the collision frequency drops with time. The thermal growth will therefore come to a halt, preventing the aggregates from growing to more than 100 μm in radius.

If the gas flow is fully laminar, so that the local velocity changes only slowly with position, the aggregates will be locked to this flow and, essentially, will not move at all relative to each other. Therefore, further growth requires the gas flow to be turbulent. In this situation, the grains will experience a rapidly fluctuating flow velocity. The smallest ones are best suited to adapt fully to these changes, while the larger ones decouple from the smallest and fastest eddies because of too long coupling times. Hence, grains of different sizes are set in relative motion with velocities characteristic of the gas turbulence. As they collide and stick, the larger ones grow further, while the smaller ones get consumed. This is *turbulent growth.*

By taking the product of τ_f and the Keplerian frequency Ω (i.e., the frequency of orbital motion), we get the so-called *Stokes number* $\mathrm{St} = \Omega\tau_f$. This expresses how efficiently the grain couples to the gas motion, when the latter changes during the orbital motion of the grain. It is an important parameter characterizing the evolution of the grains including the turbulent growth.

In its circular motion around the Sun, the gas of the solar nebula must be subject to radial forces that cancel out. These are: the central gravity due to the Sun, the outward force due to the pressure gradient of the gas, and the centrifugal force due to the orbital motion. If there was no pressure gradient, the motion would be Keplerian with a velocity given by Newton's laws. This is what big chunks of solid material would approach, if they were fully decoupled from the gas. However, the gas has a sub-Keplerian velocity due to the influence of its pressure gradient, and so the decoupled grains face a head wind that exposes them to gas drag. In turn, this drag causes an inward drift motion with a velocity

$$v_r = -\frac{2\Delta v}{\mathrm{St} + \mathrm{St}^{-1}}, \tag{7.3}$$

where Δv is the difference between the Keplerian velocity and the velocity of the gas. A good estimate for this difference is 50 m/s. It is easy to see that the *radial drift* is fastest for particles with a Stokes number of unity, for which the drift speed equals Δv.

Particles with different radii have different Stokes numbers according to Eq. (7.1) and thus different drift speeds according to Eq. (7.3). Thus, the radial drift provides a growth mechanism, as described in Sec. 7.2. From the above estimates of τ_f it is clear that during the stages of turbulent growth, when the aggregates are much less than a meter in size, the drift speed also grows with the consequence that larger aggregates overtake smaller ones and may consume them at collisions. However, another aspect of the drift is the risk of destruction by approaching the Sun.

7.1.3. *The limits of growth*

The larger the aggregates grow, the larger the collisional velocities they experience. In the range from several microns to meters, this is a direct consequence of the growth mechanisms in action. In turbulent growth the eddy velocities determine the collisional speeds — thus, a very large grain can be hit by a very small grain at a very high speed. The radial drift obviously works in a similar way. An illustration of this is found in Fig. 7.3.

The quantity plotted by means of gray shading levels and iso-contours is the speed (in meters per second) at which two particles collide at $r = 1\,\text{AU}$ near the equator of the solar nebula, according to model calculations including Brownian (thermal) motion, radial and azimuthal drift, and turbulence (Johansen *et al.* 2014). The radii of the particles are given with a logarithmic scale on the axes. The model shown is characterized by a vigorous turbulence, as seen from the relatively high α value. Although not of immediate interest here, the picture also shows what happens at very large sizes like hundreds of meters or several kilometers. Normally, such large objects would collide at very low speeds, because these are totally decoupled from the gas and are even insensitive to the drag of a $50\,\text{m/s}$ head wind. However, there are density fluctuations in the turbulent gas, which cause a fluctuating gravitational field that perturbs the orbits of km-sized objects. For this reason, the collisional speeds do not drop down to zero.

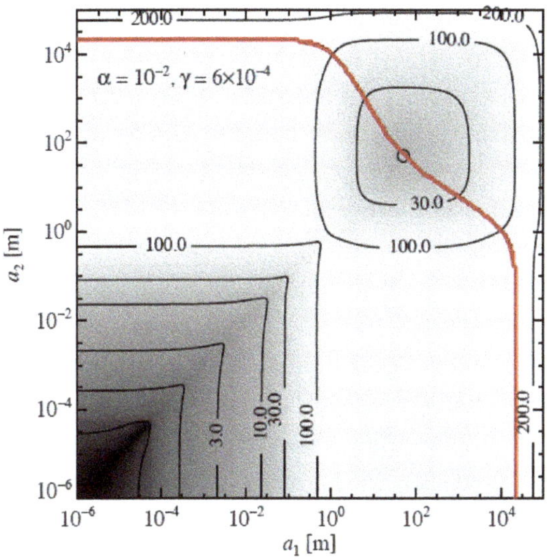

Fig. 7.3. Relative velocity of colliding particles in the midplane of the solar nebula at 1 AU from the Sun, according to a model calculation, as a function of the radii of the particles. The velocity is given by gray shading with contours indicating selected values, given in m/s. The red curve indicates the transition between velocities caused by gas drag and nebular density fluctuations. Extracted from Fig. 4 in A. Johansen *et al.*, "The Multifaceted Planetesimal Formation Process," in *Protostars and Planets VI*, edited by H. Beuther *et al.* © 2004 The Arizona Board of Regents. Reprinted by permission of University of Arizona Press.

The growth from tiny grains into larger aggregates is inexorable. If there is nothing to stop it, the aggregates will reach the sizes where the Stokes number is of order unity, and then they meet a threat of destruction by approaching the Sun, as mentioned above. If this motion is too rapid, it would be very difficult to imagine the grain growth to lead into the formation of macroscopic objects. This phenomenon is often referred to as the *meter-size barrier*, since $\mathrm{St} = 1$ in the region where terrestrial planets and asteroids are found corresponds to particle radii $R \sim 1\,\mathrm{m}$ for typical assumptions about the physical conditions.

When grains of similar or different sizes collide, numerous outcomes are possible. To begin with, only sticking, cratering and fragmentation were considered in evolution models, but experimental work has revealed a more diverse picture. Sticking is the rule at low speeds and small sizes due to molecular, van der Waals forces. Thus, accretional growth is the predestined outcome in the beginning of the evolution. As this growth proceeds, the first opposing phenomenon is bouncing, which occurs when the energy of the impact is too large to be dissipated by restructuring within the particles. Hence, the collisions tend to become elastic to some extent.

A *bouncing barrier* occurs at a typical size of about 1 mm, where the Stokes number is still much less than unity. However, it is obviously dependent on the properties of the grains. While grains of sand would bounce upon collision at a given speed, small snow flakes might not. Ice is more sticky than rock, and the porosity also plays a role. It is therefore important to clarify, which part of the solar nebula we consider. So far, most of the work has focused on the innermost part, where meteorite parent bodies and planetesimals of terrestrial planets grew, but concerning comets we are interested in regions beyond the snow line.

The next limit to growth is the *fragmentation barrier* at sizes larger than those of the bouncing barrier but small enough for the Stokes number to remain small compared to unity. This concerns collisions between particles of similar size, while an individual particle might still grow by sweeping up much smaller projectiles. For high velocities, this depends on the competition between erosion and accretion. Again, the behavior of icy aggregates is not as well characterized as that of silicate grains.

To summarize, the early growth stages of cometary material may have proceeded to the stage, when objects with Stokes numbers of ~ 0.01 or more were formed. If these would grow further until St ~ 1, the above-mentioned meter-size barrier would be encountered, although the relevant size of St ~ 1 particles was certainly larger in the more distant region where comets were to form. Due to the wide range of heliocentric distances involved and uncertainties over the typical densities of the objects in question, it is difficult to specify accurately the typical size of the most vulnerable objects. But unless some other mechanism allows these to come together into larger sizes before they drift into destruction, the growth process may be halted before objects like comet nuclei are formed.

7.2. Nucleus Formation

The first growth stages of the Sun were characterized by a very high accretion rate, a very hot inner part of the disk, and vigorous turbulence with vertical convection transporting heat from the midplane to the surface. But during this phase it was impossible for macroscopic objects to form out of the grain aggregates. The gas was actively accreting onto the forming Sun, limiting the time available for growth, and survival of grain aggregates against collisional destruction was problematic (Weidenschilling 1984). Within a million years, almost all of the solar mass had been accreted and the solar nebula had decreased in mass to the value appropriate for planetary formation (for instance, the above-described MMSN). In this situation it became possible for those large aggregates that had decoupled from the turbulent gas motions to settle toward the nebular midplane due to the vertical component of the solar gravity. This sedimentation occurred very fast and triggered a fundamental change of the prospects for the particles involved.

One should, however, remember that it is not yet possible for astrophysical theory to predict at which stage the formation of planetesimals actually took place. There is evidence from achondritic meteorites that sizeable parent bodies were formed very early, while the undifferentiated chondrites give evidence for what may possibly be a second generation of planetesimals, formed millions of years later.

If we stick to the accepted mechanism of planetesimal formation (keeping in mind that reality may have been different), we have to start from the above-described picture of sedimentation to the midplane of the solar nebula. The aggregates were now concentrated to a midplane layer that was very thin compared to the disk they came from. The finite thickness of this layer was set by the balance between sedimentation and vertical, turbulent diffusion. The latter was due to a shear between the dusty midplane layer and the surrounding gas on either side, and this shear was caused by the fact that the particles pushed the gas near the midplane closer to Keplerian rotation. Hence, there was a natural limit to the amount of dust enrichment. But the mass density of solids in the midplane layer was in fact very much enhanced.

Of course, the inward drift problem would still occur for Stokes numbers of order unity. However, the close packing of the particles and the high mass density offer ways around this problem, which may trigger the formation of full-scale planetesimals. By *planetesimals* we mean objects large enough to cause significant gravitational attraction on other planetesimals or smaller particles. Of course, whether their gravity is significant or not depends on the velocity dispersion of the population in which they exist. In any case, the appropriate radius is at least $\sim 1\,\mathrm{km}$. The formation of planetesimals beyond the snow line is a crucial step toward the formation of comet nuclei. In fact, these nuclei are either icy planetesimals or fragments thereof, as we shall see below.

7.2.1. *Gravitational collapse*

In a rotating disk such as the dusty midplane layer of the solar nebula, gravitational instability may occur for large enough density. This occurs when the particles in a local region experience a stronger gravitational potential from the neighboring field of particles than the kinetic energy available to counteract this attraction. The latter can be expressed in terms of "thermal" energy corresponding to the random motions of the particles, plus the shearing motion of the disk. If we consider the particles to be fully adapted to the quiescent gas flow, we can neglect the thermal component, and we get a condition for gravitational instability:

$$4\pi G\rho_R = 9\Omega^2, \qquad (7.4)$$

where G is the gravitational constant, ρ_R is the so-called *Roche density* marking the limit to gravitational instability, and Ω is the Keplerian frequency of the local environment in the disk. The shear of the disk is expressible in terms of a tidal force from the central mass, so that the instability criterion can also be formulated as follows. If the local density overcomes the Roche density, the solid particles are free to contract under their own gravity in spite of the tidal force, which tends to pull them apart.

It was suggested long ago (Goldreich and Ward 1973) that the sedimentation of grains toward the nebular midplane could thus trigger a gravitational instability, which would quickly lead to the formation of objects large enough to be called planetesimals. However, it was later realized that vertical, turbulent diffusion (see above) would prevent the densities to reach the necessary level for such instability. Thus, for some time the favored picture of planetesimal formation was the one of hierarchical accretion, to be described below.

The Goldreich–Ward planetesimals were formed by a global instability, but interest has recently focused on local instabilities caused by the concentration of particles in local regions of a turbulent gas disk. Such trapping can occur due to a few different mechanisms, which are schematically illustrated in Fig. 7.4 from a review by Johansen *et al.* (2014).

For the smallest eddy sizes of a kilometer or so, where the gas performs a cyclonic rotation around a local pressure minimum unaffected by the Coriolis force of disk rotation, small particles inside

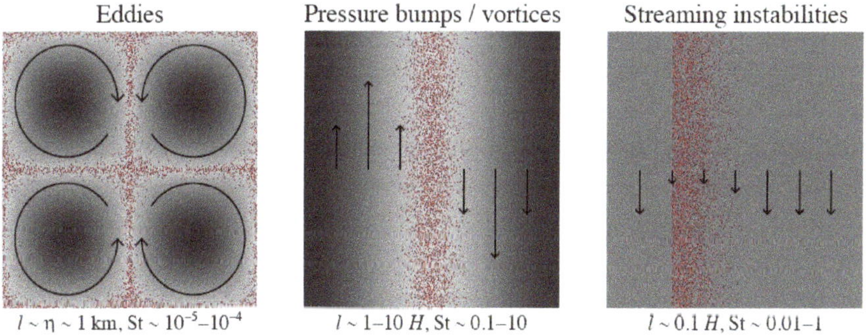

Fig. 7.4. Dust trapping mechanisms in a turbulent gas disk. The left panel shows small-scale eddies rotating around local pressure minima. The middle panel illustrates pressure maxima ("bumps") on a global scale. The right panel refers to the streaming instability, where particles play an active role in the gas dynamics. The length scales are given in terms of the disk scale height H, and the Stokes numbers refer to the orbital frequency of the whole disk. From A. Johansen *et al.*, "The Multifaceted Planetesimal Formation Process," in *Protostars and Planets VI*, edited by H. Beuther *et al.* © 2004 The Arizona Board of Regents. Reprinted by permission of University of Arizona Press.

the eddy will drift outward as long as they are big enough to have a friction time that is significant relative to the eddy turn-over time scale. This drift is most efficient for the particle size range, where the two time scales coincide, leading to a trapping of such particles into the high-pressure regions between the eddies. Of course, depending on the eddy size, different particle sizes are affected. However, this particular mechanism is not of primary concern here, because the relevant grains are very small. Since the time scales are very short compared to the Keplerian time scale of disk rotation, the Stokes numbers are much smaller than unity, and the radial drift is not a problem. The trapping between the eddies will instead speed up the accretional growth toward larger aggregates, which are still much smaller than planetesimals

Planetesimals may rather be formed under the influence of the other two phenomena illustrated in Fig. 7.4. In the case of pressure bumps (central panel), the particles pile up due to their response to the gas motion without actively participating in this dynamics, similar to the case just discussed for the small-scale eddies. On the other hand, if the streaming instability is at work, the particles play an active role in the gas dynamics thanks to mass-loading (see below).

The gas of the solar nebula was moving along circular streamlines around the gravitating center. The speed was everywhere close to Keplerian, thus decreasing with radial distance. The result was a *Keplerian shear.* Moreover, in a frame rotating with the local angular velocity of Keplerian rotation, any gas flow was deflected by the Coriolis force. This is in particular true for the velocity deficit of the sub-Keplerian gas in a region with a negative pressure gradient $\partial P/\partial r$ or the velocity excess of the super-Keplerian gas, where the pressure gradient is positive. The equilibrium structure in a rotating and shearing disk such as the solar nebula is a radial pressure bump, symmetric around the center. This is bounded by an inner, super-Keplerian flow, where particles are pushed outward by the gas and an outer, sub-Keplerian flow, where particles are dragged inward by gas drag. Hence, a pile-up occurs at the pressure bump, most notably for particles with Stokes numbers close to unity.

This mechanism was suggested by Fred Whipple (1972b) as an obstacle to the inward drift that threatens all particles with such Stokes numbers. It has been explored in many recent papers and may be counted as a strong contender for ways to trap solid particles locally in the solar nebula and thereby facilitate further growth toward planetesimals. There is also a variant on the same theme, where high-pressure, anti-cyclonic vortices develop in the disk by baroclinic instability (Klahr and Bodenheimer 2003) or strong local gradients of the turbulent viscosity (Lyra *et al.* 2008). The rotating gas velocity field around such a vortex also traps solid particles by inward drift, thus causing a local pressure bump with an enhanced solid density. In this case, the favored particles have Stokes numbers somewhat smaller than in the axisymmetric case.

Finally, the *streaming instability* means that a fully interacting system of gas and solid particles forming a rotating disk develops an instability, first found by Youdin and Goodman (2005), that may lead to very large concentrations of the particles into specific regions. This is both theoretically understood and borne out by numerous numerical simulations. The instability happens in media, where the mass density of solids is not much smaller than that of the gas, which means a high degree of dust mass-loading. This needs to be achieved by independent collection mechanisms like midplane sedimentation or pressure bumps, but given the formation of such an environment, an enormous enhancement of the solid density may result from the streaming instability (e.g., Johansen and Youdin 2007; Bai and Stone 2010) in the form of overdense dust lanes such as illustrated in Fig. 7.5.

Thus, local solid densities large enough to trigger gravitational instability have been shown to be possible via the above mechanisms. While nothing forbids particles of any size to be present in such super-dense regions, there is still a preference for those with St $\sim$ 0.01–1, since these are the ones mostly responsible for the concentration. Considering a standard model for the solar nebula and assuming a density not much less than $1\,\mathrm{g\,cm^{-3}}$ for the particles, the typical radius of the preferential particles at a distance of 5–20 AU from the Sun is of the order of several cm — i.e., pebble-sized objects.

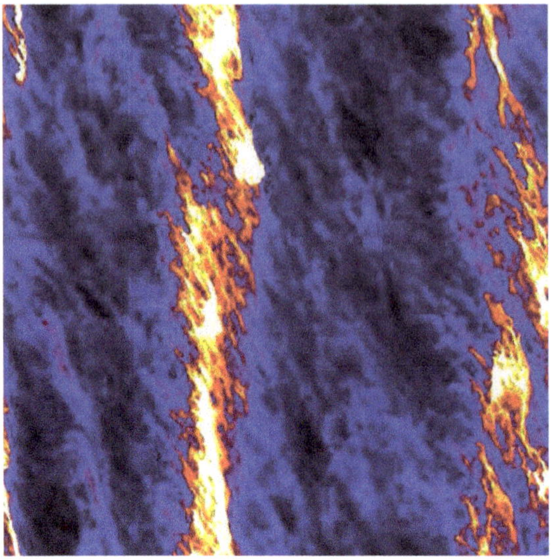

Fig. 7.5. Particle concentration in a local region of the solar nebula, illustrated by color coding. The view is from a perpendicular direction, and the outward radial and orbital directions are along the horizontal and vertical axes, respectively. Dust lanes resulting from the streaming instability are seen to extend along the direction of Keplerian rotation and to exhibit a fractal-like substructure. From A. Johansen *et al.*, "The Multifaceted Planetesimal Formation Process," in *Protostars and Planets VI*, edited by H. Beuther *et al.* © 2004 The Arizona Board of Regents. Reprinted by permission of University of Arizona Press.

Simulations of the outcome as gravitational instability sets in have shown formation of giant planetesimals on a very short time scale (Johansen *et al.* 2007) out of precursors that are often called *pebble clouds*. Typically, the step from centimeters to hundreds of kilometers is thus practically immediate (about ten orbital periods), showing that the embryos of terrestrial planets and giant planet cores could start to form out of sizeable building blocks at a very early stage. This is in line with the discovery from the dating of achondritic meteorites that large, differentiated parent bodies were present very early in the inner solar system. It also helps to understand the quick growth of the giant planet cores. Moreover, a paper by Morbidelli *et al.* (2009) with the provocative title *Asteroids were born big* argues that the main belt started out with primordial asteroids ~ 100 km or more in diameter and that the current small asteroids are essentially fragments formed much later during the collisional grinding of this initial population.

However, it is difficult to imagine the typical, km-sized comet nuclei to originate from such giant objects. In fact, as the mentioned simulations have been pursued with improved spatial resolution, it appears that the size spectrum of the primordial planetesimals is wide, possibly including objects no larger than typical comet nuclei (Wahlberg Jansson and Johansen 2014). The dynamical evolution of a large cluster of planetesimals of different sizes formed by the collapse of the same pebble swarm in the solar nebula is an interesting problem that has been addressed to some extent (e.g., Nesvorný *et al.* 2010) but still needs more attention.

7.2.2. *Collisional accretion*

A different scenario for comet formation was described by Stu Weidenschilling (1997). This starts with the thermal coagulation of micron-sized initial grains in a solar nebula, which is assumed to be inherently quiescent. There is hence no turbulent growth, and a time $\sim 10^4$ yr is needed for the production of aggregates large enough to experience sedimentation toward the midplane. Figure 7.6 shows the radial and transverse velocities of particles of different sizes with respect to the local, pressure-supported gas, as calculated for the solar nebula at $r = 30\,\text{AU}$. The solid curve gives the speed of the inward drift, and the dashed curve gives the speed of the headwind suffered by the particles.

Both the settling motion and the radial drift proceed faster for larger particles, so that the motion itself is a growth mechanism, where larger particles consume smaller ones by overtaking them. When the midplane is reached, the largest particles are cm-sized. The above-mentioned bouncing and fragmentation barriers had not been explored at the time of Weidenschilling's work and thus were not treated explicitly, although a simple recipe for collisional destruction was included.

As a result of the settling, the gas near the midplane is subject to substantial mass loading by the solid aggregates. Consequently, it gets dragged by those solids, so that its velocity deficit with respect

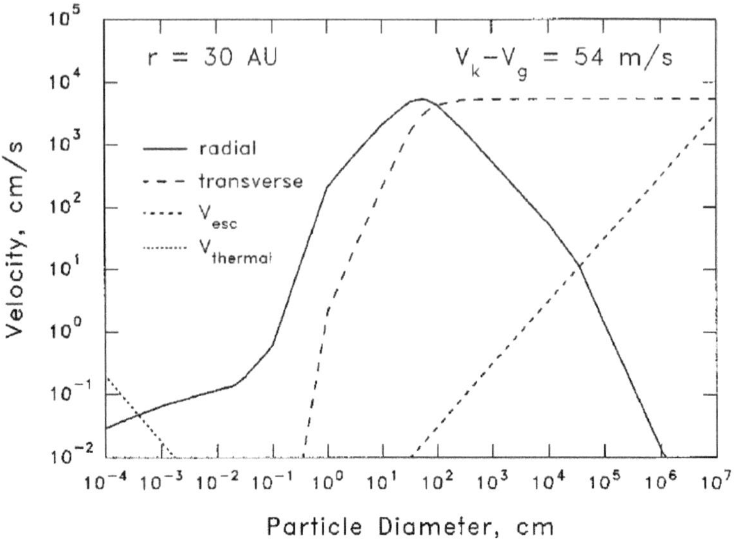

Fig. 7.6. Radial (solid curve) and transverse (dashed curve) velocities of grain aggregates relative to the ambient gas as functions of size, calculated for a heliocentric distance of $r = 30\,\mathrm{AU}$ in a quiescent solar nebula. The behavior of the radial drift speed agrees with the prediction of Eq. (7.3). The dotted curve gives the thermal velocity at a temperature of 50 K. The short-dashed curve gives the gravitational escape velocity from the surface of the particle. Reprinted from Weidenschilling, S. J., *Icarus* **127**, 290–306 (1997), with permission from Elsevier.

to Keplerian flow is decreased. Since the amount of mass loading is maximal in the midplane and tapers off in both directions, the effective gas velocity at any heliocentric distance varies with the distance from the midplane. This causes a shear in the laminar gas flow, which causes a Kelvin–Helmholtz instability, if the Reynolds number of the flow is high enough. As a result, turbulence arises, and this is assumed to have a Kolmogorov spectrum of eddy sizes with both upper and lower cutoffs.

The eddy velocities are assumed proportional to the difference between the local gas velocity and the pressure-supported value that would hold in the absence of particles. They hence decrease with distance from the midplane, but they always add a vertical component to the particle velocities. This implies a finite thickness of the particle layer, which in turn decreases with increasing particle size.

The largest objects thus grow closest to the midplane, and their effective radial drift speed remains much less than the maximum of 54 m/s indicated in Fig. 7.6, since the local gas moves at a speed closer to Keplerian due to the solid mass loading. This means that the threat of loss by thermal destruction owing to the inward motion is much reduced. As a rough estimate, the drift from $r = 30$ to 20 AU by meter-sized objects would take thousands of years. This is time enough for many objects to grow beyond the size of fastest drift, so they may slow down and continue to grow by sweeping up smaller colleagues overtaking them from the outside.

Although very high solid densities are reached within a very thin layer next to the midplane, so that the criterion for gravitational instability in terms of the Roche density in Eq. (7.4) is satisfied, the simulations by Weidenschilling did not show any collapse to occur. This is because the Roche density is only a lower limit to the true minimum density for collapse, since collapse is opposed not only by the Sun's tidal gravitational field but also by the random motions of the local particles. Here is a key point of divergence between Weidenschilling's model and those mentioned above, where the streaming instability leads to local collapse of the pebble clouds.

The growth of objects with $\mathrm{St} \gg 1$ in Weidenschilling's model involves a net mass gain by the larger objects through non-destructive sweep-up of smaller ones. As the radius increases beyond $R \sim 1$ km, gravitational focusing plays an increasing role in speeding up the growth. This is the time when planetesimals have been formed. It also means that comet nuclei may have been formed. However, the story does not end there. The time needed to produce km-sized objects at $r = 30$ AU is about 200 000 years (Weidenschilling 1997). This is comfortably short in order for giant planet cores to grow within the lifetime of the solar nebula. While the details of this growth are still unclear, we may imagine the gravitational stirring caused by very large objects to decrease the efficiency of collisional growth and to preserve a population of remnant, stray planetesimals with radii in the 1–100 km range.

7.2.3. *Towards a synthesis*

From a theoretical point of view, the earliest stages of comet formation are the best known. There is not much controversy about the following conventional wisdom. Primordial grains in the solar nebula underwent a coagulation process that produced larger and larger aggregates. These grains were micron- or submicron-sized and contained the full inventory of chemical compounds existing in comets, likely in the form of a refractory core with an icy mantle. The growth started out as a local phenomenon at all places in the disk, but the gradual decoupling of larger particles from the gas motion led to vertical settling to the midplane and radial drift. A thin, turbulent midplane layer strongly enriched in pebble-sized particles thus developed.

The composition of these pebbles, both elemental and isotopic, probably varied significantly with heliocentric distance, but only some general trends can be identified and many uncertainties remain. Two different avenues for further growth across the meter-size barrier into km-sized planetesimals have been proposed. However, these are theoretical concepts, and real comets formed in a way that is not necessarily married to either of them. One issue is whether comet nuclei are indeed primordial planetesimals or rather fragments produced when such planetesimals collide. Another issue is whether all "cometary" planetesimals formed strictly according to one of the mechanisms, or if both of them hold part of the truth.

We shall return to these issues in Secs. 7.3 and 7.4, but let us now summarize some predictions that appear to arise from the gravitational instability and collisional accretion scenarios. First, we consider the time of formation of the planetesimals. This has to be defined in relation to a reference event, and the usual reference is the formation of the calcium-aluminum-rich inclusions (CAIs) found in chondritic meteorites. The age of CAI formation is usually quoted as 4 567 Myr before present, and it marks the appearance of the first chunks of solid material in the solar nebula that we are able to trace. Since the minerals in question are extremely high-temperature condensates, CAI formation is naturally identified with the early phase of solar nebula evolution, when the accretion rate onto the

Sun was high and the aggregation of comet material had not yet started.

A certain time had to pass, before cometary planetesimals could appear independent of which mechanism we consider. This is the time needed to establish favorable conditions for the settling and rainout of solid particles, whereby the pebble-enriched midplane layer was formed. However, at this point the two mechanisms differ significantly. The streaming instability works promptly, while a few hundred thousand years are needed to grow full-size planetesimals by collisional accretion — Weidenschilling (1997) stated that this result of his simulations may be an underestimate due to the neglect of intrinsic turbulence in the gas disk.

Why should this matter? The CAIs provide evidence that the solar nebula in its early phases was "infected" with short-lived radioisotopes that likely came from stars of the solar birth cluster (see Sec. 6.2). In particular, the decay of ^{26}Al is an important heat source of chondritic material, until several half-lives have passed and the isotope is practically extinct. Small pebbles or boulders cannot store this heat, which instead escapes into space by thermal radiation, but a full-size planetesimal can. Hence, the question is if such planetesimals formed rapidly enough to experience strong heating by aluminum decay. In the case of the streaming instability, at first sight, this would almost certainly have been the case.[1] It is true that in cometary planetesimals the chondritic material was diluted by ice, but the evidence from Rosetta indicates a relatively low degree of dilution (see Sec. 2.2.3). In the collisional accretion scenario the chances of saving the cometary planetesimals from excessive heating are better, but no definitive answer can yet be given.

What does "excessive heating" mean? This primarily concerns the retention of very volatile molecules like CO and CO_2 in comet nuclei. The chances of retention are good, if the H_2O ice is amorphous and the volatiles are trapped in this ice, because the crystallization

[1]According to Anders Johansen (private communication), there may be a delayed, second generation of streaming instabilities after the solar nebula has undergone photoevaporation and the mass loading by solids has thus increased.

temperature $T_{cr} \simeq 130\,K$ is far higher than the equilibrium sublimation temperatures into vacuum.[2] We do not know, if cometary ice is really amorphous, but this can be assumed for a conservative estimate of the limits to radiogenic heating.

An exploration of those limits has been done by Dina Prialnik and co-workers. They found that crystallization can be avoided in comet nuclei with 4 km diameter, if these are formed at least $\simeq 1.5$ Myr after the CAIs (Prialnik *et al.* 1987). Some of the parameters that they used would need to be updated; for instance, a lower thermal inertia would translate into a longer delay of formation, while a smaller ice fraction in the material, or a smaller nuclear radius, has the opposite effect.

Next we consider the size distribution of planetesimals. As already mentioned, using improved spatial resolution in the streaming instability simulations, the formation of smaller and smaller planetesimals is found. Wahlberg Jansson and Johansen (2014) showed that objects with diameters of 1–10 km are probably formed along with larger ones from the same pebble swarm, although on a longer time scale. If this happens or not needs further investigation, since the collapse of the smaller objects has to occur in the presence of gravitational perturbations from the larger ones. Assuming that it does happen, a broad size distribution of cometary planetesimals is predicted, including the sizes of both typical observed comets and the largest TNOs.

As noted above, collisional accretion also leads to a wide size spectrum. However, in this case the smallest planetesimals would form first, while in the streaming instability case they would form last. In terms of morphological structure, one prediction appears to be common to the two mechanisms, namely, that the pebble-sized units that collectively started the final growth into planetesimals should be recognizeable within the comet nuclei. In the streaming instability case, these are the fundamental units out of which all planetesimals form by gravity-assisted coagulation. In

[2] Already at temperatures below 100 K amorphous ice would crystallize, if the same thermal conditions were maintained for billions of years.

Weidenschilling's picture, they collide with low velocities and may thus avoid being destroyed during the growth. Next, a minimum of the collision velocity occurs for objects with diameters of ~100 m, so that structures of this scale may also be preferentially preserved (Weidenschilling 1997).

A new model of comet nucleus formation, based on both the above mechanisms, was advanced by Davidsson *et al.* (2016). Within this scenario, the gravitational instability first forms objects with diameters from 50 to 400 km in the $r = 15$–30 AU region of the solar nebula. This process starts from pebble swarms near the gas disk midplane, where the density exceeds the Roche density by a factor 8 at $r = 30$ AU. If 7/8 of the solid mass is used for the collapse of the large objects, the rest of the pebbles remain in the absence of gravitational instability. Between 15 and 30 AU from the Sun, this means from 13% to 25% of the initial pebbles. These enter into a growth phase by collisional accretion, which leads to comet-sized planetesimals within ~ 2 Myr (somewhat longer in the outer part of the disk).

After the dispersal of the solar nebula a few Myr after CAI formation, the planetesimal disk continues to evolve dynamically and collisionally. During the time until the planetary instability of the Nice Model, which is usually assumed to be ≃400 Myr, the large objects that were formed by gravitational instability continue to evolve by collisional mergers, until toward the end of the interval several hundred very large objects — comparable to Pluto, Eris and Triton in size — have arisen. The comet-sized objects evolve through an initial phase of low-velocity mergers in a dynamically cold disk, followed by a destructive phase of mutual, high-velocity collisions after the disk has experienced some degree of viscous stirring (see Sec. 6.1.3). This is similar to the case studied by Morbidelli and Rickman (2015), which was described in Sec. 6.5.2, but the survival probability estimated by Davidsson *et al.* (2016) is higher.

7.3. The Pristine Nature of Comets

Comets in general, and comet 67P/Churyumov-Gerasimenko in particular, exhibit some properties that are consistent with the notion of comet nuclei being intact witnesses of the first stages in the origin of the planetary system. This means that, whatever happened to these nuclei after they were formed, they retain some characteristics that may be used to constrain theories about this origin. Of course, comet nuclei today do not necessarily look like they did, when they were newborn. Evolution has certainly occurred. But it is important to identify those characteristics that can indeed be used to probe the formation process. These are the ones revealing the pristine nature of comets.

7.3.1. *The structure*

First, we repeat the conclusion of Sec. 2.5.3. Nongravitational orbital perturbations of several short-period comets have been used to reach a general conclusion of bulk densities around $0.5\,\mathrm{g\,cm^{-3}}$. While the error bars on individual estimates are large, the sample seems internally consistent. Above all, there is a very good verification in the Rosetta result on comet 67P: $0.533 \pm 0.006\,\mathrm{g\,cm^{-3}}$ (Pätzold *et al.* 2016). According to these authors, there are good indications against large voids in the interior of the nucleus. The material thus has to be quite porous throughout.

Just how porous depends on the bulk chemical and mineralogical composition, and in particular the mass ratio of non-volatile to volatile substances or, in popular terms, the dust-to-ice ratio. The values reported by Marco Fulle and co-workers imply an average porosity in excess of 70%. The measurements by the CONSERT experiment, referring to some regions of the smaller lobe, have been interpreted by Ciarletti *et al.* (2015) to mean a $\simeq 15\%$ increase of porosity from the surface to 150 m depth, or alternatively, a decrease of the dust-to-ice mass ratio from ~ 4 to ~ 0.1. The porosity change seems more likely, but there might be a combination of both effects.

The high degree of porosity has implications for the collisional speeds involved in the accretion of the nucleus. According to Davidsson *et al.* (2016), impact speeds larger than 30 m/s would cause too much compaction to be compatible with the observations. On the other hand, such speeds are consistent with the formation scenario proposed in the same paper. They are also consistent with the very gentle accretion inside pebble swarms, as predicted by the streaming instability model.

In any case, only the models of comet formation in the solar nebula are able to explain the high porosity of comets. Such porosity could not have been acquired by later evolution. Moreover, if comet nuclei arose later as collisional fragments of such primordial planetesimals, this process has to keep the initial porosity more or less unaffected — something that would have to be convincingly demonstrated. Thus, the porosity is one of the critical properties showing comets to have a pristine nature.

One structural feature of comet nuclei had been suspected from visual imaging as well as radar mapping but came into focus in earnest with the Rosetta images of comet 67P. This is the bi-lobate shape with two parts of comparable size, apparently glued together by the material of the waist region. Not all nuclei are like this, but a significant fraction seem to be (Sec. 2.5.1). It is hence a feature of relevance, when we discuss the origin of the nuclei, although we cannot be sure that it is a pristine property. Assuming that it is, we may in principle have an indication about which formation process is more likely.

The criterion is that comet nuclei should be formed in such a way that km-sized objects are likely to come together at low speeds, forming composite objects with two or more parts of similar sizes. In the case of collisional accretion, we note that the same did not occur for the preceding generations of smaller objects, since each lobe in the 67P nucleus looks monolithic without any obvious binary structure. Davidsson *et al.* (2016) indeed found in their model that the cometesimals, which preceded the actual comet nuclei, were expected to be single-lobed, while pairs of such objects would later come together in low-velocity collisions, thus typically giving rise to bi-lobate nuclei. Jutzi and Asphaug (2015) made numerical simulations of such collisions, from which it seemed that the relevant, bi-lobate structures would likely arise for colliding objects of low density and strength. However, according to these authors, the simulated collisions might occur among the ejecta after a major collision as well as among primordial cometesimals.

For the scenario of pebble swarm collapse, it is necessary that similar-sized planetesimals within the same swarm may come together at least during the late stage, when the km-sized planetesimals are formed. Whether much larger, related objects like TNOs share the property of contact binarity is not known, but orbiting binaries with components of similar size are common in the cold part of the EKB (Sec. 5.4.1) though not in the hot part. From the modeling aspect, binary formation is very likely in pebble swarms among planetesimals formed with an excess of angular momentum (Nesvorný *et al.* 2010).

It thus seems possible to understand why comet nuclei may often form with bi-lobate shapes independent of which formation scenario we consider. On the other hand, the alternative scenario of major collisions, where objects like the 67P nucleus can form by mergers of similar-sized ejecta, also remains viable. This will be further discussed in Sec. 7.4.4.

7.3.2. *The molecular composition*

Basically, in this case the conclusion is rather evident. Ever since the Halley exploration, cometary material is known to be just as pristine as chondritic meteorites by imitating the elemental abundances of the Sun. It is impossible to imagine the formation of such material at any time later than, or any place different from, the solar nebula. But we may pursue the issue a bit further and ask, to what degree comets contain solid material that formed even earlier — in interstellar space or the presolar cloud.

We may divide the cometary material into three parts: the silicates and sulfides that characterize CI chondrites, the organic refractories, and the ices. As shown in Sec. 2.6.4, all these parts were combined in the individual, presolar grains, but those grains may have been thermally processed in the solar nebula. The most vulnerable part is the ices, where H_2O is a dominant molecule. The H_2O ice of presolar grains was probably amorphous and would have trapped the ambient gas molecules, possibly in large quantities (depending on the temperature). Thereby, the molecular composition of comets would naturally be similar to that observed in current star-forming regions. Current observational data indicate that this is indeed so.

However, in view of the heating expected to have occurred in the comet formation region, the survival of the amorphous ice appears problematic. The temperatures would have had to stay below $\sim$130 K to avoid crystallization. In the beginning, the temperature was relatively high, and retention of amorphous, gas-laden ice could only occur in the outskirts of the solar nebula — if at any place. Then the nebula cooled, and amorphous ice would be stable closer and closer to the Sun, but the amorphous-crystalline phase transition is irreversible, so the ice remained crystalline. During this stage, the gases that were once trapped in the ice had to coexist with H_2 in the solar nebula. This means a risk of chemical equilibration, whereby the abundant CO and CO_2 would be reduced to methane and water, and N_2 would be transformed into ammonia. Depending on how quickly the temperature fell, and how far from the Sun the comet formed, these reactions could have occurred to different degrees before the crystalline grains were able to adsorb ambient molecules and form clathrates.

Very important measurements in comet 67P were made by the ROSINA experiment, as mentioned in Sec. 2.6.1. The derived ratios of production rates for N_2/CO and Ar/N_2 were used by Mousis *et al.* (2016) to draw conclusions about the way in which these species coexist with H_2O ice in the nucleus. In Fig. 7.7 we can see that the measured ratios are consistent with trapping as guest molecules in H_2O clathrate at temperatures of about 45–50 K, while they are incompatible with either release from amorphous ice or sublimation of separate ices.

To explain this, Mousis *et al.* (2016) prefer a scenario, where the presolar H_2O ice crystallizes and then sublimates at $T > 150$ K, after which H_2O recondenses into crystalline ice and upon cooling to an equilibrium temperature of 45–50 K adsorbs the ambient gases. However, to make the fit better they speculate about an extra source of the gases in question, which could be either amorphous, gas-laden H_2O or separate supervolatile ices.

If this explanation is correct, most of the H_2O ice in comets like 67P is not presolar but a solar nebula condensation product. The

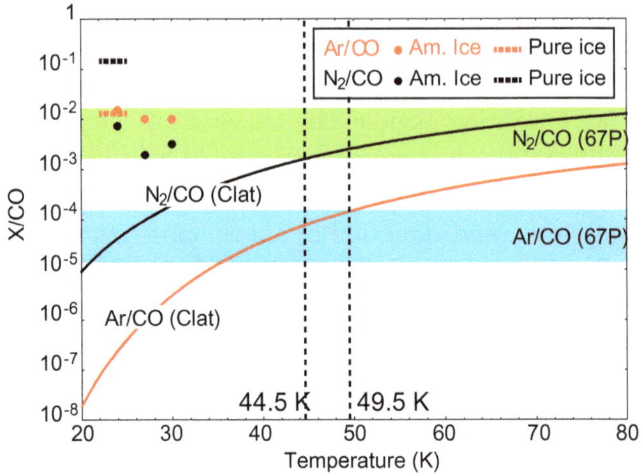

Fig. 7.7. Ranges of the N_2/CO (green) and Ar/N_2 (blue) ratios measured in comet 67P by the Rosetta/ROSINA instrument. The black and red curves show modeling results for these ratios in H_2O clathrates as functions of temperature. Dots show trapping efficiencies measured by Bar-Nun *et al.* (2007), and dashed bars show the protosolar ratios of condensed ices. From Mousis *et al.* (2016). © AAS. Reproduced with permission.

other molecules that are common to star-forming cloud cores do not come directly from the solid grains of the presolar cloud core but had to spend some time as gases in the solar nebula before being incorporated into the comet nuclei. We may say that comets are pristine though not as pristine as they might have been, and as many found it reasonable to expect.

Only the rock-forming minerals and organic substances appear to date back to the presolar cloud. However, there is solid evidence that the silicate minerals of comets represent a wide range of thermal conditions and that there is an abundant high-temperature component (Sec. 2.6.2). This has led to a widely held opinion that *radial mixing* was important in the solar nebula at an early stage, before the grain growth preceding the comets had come very far. Different processes have been discussed as responsible for this mixing, but the main implication for cometary origin is that the native refractories of comet material received an extraneous admixture of high-temperature minerals from much hotter parts of the solar nebula. This is interesting in itself but should not affect the history of the cometary volatiles.

Obviously, if all the supervolatiles were trapped in clathrates, they would not be released until the H_2O ice sublimates. It would then be hard to explain the activity in many comets way beyond the H_2O snow line. Indeed, the above-mentioned extra source is often necessary, but it may be more important in some comets than in others. The proposed scenario for ice evolution may explain how different comets can have different abundances of supervolatiles (e.g., CO) by incorporating nucleus material with different clathrate formation temperatures, different amounts of chemical processing of the gases or different amounts of pure ices. All these circumstances are likely sensitive to heliocentric distance.

An explanation of the observed fact that comets differ quite a lot in their CO abundances would certainly be attractive, but more studies are needed to fully understand the issue. The same may be said about understanding the remarkably high dust-to-ice ratio indicated by the observations of comet 67P. Here the question is, if at least partial sublimation of icy grain mantles and later recondensation may cause the comet-forming pebbles to be depleted in ice relative to the presolar grains. How this works, and what role is played by the heliocentric distance, needs to be clarified.

Finally, what about the preservation of the pristine imprint in comets in case they are collisional fragments? This issue mainly deals with the retention of supervolatiles during the impacts in question. Of course, if the impacts would deprive the fragments of all their CO (taking this molecule as a well observed example), the conclusion would be that comets cannot be collisional fragments. However, if CO largely exists in comets as clathrate, the only way to get rid of it is to vaporize the H_2O ice, and then the fragment would no longer be a comet at all. While some small-size ejecta from high-speed impacts on comet nuclei, originating from near the points of touchdown, may be ice-free, km-sized fragments cannot lose their ice in the same way. Thus, if km-sized fragments arise from impacts onto larger cometary objects, these should preserve their H_2O ice including the supervolatiles.

Those fragments might get differentiated by the energy of impact, so that the outer layers lose their supervolatile ices. But this would not prevent the 67P nucleus from having an impact origin. As long as ice exists close to the surface, there will be a certain amount of erosion by mass loss driven by outgassing. According to Sierks *et al.* (2015, supplementary material), the likely previous amount of erosion for 67P is several hundred meters, so we are probably looking deep inside the original nucleus. In case the original surface was modified, the currently observed object may not retain any memory of this. It is likely differentiated in any case, so that the supervolatile ices are found at some depth below the surface, but this depth should only reflect the recent orbital history.

7.3.3. *The orbital distribution*

As we have seen, the basic reservoirs of comets in the solar system today are the Oort Cloud, the Scattered Disk and the Edgeworth–Kuiper Belt. The latter is the only place, where comets could possibly have been formed, and it is a relatively unimportant source for the observed comets. The question from where and when the Oort Cloud and Scattered Disk were populated is thus of critical importance. The lack of any reasonable alternative to the primordial planetesimal population (possibly including collisional fragments) is thus a strong clue to the pristine nature of comets.

In this connection it is worth discussing to which extent the different reservoirs can be traced back to different regions in the solar nebula. In principle, the comet formation zone could have been very wide, if we only require H_2O ice to be condensible. It is thought that, after the initial cooling of the solar nebula, the snow line was situated not far from Jupiter's present orbit — a little inside Jupiter's formation zone. All the planetesimals that were formed beyond the snow line would be potential comet nuclei or precursors thereof.

According to current thinking, there are thus two distinct regions where comets were formed. The inner one coincides with the formation zone of the giant planets, which may have extended from about 5 to 15 AU, and the outer one covers all the larger distances out to about 32 AU. If the observed comets would sample this whole range of possible birthplaces, there is room for significant variations in chemical abundances, dust-to-ice ratio, and D/H isotopic ratio.

It has long been tempting to look for physical or chemical differences between Jupiter Family comets and the other types: long-period and Halley Type comets. To begin with, this was natural just because the orbital properties are so different that different formation regions are easy to imagine. When the Jupiter Family was realized to originate mostly from captures from the TNO population, while the other comets mostly come from the Oort Cloud, one only needed to imagine different sources for these two reservoirs. This too was easy, because the Oort Cloud was thought to have been formed by planetesimal scattering from the giant planet formation zone, while the TNOs were often identified with the Edgeworth–Kuiper Belt, which was thought to have been formed in situ. Thus, the two comet categories were thought to probe the two above-mentioned formation regions at quite different heliocentric distances. Differences in physical and chemical properties could be expected, and it was natural to look for them.

However, the picture has changed in several respects. First, after several large observational programs have been carried out, the conclusion is that the differences between Jupiter Family and other comets have basically failed to show up (A'Hearn *et al.* 2012). One exception will be mentioned below. Second, the Jupiter Family has been shown to relate more closely to the Scattered Disk than the Edgeworth–Kuiper Belt, and a common origin for the Scattered Disk and the Oort Cloud has been proposed by Brasser and Morbidelli (2013). In this picture, if carried to the extreme, all comets originate in the outer of the two formation zones, and we do not see any comets from the inner one. However, there is no reason to take such a categoric point of view. One needs to be open to the possibility of Oort Cloud comets contributing to the Jupiter Family and of planetesimals from the giant planet formation zone contributing to the Oort Cloud.

Thus, we essentially have to deal with one category of comets, but there may be large variations in nucleus composition depending on where in the wide formation zone the particular nuclei originated. Independent of this, comets come in either of two variants depending on chance (whether or not the particular comet was extracted from the young Scattered Disk into the Oort Cloud), but the variation remains within both groups.

There is, however, one chemical feature that has mainly been observed in Jupiter Family comets, as mentioned in Sec. 2.6.3. This is a depletion of carbon chain molecules (A'Hearn *et al.* 1995). Should these molecules have an interstellar origin, it is necessary for the organic component of the presolar grains of those comets to have been partially decomposed by heating in the solar nebula. This means that the depleted comets should probe the closest of all cometary formation distances — probably about 5–10 AU. However, there is a thorny issue in that such comets would be expected to be found in the Oort Cloud rather than in the Scattered Disk or the Edgeworth–Kuiper Belt.

7.4. Current Issues

Recent work in different areas and, in particular, the Rosetta exploration of comet 67P/Churyumov-Gerasimenko have brought a lot of information that has a bearing on the problem of comet formation. However, this does not mean that all issues are closed. Sometimes different results seem to be in conflict, or at least inconsistent with each other. Let us now discuss a few of the most outstanding current issues with as little bias as possible.

7.4.1. *Composition of comet material*

As described, there seems to be evidence that the observed comets provide samples of planetesimal populations formed over a wide range of heliocentric distance. On the one hand we have the carbon chain depleted comets, for which an idea (to be confirmed or refuted) is that they were formed so close to the Sun that the semi-volatile organics were decomposed and thereby lost. On the other hand there are comets, whose behavior is consistent with the ice being amorphous. One case concerning comet 29P/Schwassmann-Wachmann 1 was presented in Sec. 2.6.6, but this is not the only one. Other cases concern remote activity and outbursts, but it is difficult to argue convincingly that crystallization of amorphous ice must be the underlying mechanism.

If we accept that the super-volatiles in comet 67P do not arise from release of trapped gases during the crystallization of amorphous ice (Mousis *et al.* 2016), we can surmise that pre-solar amorphous ice did not survive in the region, where 67P was formed. If the ice even sublimed, as these authors suggested, one would have to go far beyond this formation region to find comets that formed with amorphous ice. But the formation distance of 67P is not yet known, although the high D/H ratio suggests a remote region (Altwegg *et al.* 2015). This needs further investigation, but if amorphous ice should contribute to the material in any comets, the prime candidates are those that were formed at the largest distances.

One would then think of the outskirts of the solar nebula, perhaps beyond $r = 30$ AU. If so, the place to find these comets today would be the dynamically cold part of the Edgeworth–Kuiper Belt. It is true that this population can only account for a small part of the Centaurs and Jupiter Family comets, but it may still be possible to find a few among the observed objects, and comet 29P might be one of these. However, further work will have to show if this idea is relevant or can be refuted.

Interestingly, the carbon chain depleted comets might form the opposite category of comets in terms of formation distance and still be found in the same population today. The argument goes as follows. The above idea holds that these comets were formed within the accretion zone of the giant planets and likely in its inner part. Consequently, they underwent an efficient gravitational scattering by these planets. In the absence of the solar nebula, they would have populated a primordial Oort Cloud to a large extent (Brasser *et al.* 2006). However, as shown by Brasser *et al.* (2007), the gas drag of the solar nebula changed this situation.

One main result was a size sorting of the planetesimals in question. Very large objects would migrate into a primordial scattered disk and then populate a primordial Oort Cloud under the influence of the surrounding birth cluster. This would happen in the same way as if the gas had not been present. On the other hand, small objects including comet nuclei of typical size would have their orbital evolutions dominated by the gas drag. The model for the giant planets in Brasser *et al.*'s paper is not a complete one, and we hence need to make some educated guesses as to what such a model would yield. It seems that a large fraction of the comets would evolve into nearly circular orbits inside that of Jupiter due to a strong deceleration near perihelion by the gas drag. On the other hand, many comets would also be kicked outward and experience an acceleration due to gas drag near their aphelia. This means a certain chance of dumping into low-eccentricity orbits in the trans-planetary parts of the solar nebula — perhaps including what is now seen as the dynamically cold part of the Edgeworth–Kuiper Belt.

The above is nothing but a preliminary suggestion. However, one may ask if comets from the same region — i.e., carbon chain depleted — could also end up in the Oort Cloud and return today as long-period or Halley Type comets. The preliminary answer is yes. The above-mentioned dumping had no reason to avoid implantation into the planetesimal disk of the Nice Model. Hence, the depleted kind should be present in all the cometary populations, independent of whether the planetesimal disk was dispersed very early or very late. Further modeling is needed to verify if and how this mixing of comets with different birthplaces would work.

7.4.2. *Chemical heterogeneity in comets*

Tentatively, it seems safe to regard the region beyond the giant planet orbits as the prime site of formation of the majority of comets. This includes both Jupiter Family and Centaur comets on the one hand and long-period and Halley Type comets on the other hand. However, the thermodynamic conditions varied a lot within this wide region, and this would strongly influence the efficiency of inclusion of molecules like CO and CO_2 by clathrate trapping or ice condensation. That the two mentioned species vary in abundance by a factor ten between different comets is therefore natural to expect.

If this argument is correct, the large differences between individual comets can be taken as an indication that each comet formed out of local material only. Mixing of cometesimals originating from different heliocentric distances within the same nucleus would tend to smooth out such differences. However, many papers have been produced with an aim to look for or map out chemical heterogeneities in individual comets. Even though observations of the fragments of the split comet 73P/Schwassmann-Wachmann 3 (Dello Russo *et al.* 2007) showed this comet to be chemically homogeneous, positive evidence has been reported or suggested several times. However, the uncertainties are large.

A case in point preceding the Rosetta investigation is comet 103P/Hartley 2. The structure of this nucleus, as revealed by the EPOXI mission, bears a certain resemblance to that of comet 67P. It has a bilobate shape with a morphologically peculiar waist region — smoother than the two lobes. As mentioned in Sec. 1.5.4, the smaller lobe was found to produce strong outgassing of CO_2, while the waist region was producing water. In fact, the H_2O-rich waist surface was proposed to result from fall-back of icy grains (see Sec. 4.2.1) emitted from the CO_2-active lobe (A'Hearn *et al.* 2012). These features are remarkably similar to what was found for 67P by Rosetta, as described in Sec. 4.2.1. However, there are also important differences between the two comets, so the cases are far from identical.

The deposition of H_2O-rich ice onto one region, originating from another region where CO_2 is abundant, is an evolutionary effect that may lead to an apparent chemical dichotomy without any underlying heterogeneity, as noted by A'Hearn *et al.* (2012). When it comes to comet 67P, the chemical heterogeneity of the inner coma is an observational fact, but it is still unclear whether this is due to a difference in chemistry or thermal conditions between the northern and southern hemispheres. What is certain is that there is a large difference in thermal conditions, as emphasized by Luspay–Kuti *et al.* (2015). The strongly heated, northern Hapi region was covered by redeposited material rich in H_2O ice without super-volatile ices and was thus able to outgas H_2O, while the southern, polar region producing CO and CO_2 had been strongly eroded during the previous perihelion passage and thus exposed these ices near the surface, from where they could sublime in spite of a much lower temperature. Again, an evolutionary effect is implied rather than a chemical heterogeneity.

Meanwhile, the issue deserves further investigation. As mentioned in Sec. 2.5.1, surface features on comet 67P have been interpreted as protrusions of hundred-meter sized cometesimals (Davidsson *et al.* 2016). If this is correct, the presence of building blocks of such a size begs the question, if these are chemically identical or possibly diverse. In case there is diversity, it could have an observable imprint on the coma chemistry. The diversity in question would place an interesting constraint on theories of nucleus formation.

Let us not forget that, independent of whether there is large scale homogeneity or heterogeneity in comet nuclei, the cometary material is very heterogeneous on a micro-scale. This is shown by the large abundance of high-temperature nebular condensates (like crystalline olivine) in the Stardust samples from comet 81P/Wild 2 (see Sec. 2.6.2). The Rosetta mission did not provide for such detailed analysis — however, close-up imaging of the Philae landing site (Bibring *et al.* 2015) revealed a fractured, grainy surface with large contrast between neighboring, darker and brighter units, as shown in Fig. 7.8. Two end members of this surface were identified. One is formed by agglomerates of dark, granular material, and the other consists of brighter grains and smooth-appearing blocks up to a decimeter in size. The first is likely carbon rich, while the nature of the second was not uniquely identified.

It is reasonable to assume that this randomly selected area (selection made by the uncontrolled, final landing of Philae) is typical of large parts of the 67P nucleus. If we thus extrapolate what we see to the nucleus at large, we may conclude that the accretion process involved the mixing of quite different units of small but not microscopic size, possibly originating from different places. How this is to be interpreted, and if the difference is in radial distance from the Sun or vertical distance from the nebular midplane, remains to be discovered.

Fig. 7.8. Image of a part of the local surroundings, seen from the Rosetta/Philae lander at its final location. The field of view is about a meter in size. The circle marks part of one CONSERT antenna, apparently in contact with the surface. From Bibring, J. P. *et al.* (supplementary material), *Science* **349**, aab0671 (2015). Reprinted with permission from AAAS.

7.4.3. *Long-term surface evolution*

The surface evolution of comet nuclei is not immediately relevant to the problem of comet formation, but the way in which the activity of a comet evolves has to be understood to infer the population size of the distant reservoirs from the number of observed comets and thereby to conclude about the properties of the early planetesimal disk. Moreover, concerning comet 67P/Churyumov-Gerasimenko in particular, it is of great interest to know the relationship between the present, altered surface and the original material out of which it has been formed. It may be somewhat disappointing that these are open issues with many problems remaining to be solved, but we are facing an interesting challenge.

One issue that needs to be pursued is the non-monotonous aging of comets. At present, there are only tentative indications in favor of such a picture, based on circumstantial evidence. A real understanding of the long-term behavior of comet activity and the detailed mechanisms behind this remains a goal for the future. An obvious question is then: What have we learned from Rosetta?

We have touched upon this before, but let us return to the inferences that the Rosetta results may motivate. First, what is the nature of erosion? The easiest is to say what it is not, namely, a uniform, vertical retreat of the surface by H_2O sublimation and associated dust release. But it is not so easy to describe what it really is. Scarp retreat and cliff collapse are observed phenomena that erode the nucleus in a lateral direction by removing part of a topographic high. If this is the way most of the mass loss occurs, it remains to understand why the gaseous mass loss in Jupiter Family comets correlates with orbital changes (Rickman *et al.* 1991).

However, the picture apparently cannot be so simple. While changes in insolation are certainly of importance, there is also evidence from 67P of purely mechanical weakening of the consolidated surface, which may be caused by the outgassing-related spin up of the nucleus during the perihelion passage (Hirabayashi *et al.* 2016). In 67P, this amounted to a decrease of about 20 minutes of the rotation period during the 2009 perihelion passage (Mottola *et al.* 2014). An additional decrease by about 6 minutes occurred before the perihelion passage in 2015 (Jorda *et al.* 2016). Possibly related phenomena on the nucleus have been reported by El-Maarry *et al.* (2015) in terms of surface fractures and their evolution.

One may also ask, what is the relative role of mass loss relative to mass redistribution, given that the dust/ice ratio of the nucleus is reported to be about six with most of the dust mass that leaves the nucleus being locked up in very large particles (Fulle *et al.* 2016). Moreover, much of the fall-back material is icy as described in Sec. 4.2.1. Thus, redistribution appears to be as important as mass loss for the surface evolution, and clarifying this issue will be very important.

Rosetta has taught us that the shape of a comet nucleus can be extremely different from the idealized sphere that most thermal models have considered. Furthermore, the seasonal variations on such an object govern the activity to an extent beyond what was previously thought. Then the question arises, what the consequences may be, if there is a sudden change in the axial tilt of the comet or the associated azimuthal angle. Such changes may occur in connection with close encounters with Jupiter, especially if the orbital plane is strongly perturbed. Would the activity of the comet be expected to change, and would it increase or decrease? These are currently open questions.

An interesting paper by Sosa and Fernández (2011) was quoted in Sec. 6.4.1, where their mass determinations for long-period comets were used to translate absolute total magnitudes into nuclear radii (cf. Brasser and Morbidelli 2013). Based on this translation, Sosa and Fernández found a trend for their investigated long-period comets to have very large values of the activity parameter f (see Sec. 2.4). They proposed that this may be due to sublimation of icy coma grains. However, other possibilities exist, and whatever is the nature of the investigated comets, these may constitute a biased sample. Nongravitational effects are hard to determine for one-apparition, long-period comets, and the cases of success may be those where the effects are the largest.

In any case, an interesting issue is whether there is any evidence that long-period comets evolve differently from Jupiter Family comets or that their nuclei are indeed more active. Such differences would appear to be in conflict with the similarities generally found between those classes of comets in other respects, but they are certainly possible, and the issue remains to be investigated.

7.4.4. *Are comets collisionally evolved?*

Let us return to the issue of collisional evolution, which was described in Sec. 6.5. We will focus on comet 67P, but the discussion has a bearing on all comets. The issue can be separated into two questions. First, how likely is it that comet 67P is a collisional fragment rather than an unaffected, primordial planetesimal? Second, if this were the case, would it conflict with any observations of the comet?

As mentioned in Sec. 6.5, Morbidelli and Rickman (2015) treated a case, where 67P is represented by a planetesimal that started out with its present size — modeled as a sphere with radius 2 km. This was assumed to belong to the trans-planetary planetesimal disk and to stay there until the commonly assumed time of the Nice Model planetary instability, about 400 Myr later. The authors found the chance to escape from destructive collisions to be very small. In Sec. 7.2.3 we described the formation model by Davidsson *et al.* (2016), which makes different assumptions as to the properties and evolution of the disk. In this model the survival probability is higher, but it is not overwhelming. It appears quite likely even in this model that 67P would have been destroyed, although it could reasonably be imagined to be a survivor. Thus, comets would likely be of two kinds: the survivors and the collisional remnants.

Apparently, there is hence some uncertainty over the early fate of comet 67P. However, a few special issues need further attention. One is the size distribution of the smallest objects in the planetesimal disk. The only direct information about this comes from the NASA New Horizons mission, which imaged Pluto and Charon and allowed crater counts to be made. From this crater size distribution, Singer *et al.* (2015) found a value of $\alpha \simeq 2.3$ for the cumulative size distribution index (see Sec. 2.1.3). This is quite steep, meaning that the number of small projectiles able to disrupt objects like the 67P nucleus is very large. As a consequence, destruction appears very likely. However, Singer *et al.* (2016) suggested that the disk may have had a deficit of km-sized and smaller objects relative to the above-mentioned distribution. This would translate into a smaller index at the smallest sizes, and the number of destructive projectiles might be small enough to make the 67P nucleus survive. Obviously, this issue is not yet resolved.

The other issue concerns the timing of the planet instability of the Nice Model. Morbidelli and Rickman (2015) mentioned that the late instability appeared favored by the evidence for a Late Heavy Bombardment but did not consider this to be the only possibility. A very early instability ($\sim$ 10 Myr after planetesimal formation in a turbulent disk) would mean that collisional destruction before this event is very unlikely. In a recent paper, Kaib and Chambers (2016) argued in favor of an early instability. The reason had to do with the jumping Jupiter scenario of the Nice Model (see Sec. 6.1.2).

This scenario was conceived as a solution to the problem that a slow 2:1 resonance crossing by Jupiter and Saturn would perturb the terrestrial planet system beyond recognition by making the orbits far too eccentric. Kaib and Chambers found that, while the jumping Jupiter scenario is possible, it is very unlikely for a random initial configuration of the giant planets. Out of 200 simulations of the instability phase that they performed, only one was successful in reaching the correct pattern of orbits throughout the planetary system. One way to describe this dilemma is that the giant planets could not know how to behave in order to save the relatively low eccentricities of the terrestrial planets and were thus likely to misbehave with possibly fatal consequences for the Earth and its neighbors.

The natural way to avoid this dilemma is to suppose that the instability occurred very early, before the terrestrial planets were formed. If their embryos acquired too eccentric orbits, it would not matter because the eccentricities would be damped by dynamical friction with planetesimals on a short time scale. However, in another recent paper, Toliou *et al.* (2016) investigated the effects on the asteroid belt and found that the large jumps in orbital energy by Jupiter and Saturn are in fact necessary to save the distribution of asteroid orbits in the way we observe it. This is independent of when the instability occurred. There may not be a logical explanation as to why the planets took this particular course, but given that they must have done so, the actual evolution into the planetary system we see is not overly improbable.

While the collisional evolution before the planet instability is thus very uncertain, since the timing of the latter is uncertain, a better estimate can be made of the corresponding evolution at later times. This starts with the short dispersal phase of the planetesimal disk during the migration of the ice giants and then continues with a very long stay in the Scattered Disk for the Jupiter Family comets and a much shorter one for the Oort Cloud comets. Morbidelli and Rickman

(2015) made an estimate of this evolution for Jupiter Family comets, and this was recently updated and improved by Jutzi *et al.* (2016).

These authors made numerical SPH[3] simulations of impacts onto objects with the bi-lobate shape of comet 67P, assuming strength and other physical properties, to find the critical impact energies that result in either disruption or change of shape of the target. They then simulated the physico-dynamical evolution of such objects throughout the above-mentioned time interval and computed the numbers of destructive and shape-changing impacts. From their results it appears that a destructive impact is quite likely but also the survival without any such impact. Shape-changing impacts, on the other hand, are ubiquitous. The most likely time for the last such impact experienced by 67P would be about 1 Gyr ago, and the place would have been the Scattered Disk. Hence, the nucleus as we see it today was not formed that way, and we have practically no information about the appearance of the original planetesimal.

However, this too is dependent on the size distribution of the smallest disk objects. If these are lacking to the extent that was mentioned above, 67P might have survived without any reshaping impact. Considering all these uncertainties, it is worth asking if the observed, pristine nature of 67P (see Sec. 7.3) would have had a chance to survive in the presence of the destructive or reshaping impacts. Here, Jutzi *et al.* (2016) offered diagnostic results by computing the amount of heating and compaction that would result from the impacts in question. The result is that a comet like 67P is not modified in any appreciable way.

As a consequence, comet 67P may or may not be a collisional survivor, and we would not be able to know based on its pristine properties. Comets in general may be of two kinds — remnants and survivors — and we have no way to say which ones are of which kind. The only way forward for comet 67P may be to study local surface features like the fractured wall of Hathor (Rickman *et al.* 2015) and verify if these provide morphological evidence of impacts.

[3]Smooth Particle Hydrodynamics.

7.4.5. *Planet Nine*

We finally have to discuss briefly, which consequences the existence of Planet Nine (Sec. 5.4.2) would have for the formation and evolution of the Oort Cloud. This would apparently be an ice giant planet moving in a range of heliocentric distances between about 150 and 600–2 000 AU at relatively low ecliptic latitudes. As such, it would have the potential to cause perturbations on comets.

The first thing to note is that this proposed planet would cause very small energy perturbations. In Fig. 5.2 of Sec. 5.1 we see that the typical scatter of $1/a$ caused by planetary perturbations remains roughly constant for perihelion distances from 20 to 30 AU, i.e., when the main perturbing planet changes from Uranus to Neptune. For the case of Planet Nine, we may extrapolate to $q \sim 200$ AU and estimate a similar value of the scatter. This is of order 10^{-5} AU^{-1}.

While much smaller than the perturbations experienced by observable comets, those are not completely negligible. Since experience has shown that the perturbations by Uranus and Neptune play a role in Oort Cloud dynamics and the injection of new comets, similar effects may also occur for much larger perihelion distance, if Planet Nine exists. We may also conjecture that Planet Nine would have contributed to the formation of the Oort Cloud during the early times, when the Sun belonged to its birth cluster. The detachment of cometary perihelia from the planetary system as we know it was then very efficient for small values of the semi-major axis, as discussed in Sec. 6.3.1, which leads to a concern that the Oort Cloud becomes too concentrated to act efficiently as a provider of new comets at later times. However, the scattering of comets by Planet Nine would have acted on comets that were hitherto considered to be detached and transferred these into larger orbits before the actual detachment could occur. This seems to possibly alleviate the mentioned concern.

If the ejection of Planet Nine (Sec. 5.4.2) happened within a Nice Model scenario, one may speculate that the ejection process was coeval with part of the process that populated the Oort Cloud. However, according to the results of Brasser and Morbidelli (2013), most of the Oort Cloud had not yet been formed, as the ejection took place. It seems that the emplacement of Planet Nine into the type of orbit predicted by Brown and Batygin (2016) with its large perihelion distance necessitates the action by a cluster environment, which may be an argument in favor of an early instability. However, once again it must be emphasized that Planet Nine remains a proposal in order to explain the orbital properties of TNOs, and this proposal has neither been confirmed nor refuted.

Thus, for the moment the effects of Planet Nine on comets are not yet a real issue, but they have the potential to become one in a near future.

Chapter 8

Outlooks

Comets are fascinating celestial objects and easily attract the interest of scientists as well as laymen. The scientific interest in comets today has a distinct focus on using these as witnesses of the earliest stages of our solar system and even its formation. This was the background for all the discussions in the preceding chapters of this book. In this last chapter, we will instead turn to a few other science issues, where comets play a great role too.

8.1. Comets and the Earth's Water

The origin of the Earth's water is a major issue in geophysics research. The total mass of the Earth's hydrosphere, which is dominated by the oceans, is about 1.4×10^{21} kg. In addition to this, our planet hosts an equally significant amount of water in its mantle — largely in the form of the mineral *Ringwoodite*. The mass of the mantle water is estimated to be roughly equal to, or up to, a few times larger than that of the hydrosphere.

Why is understanding the origin of this water to be regarded as a scientific issue? At less than one per mille of the mass of the planet, the water is just a marginal constituent but even so, its abundance is in a sense surprising. This is because it has been natural to imagine the Earth to be built of local building blocks, i.e., planetesimals that were formed close to the Earth's current orbit. Since this is far inside the snow line according to the usual estimates of the temperature profile of the solar nebula, it could be argued that the Earth should have no water at all.

This problem can be dealt with without invoking any late veneer. The accretion of the terrestrial planets involved so-called embryos of almost planetary size on eccentric orbits. These would have been formed on nearly circular orbits by local accretion at different distances from the Sun and, hence, each of the terrestrial planets would incorporate local material from a wide range of distances. This could either mean that some of the Earth's embryos were formed outside the snow line, or there was a transport of icy planetesimals across the snow line such that the more distant of the Earth's embryos became relatively water-rich by sweeping up such planetesimals.

Of course, in addition to this, it is possible that the ready-made terrestrial planets were still bombarded for a long time by ice-rich material. This could have been in the form of left-over projectiles of the same nature as the planetary embryos, or in the form of comets, adding a late veneer to the water budgets long after the time of planet formation. But whatever the size of this contribution, it is clear that the water that was picked up while the planets were being accreted can be termed cometary as long as it was condensed beyond the snow line. As far as we have seen in Chap. 7, such icy planetesimals cannot be distinguished from comet nuclei other than by having experienced different dynamical evolutions. Many of the comets arriving today from the trans-neptunian region or the Oort Cloud could have had essentially identical siblings that instead went into some terrestrial planet, contributing to its water budget.

However, there is an important criterion that has to be fulfilled in order to argue convincingly for a cometary origin of the Earth's water. The deuterium/hydrogen (D/H) ratio in the oceans has a well known value, referring to what is called the *Vienna Standard Mean Ocean Water* (VSMOW), and this value is 1/6420 in number of atoms. Individual comets have turned out to have their individual D/H ratios, and if the oceans are made of cometary water, the mix of comets must be compatible with the VSMOW value.

8.1.1. *Early planetesimal migration*

Let us first describe an idea that may be relevant to the transfer of icy planetesimals across the snow line into the feeding zone for the growth of terrestrial planets. We are now considering planetesimals that grew in the accretion zone of the giant planets, which we have estimated to range from about 5 to 15 AU in heliocentric distance (Sec. 7.3.3). These planetesimals included the precursors or building blocks of the giant planet cores, which were formed in a solar nebula that was dense enough to provide for their gas capture.

According to the current theories of comet formation, there may have been two generations of such planetesimals. The first arose from the pebble swarm instabilities and comprised objects with a wide size spectrum that reached up to several hundred kilometers. The second may have come along after a few million years, when the left-over pebbles aggregated into objects at least a kilometer in size (Davidsson *et al.* 2016). While this depends on the dynamics imposed by the growing, massive planets and largely remains to be clarified, observations seem to indicate the existence of km-sized comets with an origin in this region. These would be the carbon chain depleted comets, in case the reason for the depletion is heating that caused the loss of semi-volatile organics.

In Sec 7.4.1, it was suggested that the circularization of cometary orbits caused by the nebular gas drag would transfer a certain fraction of the icy planetesimals from the giant planet zone into the interior region. The efficiency of this transfer remains to be found by numerical modeling, but for the sake of argument, let us suppose that it is reasonably high. In Fig. 8.1 we illustrate an outline of the relevant dynamics. At close encounters, Jupiter's gravity scatters planetesimals between orbits with different angular momenta (Sec. 3.2), and this chaotic evolution is bidirectional. However, gas drag acts all the time, always tending to decrease the orbital eccentricity. Here, we have assumed conservation of the orbital angular momentum (see Brasser *et al.* 2007). The evolution thus follows the green curves toward the upper left.

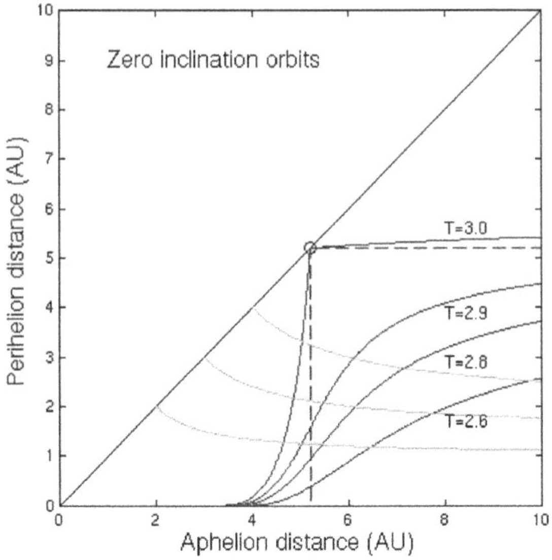

Fig. 8.1. Perihelion verses aphelion distances of planetesimal orbits in the inner part of the solar nebula after the formation of Jupiter — here, assumed to reside in a circular orbit at its current location. The planetesimals are assumed to move in Jupiter's orbital plane. Blue curves indicate the orbital evolution induced by scattering at close encounters with Jupiter for several values of the encounter speed (see Sec. 3.1). Green curves show how the gas drag is expected to affect the orbits, assuming a constant angular momentum.

Obviously, given the lapse of enough time between the jovian encounters, this may lead to decoupling from Jupiter's gravitational influence, as the aphelia are transferred to distances $Q \lesssim 4.5\,\mathrm{AU}$. At this time, for planetesimals with large enough angular momenta (and, probably, with low enough jovian encounter speeds) the circularizarion will aim at orbits beyond the current asteroid main belt. However, after the orbits become nearly circular, the objects are still subject to gas drag from the nebula and are bound to slowly spiral inward, toward the Sun.

Even though it is difficult to say without detailed modeling, what is the absolute efficiency of the described transfer, something can be said about its size dependence. The time scale of dynamical evolution by jovian encounters is of course independent of the size of the planetesimals as long as their mass is negligible compared to

Jupiter, but the time scale of the gas drag effects does depend on size and increases with the radius of the planetesimal. The efficiency of the transfer is optimal, when the two time scales coincide, and thus, there must be a preferred size of planetesimals to experience the transfer.

Specifying this size requires detailed modeling of the dynamics, but we can surmise that the planetesimals were much smaller than the asteroids initially populating the main belt. The latter are estimated to have been at least 100 km in diameter (Morbidelli *et al.* 2009). Therefore, in many cases and perhaps most cases, the icy planetesimals would drift toward the Sun much faster than the asteroids, which would remain nearly at rest. This would likely lead to low-speed collisions, whereby the outer main belt asteroids (and those beyond the current belt) were enriched in H_2O by the icy planetesimals.

All this must have happened very early, before the solar nebula was dissipated. Thus, the implantation of water into the region where terrestrial planets were formed was a very early phenomenon that was finished long before these planets arose. The described mechanism will be further discussed below, but it is not the only one. Another possibility has been proposed within the *Grand Tack* scenario (Walsh *et al.* 2011). Here, we have to step even further back in time, to the epoch when the giant planets migrated inward or outward with respect to the Sun due to their gravitational interactions with the gas disk.

In brief, the scenario looks as follows. In Sec 6.1.2 we described how Jupiter and Saturn got trapped in a 2:3 mean motion resonance, but we did not mention the detailed calculations that led Walsh and co-workers to the conclusion that the inward migration proceeded very far, before the resonance capture turned the evolution into the outward direction. At its turning point, Jupiter would have been only 1.5 AU from the Sun. The pre-existing planetesimals became scattered, and some icy ones from beyond the snow line were placed into the innermost region, from where the terrestrial planets were built. As discussed by Brasser (2013), this may provide a relevant explanation for the water on Mars as well as on the Earth.

There is even a third possibility for the early migration of icy planetesimals across the snow line. However, this would only exist in case the planet instability of the Nice Model took place very early. We have seen that there are arguments in favor of this, but the issue is open and thus, the existence of this early source of water critically depends on how it is resolved. The mechanism would be the dispersal of the trans-planetary disk by the migration of Uranus and Neptune and an ensuing gravitational scattering, which involved all the giants and left some of the disk objects temporarily in orbits penetrating into the growth region of the terrestrial planets. They could then collide with the native objects (planetesimals or planet embryos) of that region, thus bringing water to them. However, in that case the collisional velocities would be high, and much of the water would likely be lost into space. According to Levison *et al.* (2009), some disk objects would also be dynamically captured into the outer asteroid belt, which means that water-rich objects would mix in with the native objects of that region, and thus, possibly contribute to the accretion of terrestrial planets.

8.1.2. *Meteoritic and asteroidal evidence*

If water was mixed in with the early planetesimals of the inner solar system due to either of the outlined processes or a combination of them, it is worth looking for evidence in the form of water in meteorites. The great majority of the meteorites falling on the Earth are tiny fragments of main belt asteroids, knocked off their parent bodies in a relatively recent past. They come in different chemical varieties, which forms the basis of a classification. Most of them are very similar to the solar atmosphere in terms of elemental abundances, and these belong to either of a few *chondrite* classes. The elemental composition makes them especially interesting, because they can be thought to represent the pristine material that made up their parent bodies. Since different chondrite types can be tied to different groups of main belt asteroids by their spectral similarities, this provides a way to map the asteroid belt, for instance in terms of water content.

In this regard, the most interesting result is that water occurs in significant quantities only in some types of *carbonaceous chondrites.* These are spectrally linked to the *C-type asteroids*, which in turn are found to mostly inhabit the outer part of the main belt. Furthermore, the *D-type asteroids* that dominate beyond the main belt are in some regards reminiscent of the C-types. As yet, there not enough meteoritic analogues of the D-types to say with confidence, how much water these may contain.

Overall, the asteroid main belt seems rather well mixed regarding chemical properties. Thus, apparently, the ranges in heliocentric distance occupied by different chemical classes largely overlap, even though the objects probably were formed in different zones. Different materials with quite different geochemical properties can even coexist within the same asteroid. However, the trend in water abundance is quite clear. Therefore, if early implantation is the reason, it had to work selectively in terms of heliocentric distance, favoring the more distant targets. Tentatively, the first of the above transfer mechanisms seems best suited to meet this criterion, but the others are definitely not disqualified.

It is interesting to compare carbonaceous chondrites to cometary material — both similarities and differences are equally important. Let us start with the amount of water they contain. If comet 67P/Churyumov-Gerasimenko is typical of comets at large, a tentative value for the cometary H_2O abundance by mass is ~15–20%. This is not far from the largest values found in carbonaceous chondrites, which range around 10%. A possible interpretation is that the C-type asteroids from which the carbonaceous chondrites come have picked up different amounts of cometary material, from nearly nothing to a substantial part of the entire mass.

However, at this point a fundamental difference has to be noted. The carbonaceous chondrites can be said to be water-rich compared to other meteorites, but it is not a question of ice like in comets. Instead, they contain water-bearing ("hydrated") minerals like phyllosilicates. These are believed to be the result of so-called *acqueous alteration*, which means a low-temperature, thermal modification of the silicates due to their reaction with running water. The trigger may be the melting of ice in the interior of the asteroids in response to a moderate heating. Could anything like this have happened in comet nuclei as well? According to Davidsson *et al.* (2016), in large nuclei (diameters exceeding 10 km) that formed very early, the heating by decay of ^{26}Al could have induced wide-spread ice melting, judging from thermal modeling results by Merk and Prialnik (2006).

Unless the 67P nucleus is a fragment of such a large object, its small size should be a guarantee against ice melting and acqueous alteration. The Rosetta/VIRTIS spectral observations indeed showed no evidence of such alteration (Capaccioni *et al.* 2015). It thus seems that the icy material delivered into the asteroids after migration across the snow line might have been largely unaltered in the absence of melting. However, when buried in the interior of big asteroids that experienced radiogenic heating from both short- and long-lived isotopes, the formation of hydrated silicate minerals is to be expected. Of course, during this process, all the super-volatiles found in comets would have been outgassed. Indeed, hardly any trace of these has been found in carbonaceous chondrites.

The NASA *Dawn* exploration of (1) Ceres, the biggest main belt asteroid[1] and one that usually counts as a C type, has yielded new insights on its interior structure. Ceres has a moment of inertia similar to that of Jupiter's large moon Callisto. New models of its global structure (Park *et al.* 2016) show the object to be partly differentiated with a core of higher density and a lower density mantle. The latter likely involves large quantities of water, so that the global water content by mass may be around 20% by mass (McCord and Sotin 2005). Since this percentage is as high as the current estimates for comet 67P, it may not be appropriate to invoke accretion of comets as the source of Ceres' water. As mentioned, Ceres may be similar to the Galilean moons Ganymede and Callisto as well as the larger moons of Saturn. In fact, such regular satellites were formed in circumplanetary disks, where the relatively high pressures allowed the hydration of primordial CO or CO_2, yielding

[1] In fact, officially counted as a dwarf planet and so far the only such object with a main belt orbit.

water and methane. This is a favored way to explain their high water contents.

However, imagining Ceres to be an escaped, regular satellite of a giant planet raises obvious problems. The highly regular structures of the jovian, saturnian and uranian systems speak against such a violent, dynamical episode in their past. Only Neptune would thus seem to be a possible place to look for the origin of Ceres, if we do not count a hypothetical, fifth giant planet that has been ejected from the planetary system according to some variants of the Nice Model (Nesvorný and Morbidelli 2012). This scenario may be too speculative to be taken seriously at this point, and one weakness is easy to realize. There cannot have been many escaped, regular satellites at any time, and there are other, much more likely places for these to end up than the asteroid belt.

Since we should thus prefer a less exotic scenario for Ceres' origin, it is relevant to note that the modeling uncertainties allow for the possibility of a global water content that would be compatible with a cometary origin. Even so, Ceres is definitely water rich, and H_2O outgassing has been observed from its surface (Küppers *et al.* 2014) at a rate that is not far below those of active comets. It is not yet clear to which extent this is due to either ice sublimation or cryovolcanism.

We finally note that Ceres is not the only asteroid that actively produces water vapor. Another big C-type asteroid in the outer part of the main belt is (24) Themis. While H_2O outgassing has not been directly observed from this asteroid, its surface has been shown to be largely covered by water ice (Campins *et al.* 2010; Rivkin and Emery 2010). This must sublimate due to the heating by sunlight and likely be replenished from a subsurface reservoir. In addition, several asteroids are counted with a group called *Main Belt Comets*. These are main belt asteroids that have been seen to be surrounded by diffuse material in a way similar to active comets. Dave Jewitt (2012) called these "active asteroids" and concluded that there are many reasons for this apparent activity, but only in two cases could this be conclusively shown to be ice sublimation. In these two objects with cometary names 133P/Elst-Pizarro and 238P/Read, the activity is repetitive around perihelion, and this is the only part of the orbits where a substantial water sublimation flux is expected.

8.1.3. *Deuterium-to-hydrogen ratio*

So far, we have no definitive answer as to the ultimate origin of the Earth's water. Comets may have contributed by bringing water into the C-type asteroids in the outer part of the asteroid main belt, because such objects should have contributed to the build-up of the Earth at a level relevant to explain the Earth's water (Morbidelli *et al.* 2000; O'Brien *et al.* 2006). As mentioned above, it is important that the objects that supplied the water to the Earth had the proper mix of D/H ratios to explain the VSMOW value. In the following, we will use the latter as the standard, so we will express D/H ratios in units of the terrestrial ocean value.

Figure 8.2 shows a histogram of individual D/H ratios measured in different carbonaceous chondrites and illustrates that the average value is very close to 1 (with respect to VSMOW). We can also see that the molecular hydrogen of the solar nebula had D/H $\ll 1$, while the comets tend to deviate in the other sense. Such was the situation at the time (more than ten years ago) that this diagram was constructed. As mentioned in Sec. 1.5.4, the first measurements of D/H in cometary water were made in comets related to the Oort Cloud,[2] but these did indicate that comets had about twice as much deuterium as VSMOW. The general conclusion was that the Earth's water did not come from comets — however, a more prudent statement would have been that it did not come from comets like the three that had been observed, or possibly that it might have come from a category of comets that was less prone to populate the Oort Cloud.

The picture has changed, as measurements of Jupiter Family comets came along. The first was comet 103P/Hartley 2, which yielded the remarkable result of D/H ≈ 1 (Hartogh *et al.* 2011). A second one, mentioned by Ceccarelli *et al.* (2014), is 45P/Honda-Mrkos-Pajdušáková, for which the upper limit to D/H is just slightly above unity. This is important, because it shows that there

[2]These were 1P/Halley, C/1995 O1 (Hale-Bopp) and C/1996 B2 (Hyakutake).

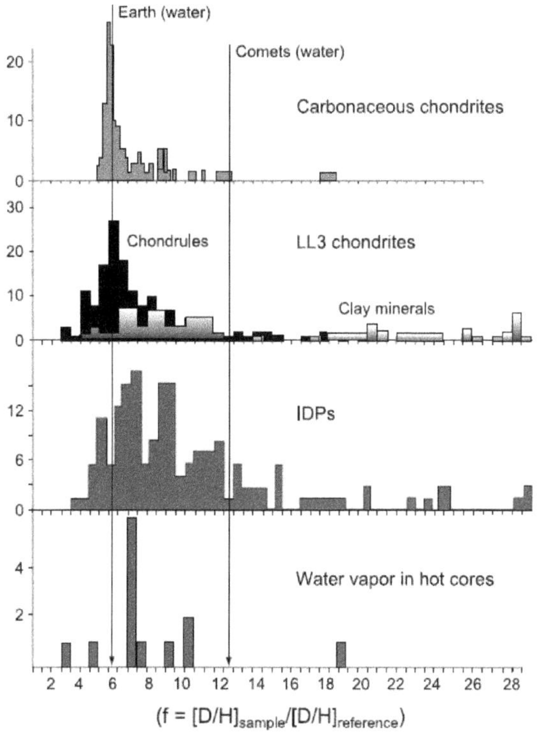

Fig. 8.2. D/H ratios measured in meteorites and interplanetary dust particles (IDPs), and observed in the hot cores of molecular clouds, where stars are being formed. The distribution of whole-rock samples from carbonaceous chondrites fits best to terrestrial water. The value indicated for comets is insignificant, being based on too few observations. From F. Robert, "Solar System Deuterium/Hydrogen Ratio," in *Meteorites and the Early Solar System II*, edited by D. S. Lauretta and H. Y. McSween Jr. © 2006 The Arizona Board of Regents. Reprinted by permission of the University of Arizona Press.

exists a type of comet that is isotopically similar to the Earth's water and may therefore be a potential source for the water in carbonaceous chondrites. The next was comet 67P/Churyumov-Gerasimenko, where the ROSINA instrument on board Rosetta measured D/H ≈ 3 (Altwegg *et al.* 2015). From this it is obvious that the Jupiter Family comets span a wide range of D/H values, and only a part of them may be relevant as contributors to the Earth's water source.

The most likely reason for the variation of D/H between different comets is isotopic exchange occurring in the solar nebula between water molecules sublimed from the pre-cometary grains and the molecular hydrogen that was prevalent in the disk. It is generally agreed that this would lead to a monotonous increase in D/H with heliocentric distance (see Geiss and Reeves 1981). We may thus conclude that Jupiter Family comets emanate from a wide range of distances in the solar nebula — certainly encompassing the whole trans-planetary disk but perhaps also including the giant planet formation zone as well, as we argued in Sec. 7.4.1. The same would be true of Oort Cloud comets too, and the bulk of data on such comets today as presented, for instance, by Altwegg *et al.* (2015), indeed shows that they scatter more or less across the whole range spanned by the Jupiter Family comets.

Hence, the acquisition of cometary water from the giant planet formation region into carbonaceous chondrites is not in conflict with the evidence on cometary D/H ratios. It thus remains as one contender for the ultimate origin of the Earth's water, even though this water may also be called asteroidal for obvious reasons (see Fig. 8.2). In the hypothetical case that comets would also have brought water directly to the Earth in the way of bombardment subsequent to the formation of our planet, it is likely that such a late veneer would be in conflict with VSMOW, unless the bombardment was insignificant. That this is the case was recently proven by Rickman *et al.* (2017b) who, as mentioned in Sec. 6.4.2, studied the bombardment of lunar and planetary surfaces that arose as a consequence of the dispersal of the trans-planetary disk within the Nice Model.

Their result concerning the delivery of water to the Earth and Mars is that this is insignificant compared to the global inventories of water on these planets. Therefore, if the dispersal was coeval with the Late Heavy Bombardment, there certainly was a late veneer but a very small one — far too small to have influenced the VSMOW in the case of the Earth. On the other hand, in case the dispersal happened very early, before the terrestrial planets and the Moon were formed, we may speculate that some delivery of water to the planetary embryos would also have occurred. This has not been investigated in detail, but even though the total collisional cross-section of the embryo population was much larger than that of the resulting planets, much less water would have been retained due to the small gravities — just like very little water would have been retained within the Moon during the LHB.

8.1.4. *The problem of the Moon-forming impact*

It should be mentioned that a problem remains concerning the Earth's water. We have seen that some of the embryos that formed our planet were likely water-rich and that they could have had the proper average of the D/H ratio. However, the last major impact in the series that formed the Earth also formed the Moon according to the most popular picture of Moon formation — the giant impact scenario. This in itself is not without problems (see Asphaug 2014), especially because it is hard to explain the great similarities between the isotope ratios of different elements in the Earth and the Moon in a scenario, where these two binary components resulted from a collision between two massive objects whose isotope ratios must have differed from each other. But in any case the question remains, what would become of the Earth's water and other volatiles during a collision that tore the planet apart.

In case much of the Earth's original water would be retained, there may be no problem. However, it has been claimed (Connelly and Bizzarro 2016) based on a study of the Earth's Pb isotopes that the Moon-forming event left our planet essentially dry. Should this be the case, we would be left with the late veneers, but these are problematic too. We have mentioned the problem of D/H in the bulk of the comets as well as the fact that far too few comets would have hit the Earth during the Late Heavy Bombardment. The only remaining option would be the tail end of the primordial bombardment, which went on for a long time after the Earth and the Moon were formed. In this case, an observational check comes from the element osmium, which contributes to the Earth's upper mantle. Since this element is highly siderophile, it went almost completely into the core, when our planet differentiated. Hence, what is now found in the upper mantle originates in the meteoritic bombardment of later times, including and dominated by the mentioned late veneer.

Drake and Righter (2002) showed that the isotope ratio $^{187}Os/^{188}Os$ of the primitive upper mantle agrees with that of ordinary chondrites — the most common meteorite type — but not with carbonaceous chondrites. The late veneer may thus be consistent with the current mix of meteorite types, where the carbonaceous chondrites are just a small fraction. The total mass of such a general, chondritic late veneer can be estimated from the measured abundance of highly siderophile elements in the mantle, and the currently favored value is 0.5% of the Earth's mass or 3×10^{22} kg (Walker 2009). It is then easy to estimate the total amount of water, and this is insignificant compared to the Earth's water inventory (see Rickman *et al.* 2017c).

In the presence of the mentioned problems, it is fair to say that the origin of the Earth's water is not known in detail. However, an argument can be made that, whatever this detailed solution may be, the water should have a strong genetic link to that of comets, albeit a part thereof that is not frequently sampled by the observed comet population.

8.2. Comets and Life on Earth

In the words of Chyba and Sagan (1997), liquid water is the *sine qua non* of life as we know it. Guided by the appearance and non-appearance of liquid water on the terrestrial planets of the solar system, the *habitable zone* is often defined to be the range of heliocentric distance from the orbit of Venus to that of Mars. However, this definition presupposes the existence of H_2O as a chemical species in the material making up the planets. From this point of view, Sec. 8.1 has already dealt with one of the fundamental aspects of the present topic.

Nonetheless, several other aspects remain to be dealt with. One is how comets may have contributed to the origin of life, and another is how life forms on the Earth's surface may have been extinguished or decimated by comet impacts from the most ancient times until the present. Let us now deal with these in turn.

8.2.1. *Prebiotic material from comets*

While comets are not considered likely as harbingers of living species from elsewhere in the Universe to the Earth, they can possibly have brought organic material in quantities that are relevant to play an important role in the supply of substances (amino acids) seen as building blocks or precursors of the first life forms. A long-standing debate concerns whether these precursors were somehow predominantly fabricated in the Earth's atmosphere or if the undeniable, exogenous supply via comets was even more important.

There are two outstanding issues involved in this debate. The putative, pre-biotic chemical factory of the early Earth's atmosphere is very sensitive to the state of oxidation, being fundamentally dependent on CH_4 as input. A methane-dominated, and therefore reducing, atmosphere has generally been considered likely for our planet, when it was young. A suggestion that energy sources (UV radiation, lightning and coronal discharges) in such an atmosphere would allow formation of more complex organic compounds out of the simplest ones (like methane) was confirmed by the famous *Miller-Urey experiment* in the early 1950s (Miller 1953). Hence, if indeed the Earth had a reducing atmosphere, large quantities of suitable molecules for the origin of life may well have been produced by this endogenous process.

One problem is that the assumption of a reducing atmosphere has been challenged. If the chemical state was at least moderately oxidizing, so that carbon existed as CO_2 rather than CH_4, the production of complex organics would be severely quenched. This would leave the stage open for an exogenous source of such organics in the form of meteoritic or cometary bombardment. However, it also needs to be established that such a source would be plentiful enough to be the primary one.

Chyba and Sagan (1997) summarized the efficiencies of organics production via different mechanisms operating in a reducing or intermediate Earth atmosphere. In the reducing case, a maximum of 1×10^{12} kg/yr was listed for the agent of suprathermal hydrogen produced by UV photolysis of hydrogen sulfide, and in the intermediate case the total of all mechanisms was much lower. If acting over

500 Myr, this would yield a maximum of 5×10^{20} kg of endogenously produced organics.

If we take the above estimate of 3×10^{22} kg for the total mass of chondritic material that fell on the Earth after core formation, and if we use 6% for the fraction of this mass contributed by carbonaceous chondrites and a rough estimate of 10% for the mass fraction of organics in such meteorites, the result is about 2×10^{20} kg of exogenously contributed organics. With these estimates, it appears that the endogenous source might be the most important one but only for a reducing atmosphere — otherwise, the exogenous source would dominate.

It is a different question, how the endogenously and exogenously derived organics differ in chemical composition and which ones are best suited from the prebiotic point of view. This is difficult to answer, but we note that amino acids have been identified in carbonaceous chondrites, the first discovery of glycine having been made in the Murchison meteorite (Kvenvolden *et al.* 1970). Glycine was also reported in the samples returned from comet 81P/Wild 2 by the Stardust spacecraft, although this discovery still needs full confirmation. Finally, it was unambiguously identified in the coma of comet 67P/Churyumov-Gerasimenko by the ROSINA instrument on board Rosetta (Altwegg *et al.* 2016).

It should be emphasized that the exogenous source of prebiotic organics mainly consists of carbonaceous chondrites swept up as part of the late veneer (subsequent to Moon formation). It is, however, of great interest that glycine has been found in a comet, since we have argued that comets could have brought water into the parent bodies of carbonaceous chondrites in the earliest days of the solar system. If they did so, they would also have brought their organics with them. Hence, the ultimate origin of the exogenous, prebiotic organics might actually be found in comets.

8.2.2. *Impact frustration of life*

It is not known, at which time life first existed on the Earth. The most ancient fossils or circumstantial evidence only set a lower limit to the age of life, which is certainly at least 3.5 Gyr and possibly as large as 3.8 Gyr. On the other hand, we can infer some properties of our planet when it was very young and some features of its environment. In particular, we know that the rate of cosmic impacts was very high. To some extent, comets were involved as projectiles. When the number of impacts is very high, the largest projectiles are bound to strike the Earth with very large energies and thus a high potential to wipe out many forms of life.

This leads to fundamental questions: Did the Earth experience impacts of such a magnitude that no life at all could survive, and did some of these occur so late that life had already started to develop? If the answer to both questions is yes, a consequence is that we cannot trace the first origin of life on Earth, since the evidence is rubbed out. But even if we imagine the second answer to be no (which would be very hard to prove), the first life may still have had to cope with extreme hardships. If this was so, one may ask if the nature of these impact-related episodes would have left any marks that could still be identified in the ancestry of terrestrial life.

Many papers have dealt with these issues, and the topic was summarized in an excellent review by Zahnle and Sleep (1997). Figure 8.3 is an updated version of their illustration. This indicates how the largest impact energy has decreased with time for the Earth and the Moon since their formation. While many details can be debated, the general trend is clear. One may formulate a unique, statistical relation between the energy of an impact (i.e., the kinetic energy of the projectile upon collision) and the depth of a hypothetical ocean that would be evaporated. The latter, in turn, is important for judging the possible consequences of the impact for life on Earth.

As stated by the authors, the Moon-forming impact must certainly have been entirely lethal, and moreover, all organic materials previously accreted by the Earth were destroyed. Therefore, the pursuit of life had to start all over by the accretion or endogenous production of fresh organics. With the recent datings of this impact at about 4.42 – 4.51 Gyr (Jacobson *et al.* 2014; Connelly and Bizzarro 2016), which are younger than most previous estimates, it seems that

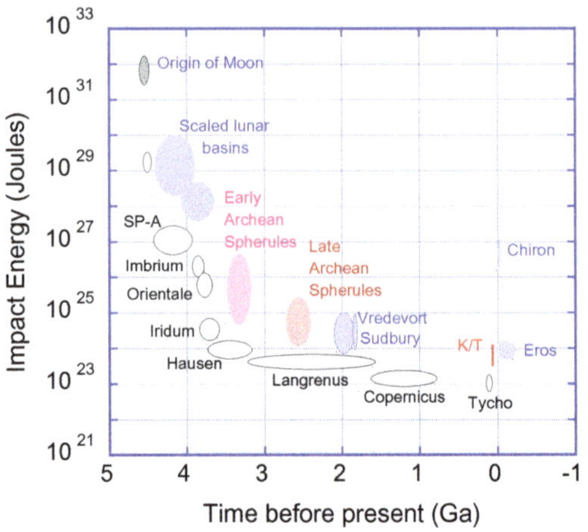

Fig. 8.3. Impact energies of the largest events for the Earth (shaded areas) and the Moon (open boxes) versus time before present. The range of uncertainty is given by the extent of the area. Reproduced with permission. Courtesy K. J. Zahnle.

this may likely have been the last fully sterilizing event. If there was a subsequent one, it probably happened so early that the Earth could not yet host an ocean.

The energy of an ocean-vaporizing impact can be estimated as about 2×10^{28} J. This would mean an average depth of 3 000 m vaporized water. Such an impact would likely be nearly fully sterilizing, and it is unclear if any such event happened during the Late Heavy Bombardment between 4.1 and 3.8 Gyr ago. Below this, at one order of magnitude less, which means about 300 m, there were certainly several impacts during that time. Zahnle and Sleep (1997) estimated that each of these impacts would allow the survival of life in a very selective fashion. Surviving organisms would be restricted to thermophiles (heat loving) and halophiles (salt loving), thriving at great depths in the ocean.

It is very likely that the Earth had an ocean at the time in question. Evidence for a very early ocean (~4.3 Gyr) from zircons and radiogenic xenon has been quoted by Drake and Righter (2002). This is close to the minimum required to solidify the crust after the Moon-forming impact, and hence, if this impact purged the Earth of its water, a large part of the current water inventory must have been re-accreted very fast. This is in line with the rapid fall-off of the impact rate due to stray planetesimals and thus presents no dilemma. Moreover, between the major impacts, the Earth's surface had enough time to cool off and allow oceans to grow by precipitation.

There is still a large uncertainty about the time that life first occurred on the Earth. However, Zahnle and Sleep (1997) made the interesting speculation that the nearly fatal impacts discussed above could have acted as a filter for terrestrial life forms. Thus, one might explain the conclusion from molecular phylogeny that life on Earth descends from sulphur-metabolizing, thermophilic bacteria (Lake 1988) without invoking an origin of life in hot environments near mid-ocean ridges. Such environments were instead the only places, where life could survive the extremely damaging effects of the major impacts that haunted our planet at the time of the Late Heavy Bombardment. As an interesting consequence, if this is true, life must have originated early enough (i.e., before the impacts in question) for the filter to operate.

What is the true story of the origin of life on Earth, time may tell. In any case, major impacts are bound to play an important role in its early development. To what degree comets were involved is less certain. We have seen that these were probably important in providing water and organics to the Earth, but the bulk of the material with which the early Earth was bombarded was not cometary — it was rather asteroidal. After all, it is of interest to estimate the cometary contribution to the major cratering events during the Late Heavy Bombardment. For the case of the Moon, Rickman *et al.* (2017b) estimated a most likely radius of about 30 km for the largest cometary projectile, and from their calculations one can also derive a corresponding radius of about 90 km for the Earth. This is close to the size of 95P/Chiron, which is one of the biggest known Centaurs.

In fact, Chiron is included in Fig. 8.3 although it has not struck the Earth and is very unlikely to do so in the future. But the energy of this potential impact is roughly correct and shows that its effect would be dramatic. Therefore, it appears that at least one of the cometary projectiles during the LHB, if the latter occurred due to a late planetary instability of the Nice Model, would have been able to make an imprint on the history of life on Earth.

8.2.3. *Comets and the impact hazard*

By the impact hazard we mean the threat to human lives and civilization posed by the fact that asteroids and comets sometimes have orbits that cross the Earth's orbit so that the minimum distance between the heliocentric ellipses (the so-called MOID[3]) is smaller than the Earth's collisional radius. The latter is the distance from the Earth's center within which a projectile must aim on approach to our planet for a collision to take place. It is larger than the Earth's actual radius due to gravitational focusing and hence depends on the speed of approach. Due to orbital evolution, the MOID of any object varies with time. It is typically large enough to avoid any risk of collision for most of the time, but occasionally the orbits do intersect and the MOID goes to zero.

Objects in Earth-crossing orbits thus pose non-zero impact risks on the average over time, and impacts do occur. Relatively recent impact craters bear witness of this, and the risk is vidimated by the frequent observations of small near-Earth asteroids (NEAs) encountering the Earth at close range. By monitoring the population of near-Earth objects (NEOs: comets and asteroids) and accurately tracing their orbits, it is possible to quantify the impact hazard in different ways. For instance, we can now say with high confidence that none of the monitored objects poses any significant impact risk during the coming century.

Another inference concerns a statistical estimate of the rate of impacts onto the Earth. Investigations of the orbital evolutions of NEOs allow to determine the average impact probability per object per Myr, and from the discoveries made at observational surveys one can estimate the total number of NEOs down to a given diameter.

[3]MOID stands for Minimum Orbit Intersection Distance.

Due to the major efforts within the *Spaceguard* project, this number is now known with a fair accuracy to be close to 1 000 with diameters $D > 1$ km (Mainzer *et al.* 2014). Combined with an average impact probability of a few times 10^{-3} per Myr, the expectancy is a few impacts per Myr.

This estimate can be compared with crater counts on the Earth. As a rough guideline, a $D = 1$ km NEO would yield a crater of 20 km diameter upon impact. However, only a minority of these craters are recognizeable because most impacts occur in oceans, and most of the rest are wiped out relatively fast by natural erosion. In spite of such uncertainties, this independent estimate of cratering rate is in good agreement with the above NEO impact rate. Further confirmation comes from the lunar maria, where the number of craters with diameters above the corresponding limit (roughly, 10 km) yields an average for the last 3 Gyr. Again, this estimate is in agreement with the others.

The cometary contribution to the current NEO population is relatively small. The majority of NEOs derive from the asteroid belt. We may therefore take the agreement between the different estimates of impact rate, which refer to very different time scales, to indicate that the cometary contribution has been small — both currently and as an average over billions of years. We can make yet another estimate based on the dynamics of Jupiter Family comets. Rickman *et al.* (2017a) used a model of the physical lifetimes to derive a steady-state capture rate for the current Jupiter Family of about 16 comets per century with $D > 1$ km, and Rickman *et al.* (2017b) derived an average collisional probability with the Earth of about 7×10^{-7} for each Jupiter Family comet. From these two numbers, we obtain an estimate of $\sim$0.1 impacts per Myr by Jupiter Family comets, confirming the impression that comets — including dormant ones masquerading as asteroids — make only a minor contribution.

Since the Jupiter Family derives mostly from the Scattered Disk and the dynamical transfer mechanism has been the same since more than 3 Gyr, the rate of such impacts can only have decreased with time due to the slow decline of the Scattered Disk population. The leakage rate of main belt asteroids into NEO orbits is not expected to have changed during the same time. However, even if the cometary contribution was thus larger in the past, due to its smallness there is no observable decrease of the cratering rate. By contrast, the arrival rate of new comets from the Oort Cloud may well have shown large fluctuations on top of an expected, slow decrease. We referred to these in Sec. 5.3 as comet showers. As a consequence, during and just after such a shower, the rate of perihelion passages of long-period and Halley Type comets would also have surged. If, indeed, there have been several major comet showers, there must have been temporary increases in the cratering rate on the Earth, but the preserved record of ancient craters is not good enough to identify those peaks.

We hence need to estimate the cratering rate caused at present by new, long-period and Halley Type comets. The current combined rate of perihelion passages of new and old long-period comets with nuclear diameters $D > 1$ km according to the conversion between absolute total magnitude and nuclear diameter by Sosa and Fernández (2011) is very small — at most one per year in Earth-crossing orbits. With the typically high inclinations, the average collision probability is very small, so the rate of impacts by such comets is much less than the one due to Jupiter Family comets. On the other hand, Halley Type comets seem to be much more important, though the observational basis for any conclusion is very poor. Reasonably reliable estimates of nuclear sizes are very few, but nonetheless it is worth noting that there is no Jupiter Family comet on a currently Earth-crossing orbit, whose nuclear radius is anywhere near that of 1P/Halley or the one tentatively estimated for 109P/Swift-Tuttle,[4] both of which have $q <$ 1 AU.

Thus, the number of large, multi-km nuclei that could collide with the Earth is as much as ten times larger for Halley Type than for Jupiter Family comets as a long-term average. On the other hand, these also have roughly ten times longer orbital periods, so

[4]This is the parent comet of the August Perseid meteor shower, which is known to provide large numbers of bolides annually. The size estimate is due to Sekanina (1981); see also Lamy *et al.* (2004).

the overall contributions to the terrestrial impact rate may be of similar magnitude. Of course, the Halley Type comets with their large inclinations and often retrograde orbits would strike the Earth at much larger velocities, so the craters would be larger for a given nuclear size. In principle, it is therefore possible that major comet showers are associated with periods of several Myr (Fernández and Ip 1987), when the Earth experiences a dramatically increased risk of a very high energy impact than under the usual conditions. Those impacts would most likely involve cometary projectiles of the Halley Type. Whether the largest impact craters on Earth were caused by asteroids or comets has been subject to some discussion, and we note that Wetherill (1989) favored a cometary origin in most cases.

In fact, there has been a great deal of discussion of cometary impacts on Earth in connection with comet showers in the literature of the last decades. A causal link between a major impact and a mass extinction of biological species was first demonstrated by Alvarez *et al.* (1980) regarding the Cretacious-Tertiary (K/T) extinction about 65 Myr ago. The size of this projectile is usually estimated at roughly 10 km. From today's perspective, it may be noted that none of the other mass extinctions identified in the geological record has been linked to such an impact with the same confidence as in the K/T case. On the other hand, even though there are certainly other events than impacts that may pose a great risk to life forms globally, large enough impacts are certainly able to do so, and more than one impact of energy comparable to the K/T event has probably occurred since the Cambrian explosion about 500 Myr ago. There is no solid evidence that precludes impacts from having been a major cause of mass extinctions.

The reason that comets came into the focus of this discussion is that a periodicity was suspected in the timing of both extinctions (Raup and Sepkoski 1984) and cratering impacts (Alvarez and Muller 1984). The reported periods were between about 25 and 30 Myr. Because neither geological triggers nor delivery of large main belt asteroids have any reason to occur periodically, suspicion naturally fell on Oort Cloud comets. However, there are two problems that effectively remove all credit from this idea.

First, the very existence of a periodicity has later been challenged by in-depth statistical analysis of the data available in the 1980's as well as a decade later (Jetsu and Pelt 2000). These authors found no real periodicities in either kind of data (biological extinctions or impact craters) and were also able to identify plausible reasons for spurious periods in the above-mentioned range to appear, depending on the technique used. Doubts on the periods had earlier been voiced, e.g., by Yabushita (1992).

Second, the physical mechanism behind the periodicity remains doubtful, to say the least. The most favored idea was to identify the period in extinctions and cratering with the half-period of the Sun's vertical oscillation with respect to the Galactic midplane during its galactocentric orbital motion (for instance, Stothers 1985; Rampino and Stothers 1986). The maxima of impact rate would then occur when the Sun passed the midplane and experienced an increased encounter rate with the more massive types of Galactic field stars. These could send large numbers of comets from the cloud into orbits with perihelia within 1 AU from the Sun. While attractive at first glance, this idea suffers from serious problems. Recent simulations of Oort Cloud dynamics (Fouchard *et al.* 2017) have shown that the stars triggering the strongest showers are not only of the massive kinds but often belong to the majority of red dwarf stars with small masses. Such stars are distributed around the Galactic midplane with a scale height exceeding the amplitude of the solar motion, so that their closest encounters would not favor the Sun's midplane passages. Other possible Oort Cloud perturbers like Giant Molecular Clouds (see Sec. 3.4) have also been discussed, but even in the midplane these are too rare to be able to upset the Oort Cloud with any regularity.

In summary, it seems likely that some mass extinctions have been caused by the impacts of large comet nuclei, possibly though not necessarily during comet showers. The most likely among the orbital classes is the Halley Type, which allows the impact velocity to be very large. Thus, even though comets do not present a major impact hazard in the short term currently, they may have been the arbiters of life forms on Earth in the remote past. But, then again, comets give and comets take. The water that the Earth obtained from its building blocks and early bombardment may ultimately derive from cometary objects. Finally, in the very beginning of life on Earth, comets may have been instrumental in providing the prebiotic substances, from which life arose.

Bibliography

A'Hearn, M. F., Schleicher, D. G., Feldman, P. D., Millis, R. L. and Thompson, D. T. (1984), *Astron. J.* **89**, 579–591.

A'Hearn, M. F., Millis, R. L., Schleicher, D. G., Osip, D. J. and Birch, P. V. (1995), *Icarus* **118**, 223–270.

A'Hearn, M. F., Belton, M. J. S., Delamere, W. A., *et al.* (2011), *Science* **332**, 1396–1400.

A'Hearn, M. F., Feaga, L. M., Keller, H. U., *et al.* (2012), *Astrophys. J.* **758**, 29.

Adams, F. C. (2010), *Annu. Rev. Astron. Astrophys.* **48**, 47–85.

Aggarwal, H. R. and Oberbeck, V. R. (1974), *Astrophys. J.* **191**, 577–588.

Alvarez, L. W., Alvarez, W., Asaro, F. and Michel, H. V. (1980), *Science* **208**, 1095–1108.

Alvarez, W. and Muller, R. A. (1984), *Nature* **308**, 718–720.

Altwegg, K., Balsiger, H., Bar-Nun, A., *et al.* (2015), *Science* **347**, 1261952.

Altwegg, K., Balsiger, H., Bar-Nun, A., *et al.* (2016), *Sci. Adv.* **2**, e1600285.

Asphaug, E. (2014), *Annu. Rev. Earth Planet. Sci.* **42**, 551–578.

Asphaug, E. and Benz, W. (1996), *Icarus* **121**, 225–248.

Bai, X.-N. and Stone, J. M. (2010), *Astrophys. J. Suppl.* **190**, 297–310.

Bailey, M. E. (1986), *Nature* **324**, 350–352.

Balsiger, H., Altwegg, K., Bar-Nun, A., *et al.* (2015), *Sci. Adv.* **1**, e1500377.

Bar-Nun, A. and Kleinfeld, I. (1989), *Icarus* **80**, 243–253.

Bar-Nun, A., Dror, J., Kochavi, E. and Laufer, D. (1987), *Phys. Rev. B* **35**, 2427–2435.

Bar-Nun, A., Notesco, G. and Owen, T. (2007), *Icarus* **190**, 655–659.

Barucci, M. A., Filacchione, G., Fornasier, S., *et al.* (2016), *Astron. Astrophys.* **595**, A102.

Batygin, K. and Brown, M. E. (2016), *Astron. J.* **151**, 22.

Belton, M. J. S., Thomas, P., Veverka, J., *et al.* (2007), *Icarus* **187**, 332–344.

Belton, M. J. S., Thomas, P., Carcich, B., *et al.* (2013), *Icarus* **222**, 477–486.
Bernstein, G. M., Trilling, D. E., Allen, R. L., *et al.* (2004), *Astron. J.* **128**, 1364–1390.
Bertaux, J.-L. (2015), *Astron. Astrophys.* **583**, A38.
Bibring, J.-P., Langevin, Y., Carter, J., *et al.* (2015), *Science* **349**, aab0671.
Bieler, A., Altwegg, K., Balsiger, H., *et al.* (2015), *Icarus* **583**, A7.
Bieler, A., Altwegg, K., Balsiger, H., *et al.* (2015), *Nature* **526**, 678–681.
Biermann, L. (1978), in *Astronomical Papers Dedicated to Bengt Strömgren* (eds. A. Reiz and T. Andersen), Copenhagen: Copenhagen Observatory, pp. 327–336.
Biver, N., Bockelée-Morvan, D., Colom, P., *et al.* (2002), *Earth, Moon, and Planets* **90**, 5–14.
Bockelée-Morvan, D., Crovisier, J., Mumma, M. J. and Weaver, H. A. (2004), in *Comets II* (eds. M. C. Festou, H. U. Keller and H. A. Weaver), Tucson: Univ. Arizona Press, pp. 391–423.
Bodewits, D., Lara, L. M., A'Hearn, M. F., *et al.* (2016), *Astron. J.* **152**, 130.
Boehnhardt, H. (2004), in *Comets II* (eds. M. C. Festou, H. U. Keller and H. A. Weaver), Tucson: Univ. Arizona Press, pp. 301–316.
Bowell, E. and Muinonen, K. (1994), in *Hazards due to Comets and Asteroids* (eds. T. Gehrels, M. S. Matthews and A. Schumann), Tucson: Univ. Arizona Press, pp. 149–197.
Brasser, R. (2013), *Space Sci. Rev.* **174**, 11–25.
Brasser, R. and Morbidelli, A. (2013), *Icarus* **225**, 40–49.
Brasser, R., Duncan, M. J. and Levison, H. F. (2006), *Icarus* **184**, 59–82.
Brasser, R., Duncan, M. J. and Levison, H. F. (2007), *Icarus* **191**, 413–433.
Brown, M. E. and Batygin, K. (2016), *Astrophys. J. Lett.* **824**, L23.
Brownlee, D. E., Horz, F., Newburn, R. L., *et al.* (2004), *Science* **304**, 1764–1769.
Budzien, S. A. and Feldman, P. D. (1992), *Icarus* **99**, 143–152.
Byl, J. (1983), *Moon and Planets* **29**, 121–137.
Campins, H., Hargrove, K., Pinilla-Alonso, N., *et al.* (2010), *Nature* **464**, 1320–1321.
Capaccioni, F., Coradini, A., Filacchione, G., *et al.* (2015), *Science* **347**, aaa0628.
Carusi, A. and Valsecchi, G. B. (1979), in *Asteroids* (ed. T. Gehrels), Tucson: Univ. Arizona Press, pp. 391–416.
Carusi, A. and Valsecchi, G. B. (1987), in *Interplanetary Matter* (eds. Z. Ceplecha and P. Pecina), Praha: Astron. ústav CSAV, pp. 21–28.
Carusi, A., Kresák, L., Perozzi, E. and Valsecchi, G. B. (1987), *Astron. Astrophys.* **187**, 899–905.

Carusi, A., Kresák, L., Perozzi, E. and Valsecchi, G. B. (1995), *Atlas of Dynamical Evolutions of Short-Period Comets* (http://web.archive.org/web/19970731012601/ http://www.ias.fra.cnr.it/ias-home/comet/catalog.html)

Ceccarelli, C., Caselli, P., Bockelée-Morvan, D., *et al.* (2014), in *Protostars and Planets VI* (eds. H. Beuther *et al.*), Tucson: Univ. Arizona Press, pp. 859–882.

Charnoz, S. and Morbidelli, A. (2007), *Icarus* **188**, 468–480.

Chen, J. and Jewitt, D. (1994), *Icarus* **108**, 265–271.

Chyba, C. F. and Sagan, C. (1997), in *Comets and the Origin and Evolution of Life* (eds. P. J. Thomas, C. F. Chyba and C. P. McKay), New York: Springer-Verlag, pp. 147–173.

Ciarletti, V., Levasseur-Regourd, A. C., Lasue, J., *et al.* (2015), *Astron. Astrophys.* **583**, A40.

Connelly, J. N. and Bizzarro, M. (2016), *Earth Planet. Sci. Lett.* **452**, 36–43.

Davidsson, B. J. R. and Gutiérrez, P. J. (2004), *Icarus* **168**, 392–408.

Davidsson, B. J. R. and Gutiérrez, P. J. (2005), *Icarus* **176**, 453–477.

Davidsson, B. J. R. and Gutiérrez, P. J. (2006), *Icarus* **180**, 224–242.

Davidsson, B. J. R., Gutiérrez, P. J. and Rickman, H. (2007), *Icarus* **187**, 306–320.

Davidsson, B. J. R., Gutiérrez, P. J., Groussin, O., *et al.* (2013), *Icarus* **224**, 154–171.

Davidsson, B. J. R., Rickman, H., Bandfield, J. L., *et al.* (2015), *Icarus* **252**, 1–21.

Davidsson, B. J. R., Sierks, H., Güttler, C., *et al.* (2016), *Astron. Astrophys.* **592**, A63.

Davis, D. R. and Farinella, P. (1997), *Icarus* **125**, 50–60.

Dello Russo, N., Vervack, R. J., Weaver, H. A., *et al.* (2007), *Nature* **448**, 172–175.

Delsemme, A. H. (1982), in *Comets* (ed. L. L. Wilkening), Tucson: Univ. Arizona Press, pp. 85–130.

Delsemme, A. H. (1987), *Astron. Astrophys.* **187**, 913–918.

DeMeo, F. and Binzel, R. P. (2008), *Icarus* **194**, 436–449.

Denning, W. F. (1928), *J. Brit. Astron. Assoc.* **38**, 302.

De Sanctis, M. C., Capaccioni, F., Ciarniello, M., *et al.* (2015), *Nature* **525**, 500–503.

Di Sisto, R. P., Fernández, J. A. and Brunini, A. (2009), *Icarus* **203**, 140–154.

Drake, M. J. and Righter, K. (2002), *Nature* **416**, 39–44.

Duncan, M. J. and Levison, H. F. (1997), *Science* **276**, 1670–1672.

Duncan, M., Quinn, T. and Tremaine, S. (1987), *Astron. J.* **94**, 1330–1338.

Duncan, M., Quinn, T. and Tremaine, S. (1988), *Astrophys. J. Lett.* **328**, L69–L73.

Duncan, M. J., Levison, H. F. and Dones, L. (2004), in *Comets II* (eds. M. C. Festou, H. U. Keller and H. A. Weaver), Tucson: Univ. Arizona Press, pp. 193–204.

Dybczyński, P. A. (1994), *Celest. Mech and Dynam. Astron.* **58**, 139–150.

Dybczyński, P. A. (2002), *Astron. Astrophys.* **383**, 1049–1053.

Dybczyński, P. A. and Królikowska, M. (2011), *Mon. Not. R. Astron. Soc.* **416**. 51–69.

Edgeworth, K. E. (1949), *Mon. Not. R. Astron. Soc.* **109**, 600–609.

Eggers, S. and Woolfson, M. M. (1996), *Mon. Not. R. Astron. Soc.* **282**, 13–18.

El-Maarry, M. R., Thomas, N., Gracia-Berná, *et al.* (2015), *Geophys. Res. Lett.* **42**, 5170–5178.

Elsila, J. E., Glavin, D. P. and Dworkin, J. P. (2009), *Meteor. Planet. Sci.* **44**, 1323–1330.

Everhart, E. (1972), *Astrophys. Lett.* **10**, 131–135.

Everhart, E. (1977), in *Comets, Asteroids, Meteorites: Interrelations, Evolution, and Origins* (ed. A. H. Delsemme), Toledo, Oh: Univ. of Toledo, pp. 99–104.

Fernández, J. A. (1997), *Icarus* **129**, 106–119.

Fernández, J. A. and Ip, W.-H. (1983a), *Icarus* **54**, 377–387.

Fernández, J. A. and Ip, W.-H. (1983b), in *Asteroids, Comets, Meteors* (eds. C.-I. Lagerkvist and H. Rickman), Uppsala: Astronomiska Observatoriet, pp. 387–390.

Fernández, J. A. and Ip, W.-H. (1987), *Icarus* **71**, 46–56.

Fernández, J. A., Rickman, H. and Kamél, L. (1992), in *Periodic Comets* (eds. J. A. Fernández and H. Rickman), Montevideo: Univ. de la Republica, Dept. de Publicaciones, pp. 143–158.

Fernández, J. A., Tancredi, G., Rickman, H. and Licandro, J. (1999), *Astron. Astrophys.* **352**, 327–340.

Fernández, Y. R., Jewitt, D. C. and Sheppard, S. S. (2005), *Astron. J.* **130**, 308–318.

Fernández, Y. R., Kelley, M. S., Lamy, P. L., *et al.* (2013), *Icarus* **226**, 1138–1170.

Festou, M., Rickman, H. and Kamél, L. (1990), *Nature* **345**, 235–238.

Festou, M. C., Rickman, H. and West, R. M. (1993), *Astron. Astrophys. Rev.* **5**, 37–163.

Filacchione, G., De Sanctis, M. C., Capaccioni, F., *et al.* (2016a), *Nature* **529**, 368–372.

Filacchione, G., Raponi, A., Capaccioni, F., *et al.* (2016b), *Science* **354**, 1563–1566.

Fink, U. (2009), *Icarus* **201**, 311–334.

Fornasier, S., Mottola, S., Keller, H. U., *et al.* (2016), *Science* **354**, 1566–1570.

Fouchard, M., Froeschlé, Ch., Rickman, H. and Valsecchi, G. B. (2011a), *Icarus* **214**, 334–347.

Fouchard, M., Rickman, H., Froeschlé, Ch. and Valsecchi, G. B. (2011b), *Astron. Astrophys.* **535**, A86.

Fouchard, M., Rickman, H., Froeschlé, Ch. and Valsecchi, G. B. (2013), *Icarus* **222**, 20–31.

Fouchard, M., Rickman, H., Froeschlé, Ch. and Valsecchi, G. B. (2014a), *Icarus* **231**, 110–121.

Fouchard, M., Rickman, H., Froeschlé, Ch. and Valsecchi, G. B. (2014b), *Icarus* **231**, 99–109.

Fouchard, M., Rickman, H., Froeschlé, Ch. and Valsecchi, G. B. (2017), *Icarus*, **292**, 218–233.

Fougere, N., Altwegg, K., Berthelier, J.-J., *et al.* (2016), *Astron. Astrophys.* **588**, A134.

Francis, P. J. (2005), *Astrophys. J.* **635**, 1348–1361.

Fraser, W. C., Brown, M. E., Morbidelli, A., Parker, A. and Batygin, K. (2014), *Astrophys. J.* **782**, 100.

Froeschlé, C. and Rickman, H. (1986), *Astron. Astrophys.* **170**, 145–160.

Froeschlé, Cl., Klinger, J. and Rickman, H. (1983), in *Asteroids, Comets, Meteors* (eds. C.-I. Lagerkvist and H. Rickman), Uppsala: Astronomiska Observatoriet, pp. 215–224.

Fulle, M. (2004), in *Comets II* (eds. M. C. Festou, H. U. Keller and H. A. Weaver), Tucson: Univ. Arizona Press, pp. 565–575.

Fulle, M., Marzari, F., Della Corte, V., *et al.* (2016), *Astrophys. J.* **821**, 19.

Gallardo, T. (2006), *Icarus* **181**, 205–217.

Geiss, J. and Reeves, H. (1981), *Astron. Astrophys.* **93**, 189–199.

Gladman, B., Kavelaars, J. J., Petit, J.-M., *et al.* (2001), *Astron. J.* **122**, 1051–1066.

Goldreich, P. and Ward, W. R. (1973), *Astrophys. J.* **183**, 1051–1062.

Gomes, R., Levison, H. F., Tsiganis, K. and Morbidelli, A. (2005), *Nature* **435**, 466–469.

Gonczi, R., Rickman, H. and Froeschlé, C. (1992), *Mon. Not. R. Astron. Soc.* **254**, 627–634.

Gounelle, M. and Meynet, G. (2012), *Astron. Astrophys.* **545**, A4.

Greenberg, J. M. and Hage, J. I. (1990), *Astrophys. J.* **361**, 260–274.

Gunn, J. E., Knapp, G. R. and Tremaine, S. D. (1979), *Astron. J.* **84**, 1181–1188.

Gunnarsson, M., Rickman, H., Festou, M. C., Winnberg, A. and Tancredi, G. (2002), *Icarus* **157**, 309–322.

Gustafsson, B., Church, R. P., Davies, M. B. and Rickman, H. (2016), *Astron. Astrophys.* **593**, A85.

Guzzo, M. and Lega, E. (2015), *Astron. Astrophys.* **579**, A79.

Hanner, M. S., Lynch, D. K. and Russell, R. W. (1994), *Astrophys. J.* **425**, 274–285.

Harmon, J. K., Nolan, M. C., Ostro, S. J. and Campbell, D. B. (2004), in *Comets II* (eds. M. C. Festou, H. U. Keller and H. A. Weaver), Tucson: Univ. Arizona Press, pp. 265–279.

Harmon, J. K., Nolan, M. C., Giorgini, J. D. and Howell, E. S. (2010), *Icarus* **207**, 499–502.

Hartmann, L. (2008), *Phys. Scripta* **130**, 014012.

Hartogh, P., Lis, D. C., Bockelée-Morvan, D., *et al.* (2011), *Nature* **478**, 218–220.

Hässig, M., Altwegg, K., Balsiger, H., *et al.* (2015), *Science* **347**, aaa0276.

Heisler, J. and Tremaine, S. (1986), *Icarus* **65**, 13–26.

Hills, J. G. (1981), *Astron. J.* **86**, 1730–1740.

Hirabayashi, M., Scheeres, D. J., Chesley, S. R., *et al.* (2016), *Nature* **534**, 352–355.

Holmberg, J. and Flynn, C. (2000), *Mon. Not. R. Astron. Soc.* **313**, 209–216.

Hut, P. and Tremaine, S. (1985), *Astron. J.* **90**, 1548–1557.

Jacobson, S. A., Morbidelli, A., Raymond, S. N., *et al.* (2014), *Nature* **508**, 84–87.

Jetsu, L. and Pelt, J. (2000), *Astron. Astrophys.* **353**, 409–418.

Jewitt, D. (2012), *Astron. J.* **143**, 66.

Johansen, A., Oishi, J. S., Mac Low, M.-M., *et al.* (2007), *Nature*, **448**, 1022–1025.

Johansen, A., Blum, J., Tanaka, H., *et al.* (2014), in *Protostars and Planets VI* (eds. H. Beuther *et al.*), Tucson: Univ. Arizona Press, pp. 547–570.

Jorda, L., Crovisier, J. and Green, D. W. E. (1992), in *Asteroids, Comets, Meteors 1991* (eds. Harris, A. W. and Bowell, E.), Houston: Lunar Planet. Inst., pp. 285–288.

Jorda, L., Gaskell, R., Capanna, C., *et al.* (2016), *Icarus* **277**, 257–278.

Joss, P. C. (1973), *Astron. Astrophys.* **25**, 271–273.

Jutzi, M. and Asphaug, E. (2015), *Science* **348**, 1355–1358.

Jutzi, M., Benz, W., Toliou, A., Morbidelli, A. and Brasser, R. (2017), *Astron. Astrophys.* **597**, A61.

Kaib, N. A. and Quinn, T. (2009), *Science* **325**, 1234–1236.

Kaib, N. A. and Chambers, J. E. (2016), *Mon. Not. R. Astron. Soc.* **455**, 3561–3569.

Kamél, L. (1991), *Icarus* **93**, 226–245.

Kazimirchak-Polonskaya, E. I. (1972), in *The Motion, Evolution of Orbits, and Origin of Comets* (eds. G. A. Chebotarev, E. I. Kazimirchak-Polonskaya and B. G. Marsden), Dordrecht: Reidel, pp. 373–397.
Keller, H. U., Mottola, S., Davidsson, B., *et al.* (2015), *Astron. Astrophys.* **583**, A34.
Keller, H. U., Hviid, S. F., Mottola, S., *et al.* (2016), *AAS/DPS 48th Meeting*, Abstract id. 201.03.
Kissel, J., Brownlee, D. E., Büchler, K., *et al.* (1986), *Nature* **321**, 336–337.
Klahr, H. H. and Bodenheimer, P. (2003), *Astrophys. J.* **582**, 869–892.
Kofman, W., Herique, A., Barbin, Y., *et al.* (2015), *Science* **349**, aab0639.
Kouchi, A., Yamamoto, T., Kozasa, T., Kuroda, T. and Greenberg, J. M. (1994), *Astron. Astrophys.* **290**, 1009–1018.
Kozai, Y. (1962), *Astron. J.* **67**, 591–598.
Kresák, L. (1974), *Bull. Astron. Inst. Czech.* **25**, 293–304.
Kresák, L. (1987), *Astron. Astrophys.* **187**, 906–908.
Królikowska, M. (2006), *Acta Astron.* **56**, 385–412.
Kührt, E. and Keller, H. U. (1994), *Icarus* **109**, 121–132.
Küppers, M., O'Rourke, L., Bockelée-Morvan, D., *et al.* (2014), *Nature* **505**, 525–527.
Kuiper, G. P. (1951), in *Astrophysics* (ed. J. A. Hynek), New York: McGraw-Hill, pp. 357–424.
Kvenvolden, K., Lawless, J., Pering, K., *et al.* (1970), *Nature* **228**, 923–926.
Lada, C. J. and Lada, E. A. (2003), *Annu. Rev. Astron. Astrophys.* **41**, 57–115.
Lake, J. A. (1988), *Nature* **331**, 184–186.
Lamy, P. L., Tóth, I., Fernández, Y. R. and Weaver, H. A. (2004), in *Comets II* (eds. M. C. Festou, H. U. Keller and H. A. Weaver), Tucson: Univ. Arizona Press, pp. 223–264.
Lamy, P. L., Tóth, I., Jorda, L., *et al.* (2008), *Bull. Amer. Astron. Soc.* **40**, 393.
Lara, L. M., Lowry, S., Vincent, J.-B., *et al.* (2015), *Astron. Astrophys.* **583**, A9.
Lee, T., Papanastassiou, D. A. and Wasserburg, G. J. (1976), *Geophys. Res. Lett.* **3**, 41–44.
Levison, H. F. (1996), in *Completing the Inventory of the Solar System* (eds. T. W. Rettig and J. M. Hahn), San Francisko: Astron. Soc. of the Pacific, pp. 173–191.
Levison, H. F. and Duncan, M. J. (1994), *Icarus* **108**, 18–36.
Levison, H. F. and Duncan, M. J. (1997), *Icarus* **127**, 13–32.
Levison, H. F., Morbidelli, A., Dones, L., *et al.* (2002), *Science* **296**, 2212–2215.

Levison, H. F., Terrell, D., Wiegert, P. A., Dones, L. and Duncan, M. J. (2006), *Icarus* **182**, 161–168.
Levison, H. F., Bottke, W. F., Gounelle, M., *et al.* (2009), *Nature* **460**, 364–366.
Levison, H. F., Duncan, M. J., Brasser, R. and Kaufmann, D. E. (2010), *Science* **329**, 187–190.
Levison, H. F., Morbidelli, A., Tsiganis, K., Nesvorný, D. and Gomes, R. (2011), *Astron. J.* **142**, 152.
Licandro, J., Tancredi, G., Lindgren, M., Rickman, H. and Gil Hutton, R. (2000), *Icarus* **147**, 161–179.
Lin, Z.-Y., Ip, W.-H., Lai, I.-L.-, *et al.* (2015), *Astron. Astrophys.* **583**, A11.
Lisse, C. M., Fernández, Y. R., Reach, W. T., *et al.* (2009), *Publ. Astron. Soc. Pacific* **121**, 968–975.
Luspay-Kuti, A., Hässig, M., Fuselier, S. A., *et al.* (2015), *Astron. Astrophys.* **583**, A4.
Lyra, W., Johansen, A., Klahr, H. and Piskunov, N. (2008), *Astron. Astrophys.* **491**, L41–L44.
Mainzer, A., Bauer, J., Grav, T., *et al.* (2014), *Astrophys. J.* **784**, 110.
Malmberg, D., de Angeli, F., Davies, M. B., *et al.* (2007), *Mon. Not. R. Astron. Soc.* **378**, 1207–1216.
Marsden, B. G., Sekanina, Z. and Yeomans, D. K. (1973), *Astron. J.* **78**, 211–225.
Marsden, B. G. and Williams, G. V. (2005), *Catalogue of Cometary Orbits*, 6th edn., Cambridge, Ma: IAU Central Bureau for Astronomical Telegrams.
Masset, F. and Snellgrove, M. (2001), *Mon. Not. R. Astron. Soc.* **320**, L55–L59.
Massironi, M., Simioni, E., Marzari, F., *et al.* (2015), *Nature* **526**, 402–405.
McCord, T. B. and Sotin, C. (2005), *J. Geophys. Res.* **110**, E05009.
McCuskey, S. W. (1963), *Introduction to Celestial Mechanics*, Reading, Ma/Palo Alto/London: Addison-Wesley Publ. Co.
Merényi, E., Földy, L., Szegö, K., Tóth, I. and Kondor, A. (1990), *Icarus* **86**, 9–20.
Merk, R. and Prialnik, D. (2006), *Icarus* **183**, 283–295.
Milani, A., Nobili, A. M. and Carpino, M. (1989), *Icarus* **82**, 200–217.
Miller, S. L. (1953), *Science* **117**, 528–529.
Mommert, M., Hora, J. L., Harris, A. W., *et al.* (2014), *Astrophys. J.* **781**, 25.
Morbidelli, A. and Levison, H. F. (2004), *Astron. J.* **128**, 2564–2576.
Morbidelli, A. and Crida, A. (2007), *Icarus* **191**, 158–171.
Morbidelli, A. and Rickman, H. (2015), *Astron. Astrophys.* **583**, A43.

Morbidelli, A., Chambers, J., Lunine, J. I., *et al.* (2000), *Meteor. Planet. Sci.* **35**, 1309–1320.
Morbidelli, A., Levison, H. F., Tsiganis, K. and Gomes, R. (2005), *Nature* **435**, 462–465.
Morbidelli, A., Bottke, W. F., Nesvorný, D. and Levison, H. F. (2009), *Icarus* **204**, 558–573.
Morbidelli, A., Brasser, R., Gomes, R., Levison, H. F. and Tsiganis, K. (2010), *Astron. J.* **140**, 1391–1401.
Mottola, S., Lowry, S., Snodgrass, C., *et al.* (2014), *Astron. Astrophys.* **569**, L2.
Mousis, O., Lunine, J. I., Luspay-Kuti, A., *et al.* (2016), *Astrophys. J. Lett.* **819**, L33.
Mumma, M. J. and Charnley, S. B. (2011), *Annu. Rev. Astron. Astrophys.* **49**, 471–524.
Nesvorný, D. and Morbidelli, A. (2012), *Astron. J.* **144**, 117.
Nesvorný, D., Youdin, A. N. and Richardson, D. C. (2010), *Astron. J.* **140**, 785–793.
Newton, H. A. (1893), *Mem. Nat. Acad. Sci.* (Washington) **6**, 7.
Nordlander, T., Rickman, H. and Gustafsson, B. (2017), *Astron. Astrophys.* **603**, A112.
O'Brien, D. P., Morbidelli, A. and Levison, H. F. (2006), *Icarus* **184**, 39–58.
Önehag, A., Korn, A., Gustafsson, B., Stempels, E. and Vandenberg, D. A. (2011), *Astron. Astrophys.* **528**, A85.
Önehag, A., Gustafsson, B. and Korn, A. (2014), *Astron. Astrophys.* **562**, A102.
Oort, J. H. (1950), *Bull. Astron. Inst. Netherl.* **11**, 91–110.
Ootsubo, T., Kawakita, H., Hamada, S., *et al.* (2012), *Astrophys. J.* **752**, 15.
Öpik, E. J. (1951), *Proc. R. Irish Acad. Sect. A* **54**, 165–199.
Park, R. S., Konopliv, A. S., Bills, B. G., *et al.* (2016), *Nature* **537**, 515–517.
Pätzold, M., Andert, T., Hahn, M., *et al.* (2016), *Nature* **530**, 63–65.
Pommerol, A., Thomas, N., El-Maarry, M. R., *et al.* (2015), *Astron. Astrophys.* **583**, A25.
Porubčan, V., Kornoš, L. and Williams, I. P. (2006), *Contrib. Astron. Obs. Skalnaté Pleso* **36**, 103–117.
Prialnik, D. and Bar-Nun, A. (1987), *Astrophys. J.* **313**, 93–905.
Prialnik, D., Bar-Nun, A. and Podolak, M. (1987), *Astrophys. J.* **319**, 993–1002.
Quinn, T., Tremaine, S. and Duncan, M. (1990), *Astrophys. J.* **355**, 667–679.
Quirico, E., Moroz, L. V., Schmitt, B., *et al.* (2016), *Icarus* **272**, 32–47.

Rampino, M. R. and Stothers, R. B. (1986), in *The Galaxy and the Solar System* (eds. R. Smoluchowski, J. N. Bahcall and M. S. Matthews), Tucson: Univ. Arizona Press, pp. 241–259.

Raup, D. M. and Sepkoski, J. J. (1984), *Proc. Natl. Acad. Sci. (USA)* **81**, 801–805.

Richardson, J. E., Melosh, H. J., Lisse, C. M. and Carcich, B. (2007), *Icarus* **190**, 357–390.

Rickman, H. (1986), *In Comet Nucleus Sample Return Mission* (ed. O. Melita), ESA SP-249, Noordwijk, pp. 195–205.

Rickman, H. (1989), *Adv. Space Res.* **9**, (3)59–(3)71.

Rickman, H. (2010), in *Dynamics of Small Solar System Bodies and Exoplanets* (eds. J. Souchay and R. Dvorak), Lecture Notes in Physics **790**, Berlin/Heidelberg: Springer, pp. 341–399.

Rickman, H. (2014), *Meteor. Planet. Sci.* **49**, 8–20.

Rickman, H. and Froeschlé, C. (1983a), in *Cometary Exploration* (ed. T. I. Gombosi), Budapest: Akademiai Kiado, pp. 75–84.

Rickman, H. and Froeschlé, C. (1983b), in *Cometary Exploration* (ed. T. I. Gombosi), Budapest: Akademiai Kiado, pp. 109–126.

Rickman, H., Fernández, J. A. and Gustafson, B. Å. S. (1990), *Astron. Astrophys.* **237**, 524–535.

Rickman, H., Kamél, L., Froeschlé, C. and Festou, M. C. (1991), *Astron. J.* **102**, 1446–1463.

Rickman, H., Froeschlé, Ch., Froeschlé, Cl. and Valsecchi, G. B. (2004), *Astron. Astrophys.* **428**, 673–681.

Rickman, H., Fouchard, M., Valsecchi, G. B. and Froeschlé, Ch. (2005), *Earth, Moon, and Planets* **97**, 411–434.

Rickman, H., Fouchard, M., Froeschlé, Ch. and Valsecchi, G. B. (2008), *Celest. Mech, Dynam. Astron.* **102**, 111–132.

Rickman, H., Wiśniowski, T., Wajer, P., Gabryszewski, R. and Valsecchi, G. B. (2014), *Astron. Astrophys.* **569**, A47.

Rickman, H., Marchi, S., A'Hearn, M. F., *et al.* (2015), *Astron. Astrophys.* **583**, A44.

Rickman, H., Gabryszewski, R., Wajer, P., *et al.* (2017a), *Astron. Astrophys.* **598**, A110.

Rickman, H., Wiśniowski, T., Gabryszewski, R., *et al.* (2017b), *Astron. Astrophys.* **598**, A67.

Rickman, H., Błęcka, M., Gurgurewicz, J., *et al.* (2017c), *Space Sci. Rev.*, submitted.

Rivkin, A. S. and Emery, J. P. (2010), *Nature* **464**, 1322–1323.

Robert, F. (2006), in *Meteorites and the Early Solar System II* (eds. D. S. Lauretta and H. Y. McSween Jr.), Tucson: Univ. Arizona Press, pp. 341–351.

Roig, F., Nesvorný, D. and Ferraz-Mello, S. (2002), *Mon. Not. R. Astron. Soc.* **335**, 417–431.
Rotundi, A., Sierks, H., Della Corte, V., *et al.* (2015), *Science* **347**, aaa3905.
Rubin, M., Altwegg, K., van Dieshoeck, E. F. and Schwehm, G. (2015a), *Astrophys. J. Lett.* **815**, L11.
Rubin, M., Altwegg, K., Balsiger, H., *et al.* (2015b), *Science* **348**, 232–235.
Safronov, V. S. (1977), in *Comets, Asteroids, Meteorites: Interrelations, Evolution, and Origins* (ed. A. H. Delsemme), Toledo, Oh: Univ. of Toledo, pp. 483–484.
Sandford, S. A., Aléon, J., O'D. Alexander, C. M., *et al.* (2006), *Science* **314**, 1720–1724.
Schmitt, B., Espinasse, S., Grim, R. J. A., Greenberg, J. M. and Klinger, J. (1989), in *Physics and Mechanics of Cometary Materials* (eds. J. J. Hunt and T. D. Guyenne), ESA SP-302, pp. 65–69.
Sekanina, Z. (1981), *Astron. J.* **86**, 1741–1773.
Sekanina, Z. (1982), in *Comets* (ed. L. L. Wilkening), Tucson: Univ. Arizona Press, pp. 251–287.
Sekanina, Z. and Yeomans, D. K. (1985), *Astron. J.* **90**, 2335–2352.
Sierks, H., Barbieri, C., Lamy, P. L., *et al.* (2015), *Science* **347**, aaa1044.
Singer, K. N., Schenk, P. M., Robbins, S. J., *et al.* (2015), in *AAS/Division for Planetary Sciences Meeting Abstracts* **47**, 102.02.
Singer, K. N., McKinnon, W. B., Greenstreet, S., *et al.* (2016), in *AAS/Division for Planetary Sciences Meeting Abstracts* **48**, 213.12.
Skorov, Yu. V. and Rickman, H. (1999), *Planet. Space Sci.* **47**, 935–949.
Sosa, A. and Fernández, J. A. (2011), *Mon. Not. R. Astron. Soc.* **416**, 767–782.
Sridhar, S. and Tremaine, S. (1992), *Icarus* **95**, 86–99.
Stern, S. A. (1995), *Astron. J.* **110**, 856–868.
Stern, S. A. and Weissman, P. R. (2001), *Nature* **409**, 589–591.
Stothers, R. B. (1985), *Nature* **317**, 338–341.
Stuart, J. S. and Binzel, R. P. (2004), *Icarus* **170**, 295–311.
Sunshine, J. M., A'Hearn, M. F., Groussin, O., *et al.* (2006), *Science* **311**, 1453–1455.
Sykes, M. V. and Walker, R. G. (1992), *Icarus* **95**, 180–210.
Szutowicz, S. and Rickman, H. (2006), *Icarus* **185**, 223–243.
Tancredi, G., Lindgren, M. and Rickman, H. (1990), *Astron. Astrophys.* **239**, 375–380.
Tancredi, G., Fernández, J. A., Rickman, H. and Licandro, J. (2000), *Astron. Astrophys. Suppl. Ser.* **146**, 73–90.
Tancredi, G., Fernández, J. A., Rickman, H. and Licandro, J. (2006), *Icarus* **182**, 527–549.
Thomas, N., Sierks, H., Barbieri, C., *et al.* (2015a), *Science* **347**, aaa0440.

Thomas, N., Davidsson, B., El-Maarry, M. R., *et al.* (2015b), *Astron. Astrophys.* **583**, A17.
Thomas, P. C., Veverka, J., Belton, M. J. S., *et al.* (2007), *Icarus* **191**, 51–62.
Thomas, P., A'Hearn, M. F., Belton, M. J. S., *et al.* (2013), *Icarus* **222**, 453–466.
Toliou, A., Morbidelli, A. and Tsiganis, K. (2016), *Astron. Astrophys.* **592**, A72.
Trujillo, C. A. and Sheppard, S. S. (2014), *Nature* **507**, 471–474.
Tsiganis, K., Gomes, R., Morbidelli, A. and Levison, H. F. (2005), *Nature* **435**, 459–461.
Valsecchi, G. B., Milani, A., Gronchi, G. F. and Chesley, S. R. (2003), *Astron. Astrophys.* **408**, 1179–1196.
van Woerkom, A. J. J. (1948), *Bull. Astron. Inst. Netherl.* **10**, 445–472.
Veverka, J., Klaasen, K., A'Hearn, M. F., *et al.* (2013), *Icarus* **222**, 424–435.
Vincent, J.-B., Lara, L. M., Tozzi, G. P., Lin, Z.-Y. and Sierks, H. (2013), *Astron. Astrophys.* **549**, A121.
Vincent, J.-B., Bodewits, D., Besse, S., *et al.* (2015), *Nature* **523**, 63–66.
Vincent, J.-B., Oklay, N., Pajola, M., *et al.* (2016), *Astron. Astrophys.* **587**, A14.
Wahlberg Jansson, K. and Johansen, A. (2014), *Astron. Astrophys.* **570**, A47.
Walker, R. J. (2009), *Chemie Erde Geochem.* **69**, 101–125.
Walsh, K. J., Morbidelli, A., Raymond, S. N., O'Brien, D. P. and Mandell, A. M. (2011), *Nature* **475**, 206–209.
Weidenschilling, S. J. (1984), *Icarus* **60**, 553–567.
Weidenschilling, S. J. (1997), *Icarus* **127**, 290–306.
Weissman, P. R. (1980), *Astron. Astrophys.* **85**, 191–196.
Weissman, P. R. (1986), in *20th ESLAB Symposium on the Exploration of Halley's Comet* (eds. B. Battrick, E. J. Rolfe and R. Reinhard), ESA SP-250, pp. 517–522.
Weissman, P. R. (2011), *Bull. Amer. Astron. Soc.* **43**, Abstract 1.04.
Weissman, P. R. and Kieffer, H. H. (1981), *Icarus* **47**, 302–311.
Wetherill, G. W. (1989), *Meteoritics* **24**, 15–22.
Whipple, F. L. (1940), *Proc. Amer. Phil. Soc.* **111**, 375.
Whipple, F. L. (1950), *Astrophys. J.* **111**, 375–394.
Whipple, F. L. (1972a), in *The Motion, Evolution of Orbits, and Origin of Comets* (eds. G. A. Chebotarev, E. I. Kazimirchak-Polonskaya and B. G. Marsden), Dordrecht: Reidel, pp. 401–408.
Whipple, F. L. (1972b), in *From Plasma to Planet* (ed. A. Elvius), Stockholm: Almqvist and Wiksell, pp. 211–230.

Wiśniowski, T. and Rickman, H. (2013), *Acta Astron.* **63**, 293–307.
Wooden, D. H., Harker, D. E., Woodward, C. E., *et al.* (1999), *Astrophys. J.* **517**, 1034–1058.
Yabushita, S. (1992), *Celest. Mech. Dynam. Astron.* **54**, 161–178.
Yeomans, D. K., Giorgini, J. D. and Chesley, S. R. (2005), *Space Sci. Rev.* **117**, 123–135.
Youdin, A. N. and Goodman, J. (2005), *Astrophys. J.* **620**, 459–469.
Zahnle, K. J. and Sleep, N. H. (1997), in *Comets and the Origin and Evolution of Life* (eds. P. J. Thomas, C. F. Chyba and C. P. McKay), New York: Springer-Verlag, pp. 175–208.
Zolensky, M. E., Zega, T. J., Yano, H., *et al.* (2006), *Science* **314**, 1735–1739.

www.ingramcontent.com/pod-product-compliance
Ingram Content Group UK Ltd.
Pitfield, Milton Keynes, MK11 3LW, UK
UKHW020350150926
8624IPUK00004B/99

9 789813 222571